PERSPECTIVES ON CORAL REEFS

PERSPECTIVES ON CORAL REEFS

Edited by

D. J. BARNES

Published for
The Australian Institute of Marine Science
by Brian Clouston Publisher

Australian Institute of Marine Science Contribution Number 200
Published for
The Australian Institute of Marine Science
by Brian Clouston Publisher

First published for the Australian Institute of Marine Science by Brian Clouston, P.O. Box 531, Manuka, A.C.T. 2603 Australia

ISBN 0 642 89585 6

Typeset by Canberra Times Print

Printed in Hong Kong

Cover design by Kirsty Morrison

Contents

Introduction

The suggestion for this collection of reviews came at a workshop on coral reefs held at the Australian Institute of Marine Science in August 1979. The workshop was primarily for Australian researchers, although several overseas scientists attended. In all, about 70 coral reef workers were present. Australian workers now make up a significant proportion of the scientists working on coral reefs. The size of Australia and the size of the Great Barrier Reef limit the opportunities for contacts between reef scientists. Moreover, while the various disciplines involved in reef science are becoming more and more interlocked, the workers in any one discipline are less and less able to keep up with the major developments in other fields. It was felt that a general meeting of marine scientists with interests in coral reefs would help to remedy this situation. The Australian Institute of Marine Science, a national facility, undertook to organize the meeting and to provide assistance, where necessary and as much as was possible, to ensure that most Australian reef workers could attend.

The workshop was split into three major areas which were presented consecutively: reef geology, reef growth and reef biology. By the conclusion of the workshop there was a general feeling that a combined meeting of (mainly) geologists and biologists had provided each with fresh views. There was also general agreement that scientific investigation of coral reefs would greatly benefit from a combined approach among the disciplines involved. It was proposed that a series of fairly basic reviews would provide a straightforward means by which a specialist in one field could discover what was going on in other fields. Following from this, a number of workshop participants agreed to provide reviews. Most agreed to provide a fairly brief review of this specialist area. Certain workshop participants agreed to provide a more extensive cover of general subjects. This volume is the collection of those reviews.

The task of pulling the material together for this book was far more difficult and laborious than I thought possible. I would like to thank members of A.I.M.S.'s staff who assisted, particularly my colleagues who reviewed the manuscripts. I am grateful to Mr T. Cox, Mr M. Devereux and Mrs W. Ellery for their sterling work in checking and correcting references. Mr Cox and Mr Devereux spent many hours with me checking the various drafts, and Mr Cox and Mrs D. Smith entered and corrected the manuscript in the computer.

D. J. Barnes

1 Carbon and Oxygen Isotope Probes of Reef Environment Histories

Paul Aharon and John Chappell

Research School of Earth Sciences and Research School of Pacific Studies, Australian
National University, P.O. Box 4, Canberra 2600.

INTRODUCTION

Carbon and oxygen occur in several isotopic forms. In addition to the common forms, ^{12}C and ^{16}O, the stable isotopes ^{13}C and ^{18}O occur at low but significant levels. The ratios of $^{13}C/^{12}C$ and $^{18}O/^{16}O$ differ between the various compartments of the global carbon and oxygen systems. For instance, the ratio of $^{13}C/^{12}C$ is different in atmospheric CO_2 from that in precipitated calcium carbonate; the ratio of $^{18}O/^{16}O$ in seawater is different from that in snow.

Variations in the ratios are also found within compartments of the global systems. For instance, stable isotope ratios for both carbon and oxygen vary between the annual rings found in certain trees. These variations occur for several reasons, including environmental effects on the processes which cause ratios to become altered. Organic and inorganic materials containing carbon and oxygen provide a stable isotope record of the conditions under which they were synthesised or deposited. Materials that have accumulated on the geological time scale can provide information about past environments and their changes. Thus, over the past thirty years, analyses of the $^{18}O/^{16}O$ record in micro-organisms preserved in deep sea sediments have provided a history of global ice volume and oceanic temperature (e.g., Emiliani 1955; Shackleton & Opdyke 1973). Results have contributed greatly to our understanding of the Quaternary Era, although temporal resolution in these records is about at the level of a 5000 year moving average.

More recently, interest has turned towards records decipherable at the annual level, especially where direct comparison is possible with recorded climatic patterns. Materials of particular interest include growth-banded corals and molluscs, and wood from tree ring sequences. Once a correlation is proven between isotopic and environmental variations, the prospect is exciting for palaeoclimatology and other studies, because tree rings can provide continuous annual records extending back several thousand years, and coral reef materials can provide higher resolution 'windows', each of several decades length, to much greater depths into the past.

Although more work at a very detailed level has been done on tree rings (e.g., Tans 1978; Wilson & Grinsted 1977) than on reefal organisms, the work which we describe shows that reef environment histories can be probed for temperature and salinity with greater confidence and resolution than so far achieved for temperature and rainfall from isotopic analysis of tree growth bands.

1

CARBON AND OXYGEN ISOTOPES: PRINCIPLES AND MEASUREMENT

The stable isotopes ^{13}C and ^{18}O occur at about 1% and 0.2% respectively, of the natural levels of the common isotopes of carbon and oxygen; ^{12}C and ^{16}O (in water, for example, the relative atomic abundance of oxygen is approximately 33%, of which about 99.8% is ^{16}O and 0.2% is ^{18}O). Ratios of $^{18}O/^{16}O$ and $^{13}C/^{12}C$ in natural carbon and oxygen-bearing compounds vary with chemical pathway by which the compounds were formed and the conditions under which the reaction occurred, including temperature. Alteration of ratios by physical, chemical or biological processes is called isotopic fractionation.

In this paper, and in general unless stated otherwise, the isotopic ratios are given as relative deviations from the ratio present in a standard. The deviations are given in parts per thousand (δ‰).

$$\delta‰ = 1000 \times (ratio \; sample - ratio \; standard)/(ratio \; standard).$$

The δ‰ value is thus a standardized indication of the ratio of $^{18}O/^{16}O$ or $^{13}C/^{12}C$ present in a sample. The ratio of $^{18}O/^{16}O$ in water is measured with reference to the ratio present in standard mean ocean water (SMOW standard, distributed by the International Atomic Energy Agency, Vienna). For carbonate materials, the ratios of $^{18}O/^{16}O$ and $^{13}C/^{12}C$ are measured with reference to the ratios present in the PDB standard (defined from a Cretaceous belemnite by Epstein *et al.* 1953). The relationship for oxygen ratios between the standards is (Friedman & O'Neil 1977):

$$\delta^{18}O(SMOW) = 1.03086 \times \delta^{18}O(PDB) + 30.86.$$

Isotope ratios are measured with a mass spectrometer, usually from CO_2 gas prepared by means appropriate to the sample. Full accounts are given by Bowen (1966) and Hoefs (1973); details for water samples are given by Epstein & Mayeda (1953), and details for total dissolved carbon in water are given by Aharon (1980). For work to be regarded as satisfactory, mass spectrometric error for a single determination should be around ± 0.05‰ and reproducibility between sample replicates should be better than ± 0.1‰

ELEMENTS OF OXYGEN AND CARBON ISOTOPES IN THE GLOBAL SYSTEMS

Oxygen isotope ratios on a global scale are principally affected by fractionation processes in the water cycle (fig. 1), so that $\delta^{18}O$ in precipitation is substantially lower than in sea water, especially at high latitudes and altitudes (Dansgaard 1964). $\delta^{18}O$ values in materials of organic origin may differ substantially from local aqueous or atmospheric reservoirs due to fractionation during synthesis. Owing to the isotopic difference between snow and sea water, the mean $\delta^{18}O$ value in the oceans varied throughout Pleistocene times due to growth and decay of the icecaps (implicit in fig. 1). At times of glacial maxima, when about 4% of the present oceans was locked up in northern ice sheets, sea water was about 1.2‰ heavier in $\delta^{18}O$ than it is today (Shackleton & Opdyke 1973).

Gross changes in the global biomass or amounts of biologically-derived materials (as from burning fossil fuels) have little effect on $\delta^{18}O$ values for global reservoirs, because these materials contain only a small portion of the global oxygen. However, this is not the case for $\delta^{13}C$ values of global carbon reservoirs. A large portion of global carbon is contained in organic materials, and these have a $\delta^{13}C$ value about 24‰ lower than other reservoirs. Burning of fossil fuels has increased atmospheric CO_2 and significantly affected $\delta^{13}C$ values for global carbon reservoirs (fig. 2) and the $\delta^{13}C$ values of tissues and skeleton of organisms living in the reservoirs (Tans 1978; Broecker *et al.* 1979).

Variations in $\delta^{13}C$ values between the principal global reservoirs of the carbon cycle are more complex than variations in $\delta^{18}O$ values associated with global water cycle. Carbon

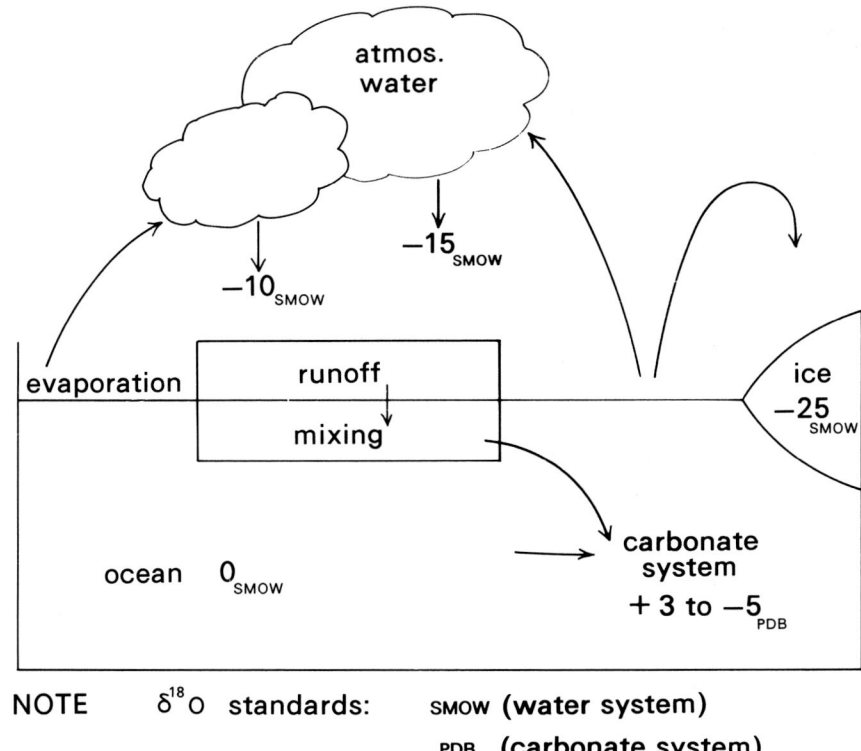

Figure 1   Simplified oxygen isotope cycles between ocean, atmosphere, and glacial ice

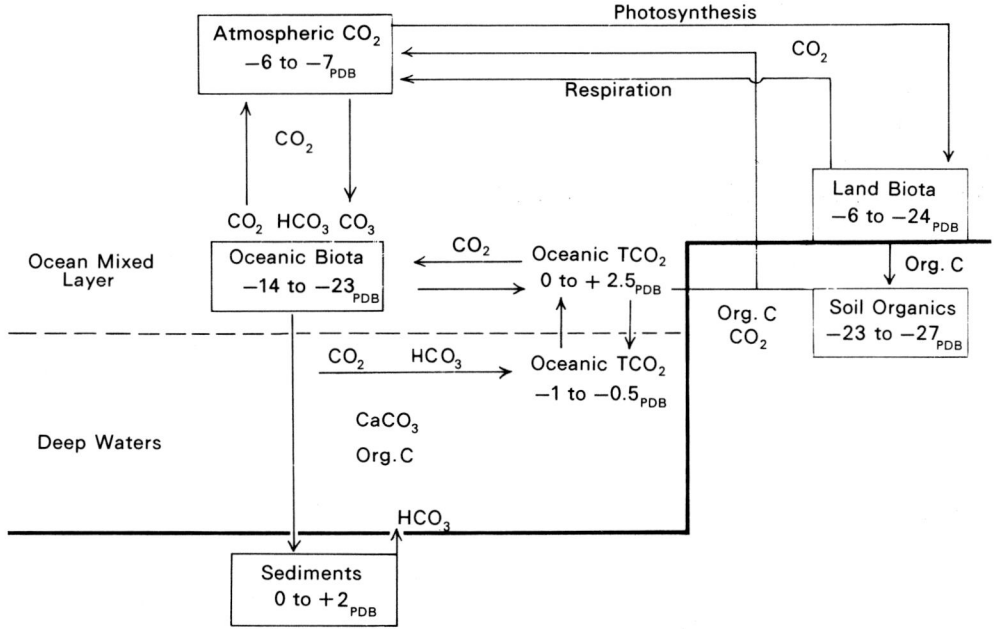

Figure 2   Natural carbon cycles and isotopic levels

dioxide is exchanged between the atmosphere, the oceans and the biosphere (fig. 2). Exchange between surface waters and deep waters in the oceans is mainly via the carbonate chemical system. The $^{13}C/^{12}C$ ratio in ocean surface waters is affected by factors, such as temperature, which alter chemical exchange coefficients (MacIntyre 1979) and factors which affect transport of ocean masses and atmospheric circulation (Shackleton 1977).

In certain important reactions, temperature does not have the same effect on fractionation of oxygen isotopes as it has on carbon isotopes. Slow precipitation of calcium carbonate is one such reaction; consequently, comparison of $\delta^{13}C$ and $\delta^{18}O$ variations provides a means for examining past and present coral reef processes.

At its simplest, precipitation of calcium carbonate is represented by the equation:

$$Ca^{2+} + 2HCO_3^- = CaCO_3 + H_2O + CO_2.$$

The carbon isotope composition of the precipitated $CaCO_3$ is determined by that of the dissolved bicarbonate and by the fractionation factor involved in the precipitation reaction: $(\lessgtr^{13}_{HCO_3^-}(CaCO_3))$. Thus,

$$\delta^{13}C(CaCO_3) = \delta^{13}C(HCO_3^-) + \lessgtr^{13}_{HCO_3^-}(CaCO_3).$$

The $\delta^{18}O$ value of the $CaCO_3$ is determined by that of the water and the respective fractionation factor:

$$\delta^{18}O(CaCO_3) = \delta^{18}O(H_2O) + \lessgtr^{18}_{H_2O}(CaCO_3).$$

The fractionation factors have different and opposite sensitivities to temperature. Whereas the $^{18}O/^{16}O$ fractionation between the calcium carbonate and water $(\lessgtr^{18}_{H_2O}(CaCO_3))$ is -0.23‰ °C^{-1} (Epstein *et al.* 1953), the $^{13}C/^{12}C$ fractionation between the same calcium carbonate and the dissolved bicarbonate $(\lessgtr^{13}_{HCO_3^-}(CaCO_3))$ is only +0.035‰ °C^{-1} (Emrich *et al.* 1970; Mook *et al.* 1974).

These different responses of stable isotopes to temperature provide one means by which environmental temperature can be separated from other factors (see table 1). Other strategies for separating environmental factors, both global and local, are discussed below.

CARBON AND OXYGEN ISOTOPES IN REEFS: CLAMS AND CORALS

Long-lived organisms with marked skeletal growth bands are the most useful for analysis of past environment changes using stable isotopes. Organisms need to be sufficiently abundant throughout the various reef zones to allow detailed analysis of reef environment changes. Certain types of scleractinian coral and various species of *Tridacna,* the giant clam, meet these criteria. *Tridacna* spp. have the important advantage that their skeletons are less subject to diagenetic alteration, which can alter isotopic composition. By using radiocarbon (^{14}C) as a tracer of chemical exchange, Chappell & Polach (1972) showed that skeleton of *Tridacna gigas* in Pleistocene reefs can remain isolated from exchange with the groundwater, while corals recrystallize and exchange far more readily.

In establishing an environment probe based on stable isotopes in skeletal material, it is first necessary to determine the relationship between isotope ratios in skeletons and the ratios in the parent reservoir. As $\delta^{18}O$ values are more altered by changes in environmental temperature than $\delta^{13}C$ values, they provide the better example. Corals and Tridacnids both form $CaCO_3$ exoskeletons of aragonite. The theoretical (thermodynamic equilibrium) value for aragonite precipitating in water having $\delta^{18}O = 0.0$‰ SMOW at 25°C is -1.29‰ (Tarutani *et al.* 1969). *Tridacna* aragonite approaches this figure more

Perturbation type	$^{18}O/^{16}O$ in $CaCO_3$	$^{13}C/^{12}C$ in $CaCO_3$	Range in coral reef environment
Sea surface warming	falls	(rises)	up to 1‰ in $\delta^{18}O$
Fresh water runoff dilutes ocean	falls	falls	up to 0.5‰
Evaporation	(rises)	—	—
Ocean upwelling	—	falls	up to −0.7‰*
Major glaciation	rises	—	up to 1.2‰ in Pleistocene
Biomass productivity increase	—	rises	up to 2‰ in a lagoon
Organic matter oxidation	—	falls	up to 2‰ in a lagoon
Fossil fuel burning	—	(falls)	?

* Killingley & Berger (1979)
(falls) or (rises): small effect

Table 1 Effects of environmental perturbations on stable isotopes, with emphasis on the coral reef environment

closely than corals, and corals show considerably greater variation than *Tridacna*. Significant departures from equilibrium isotopic values are likewise noted for $\delta^{13}C$ values in corals but not in *Tridacna*.

Values of $\delta^{18}O$ and $\delta^{13}C$ in coralline aragonite vary much more widely than in the surrounding seawater. Measurements from corals show a $\delta^{18}O$ range of 4‰ and a $\delta^{18}C$ range of 13‰ (Weber & Woodhead 1970; Land *et al.* 1975). *Tridacna*, on the other hand, shows a consistent temperature-dependent variation of $\delta^{18}O$ of 0.23‰.°C^{-1}, agreeing with the theoretical prediction of Urey (in Garlick 1970), and a value of −0.84 ±0.05‰ at 25°C. These values are based on Tridacnid samples from the Great Barrier Reef province, northern New Guinea, and the Red Sea are summarized in table 2 (Aharon 1980). The $\delta^{13}C$ range in the same set of Tridacnid samples is 0.76% (table 2). Typical isotopic values from corals (from Weber & Woodhead 1972a,b; Erez 1978) are compared with *Tridacna* results in figure 3. Clearly, Tridacnids offer a superior probe, closely matching isotopic values for aragonite precipitated in equilibrium with the surrounding water (fig. 3). However, the possibility of using corals as a probe must not be dismissed. Careful work with a single species taken from similar sites across a range of reefs with different mean annual temperature can show a good relationship between $\delta^{18}O$ and temperature. The best available example is for the hermatypic coral *Galaxea* (Weber 1977), which is compared with the *Tridacna* results in figure 4. Although the data are offset from the theoretical values for aragonite ($\delta^{18}O$ at 25°C = −4.67‰) and have a different temperature dependancy (−0.17‰ °C^{-1}), they do show a good trend.

Locality	Mean annual water temp.	$\delta^{18}O$ seawater ‰ SMOW	$\delta^{18}O$ Tridacna ‰ PDB	$\delta^{13}C$ Tridacna ‰ PDB	No. of analysed specimens
Huon Peninsula, New Guinea (6°S, 147°36'E)	27.9±0.2	−0.07	−1.64±0.22	+2.28±0.25	10
Raine Island[1] (11°35'S, 144°02'E)	27.2±0.5	+0.20	−1.03±0.02	+3.04±0.02	1
North Great Barrier Reef[1] (14°28'S to 14°44'S)	26.5±0.7	+0.20	−1.22 ±0.24	+2.62±0.22	12
Low Isles[1] (16°24'S, 145°33'E)	25.7±0.7	−0.02	−0.76±0.04	+2.37±0.02	1
Lady Elliot Reef[2] (24°07'S, 152°46'E)	24.4±0.8	+0.40	−0.36±0.05	+2.40±0.02	2
Huweiba, Red Sea (28°58'N, 34°38'E)	22.9±0.7	+1.33	+0.79±0.03	+2.65±0.05	2

[1] Recent *Tridacna* sp. collected by Great Barrier Reef Expedition, 1973
[2] *Tridacna maxima* analysed by Dr P. Flood (pers. comm.)
Other sources: Great Barrier Reef environmental data from Pickard *et al.* (1977); water salinity translated into δ^{18} O by relationship given by Craig and Gordon (1965). Data from Nuweiba from Aharon (1974); other data from Aharon (1980)

Table 2   Oxygen and Carbon isotopes from seawater and *Tridacna* aragonite, from various Indo-Pacific localities

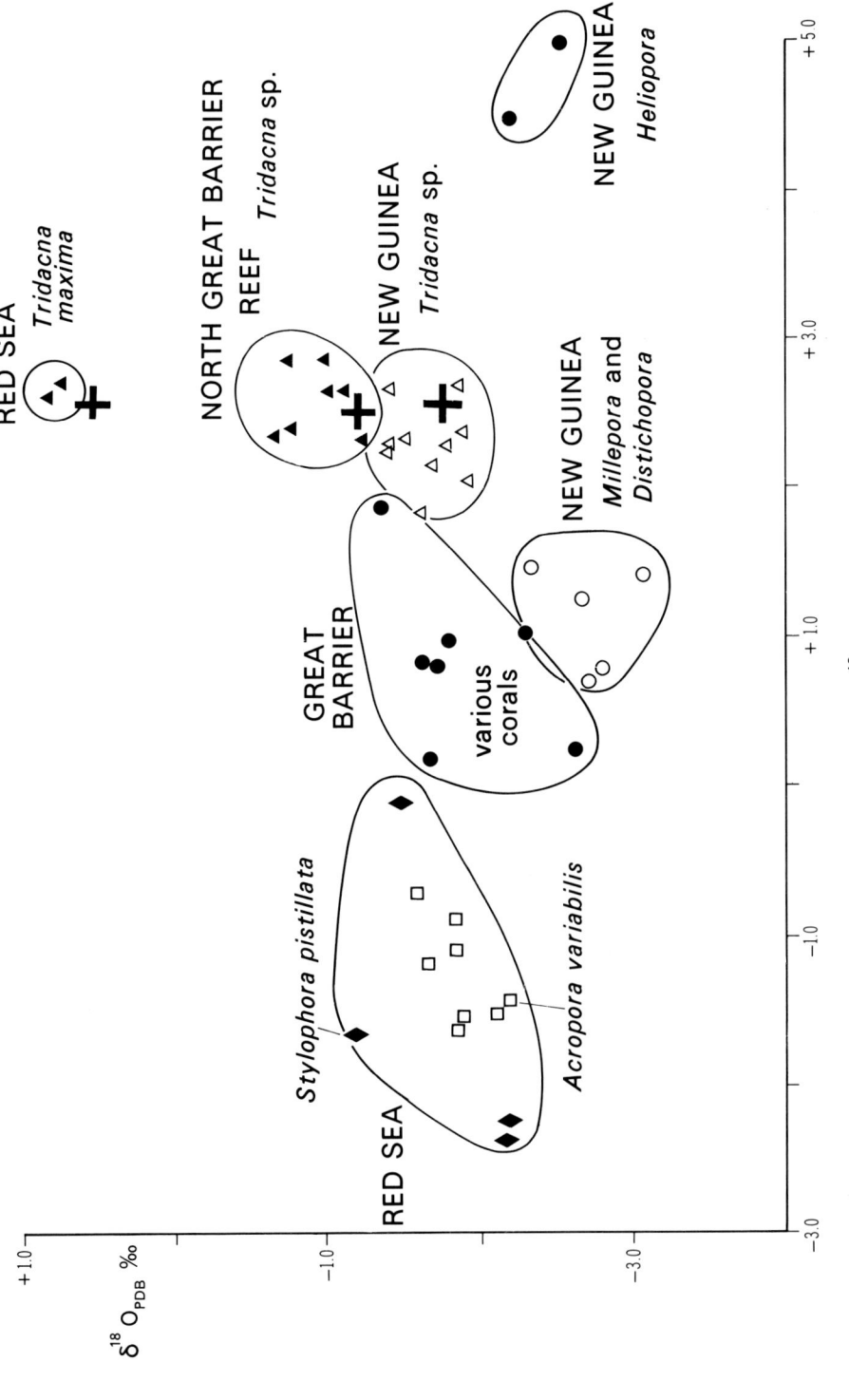

Figure 3 Oxygen and carbon isotope relationships in *Tridacna* and various corals from the same environments. Coral data from Weber & Woodhead (1972a,b) and Erez (1978). Heavy crosses represent isotopic equilibrium values for aragonite precipitated in the same environments as the samples

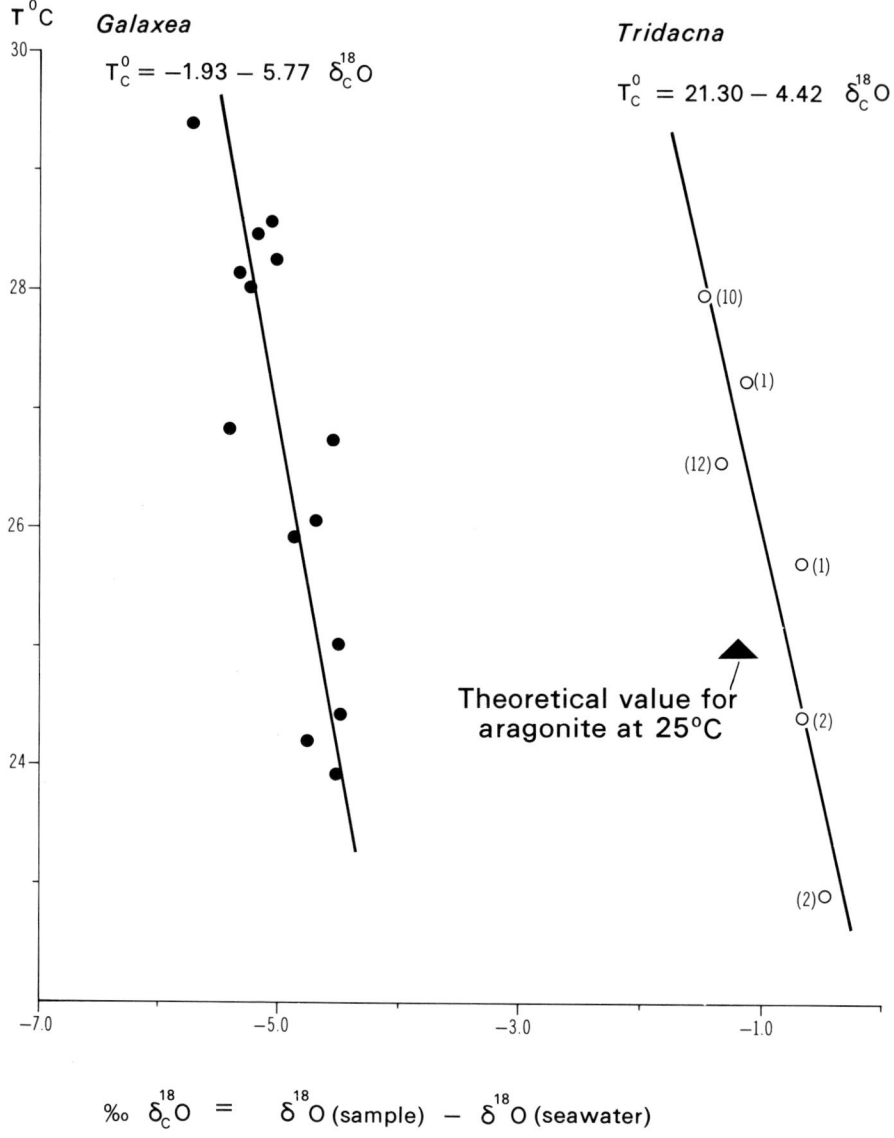

$$\% \ \delta_c^{18}O \ = \ \delta^{18}O \text{ (sample)} \ - \ \delta^{18}O \text{ (seawater)}$$

Figure 4 ^{18}O thermometer for *Tridacna* aragonite. Bracketed figures are numbers of specimens at each averaged point. Regression line fitted by cubic method (York 1969). Also shown are results for the hermatypic coral *Galaxea* (Weber 1977)

The main problem with hermatypic corals is that their $\delta^{18}O$ and $\delta^{13}C$ values vary with growth position on a reef, as demonstrated by results from Weber & Woodhead (1970, 1972a,b) and Land *et al.* (1975). This is biologically interesting when compared with the regular near-equilibrium behaviour of *Tridacna*. *Tridacna* is distinctive amongst molluscs in having symbiotic zooxanthellae in its tissues. These zooxanthellae are possibly instrumental in the high rate of calcification in *Tridacna* which well exceeds other

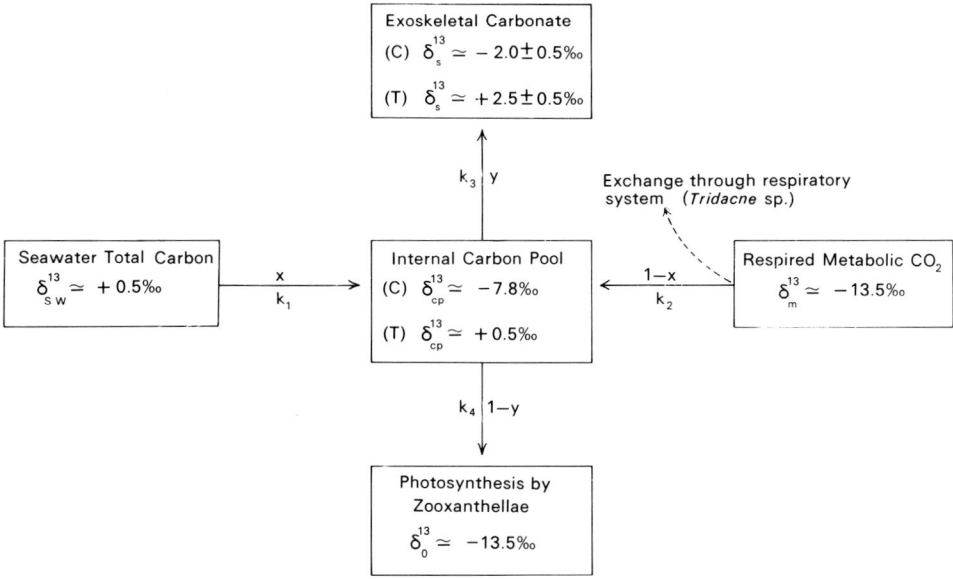

Figure 5   Carbon exchanges suggested by Goreau (1977) leading to precipitation of skeletal carbonate in corals via an internal pool; and a suggested modification for *Tridacna*. $\delta^{13}C$ values prefixed by (C) refer to corals; those prefixed by (T), to *Tridacna*. Note transport rate differences: $k_1(T) >> k_1(C)$, $k_4(T) << k_4(C)$, and $k_3(T) \simeq 3 k_3(C)$

molluscs. Thus, *Tridacna* has a superficial resemblance to the hermatypic corals amongst which it lives. The differences between carbon isotope behaviour in clams and in corals shed light on mechanisms of calcification. Goreau (1977) suggested a carbon exchange model for corals (fig. 5) in which skeletal carbonate is drawn from a central pool made up of two sources: seawater bicarbonate-carbonate ($\delta^{13}C = +0.5\%$) and metabolic CO_2 ($\delta^{13}C = -13.5\%$). The model assumes that the fraction of inorganic carbon from seawater used in skeletal precipitation is about the same as the fraction used in photosynthesis (for each molecule of CO_2 fixed in photosynthesis, one molecule of $CaCO_3$ is precipitated). A simple input-output material balance shows that roughly 60% of the pool of inorganic carbon within the coral tissues is derived from respiration and 40% from seawater, leading to $\delta^{13}C$ values in the internal pool of approximately -7.8%. The relatively large carbon isotopic fractionation induced by the zooxanthellae extracting CO_2 from the central pool and the fractionation between the dissolved carbon and the deposited $CaCO_3$ results in the skeletal $\delta^{13}C$ values around -2.5%. *Tridacna*, on the other hand, has skeletal $\delta^{13}C$ values around $+2.5\%$, which is close to isotopic equilibrium after fractionation is allowed for. This suggests that the zooxanthellae either play no significant role in *Tridacna* calcification, perhaps because their relative mass is very small compared with that of the host *Tridacna* tissue (Yonge 1936, 1975), or do not contribute to the internal carbon pool (i.e. the alternative exchange shown at the right of figure 5).

DETAILED TESTING OF THE *TRIDACNA* PROBE: SEASONAL VARIATIONS

The most interesting test of an isotopic probe based on a growth-banded organism lies in its seasonal signal. Wilson & Grinsted (1975, 1977) have shown it should be possible to obtain seasonal temperature and rainfall data from tree rings using $\delta^{18}O$, $\delta^{13}C$ and δD (the deuterium/hydrogen isotope ratio), although interpretation is complicated by several factors. Although coral skeletons usually display relatively wide annual bands around 1 cm wide (Buddemeier 1974; Fairbanks & Dodge 1979), the annual growth bands in Tridacnids are more finely subdivided than those in coral. A typical section through a *Tridacna gigas* shell (fig. 6) shows two regions, one of fine internal layers deposited by the inner mantle fold under the pallial line, and the other of broader growth bands, outside the pallial line. Both regions can be resolved microscopically into daily micro-bands, about 7 μm thick for the internal layers, and about 30 μm thick for the external layers. Annual banding in both internal and external regions is indicated by broad light and dark zones. A specimen collected live at Huon Peninsula, New Guinea, in October 1977 shows that the light-coloured zones correspond to the austral summer and the dark zones relate to the winter (Aharon 1980). A sequence of 17 $\delta^{13}C$ and $\delta^{18}O$ determinations across 5 years of growth from this specimen is plotted in figure 7 and shows a clear annual signal.

The following points emerge.

1. Variations of $\delta^{13}C$ and $\delta^{18}O$ are positively correlated (in figure 7, $\delta^{18}O$ is plotted in the opposite sense to $\delta^{13}C$; this is in accordance with the convention in geologic literature, where $\delta^{18}O$ is plotted negative-upwards). The variations do not simply reflect seasonal temperatures because the temperature sensitivity of $\delta^{13}O$ values is both small and in the opposite direction of $\delta^{18}O$ values. In any case, annual seawater temperature range in the Huon Peninsula area is only 1.5°C (Aharon 1980) and could account for no more than 0.35‰ of the observed 0.75‰ variation in $\delta^{18}O$.

2. North-eastern Huon Peninsula is subject to strongly seasonal rainfall. High run-off in the austral summer strongly dilutes coastal sea water with fresh water, which has a $\delta^{18}O$ value around -9‰ SMOW. The summer growth bands have increased negative $\delta^{18}O$ values which correspond to this dilution.

3. $\delta^{13}C$ values for the local water column become more positive as productivity increases (confirmed by water column measurements taken through the diurnal cycle; Aharon 1980). We suggest the increase of $\delta^{13}C$ in the winter growth bands (fig. 7) is due to greater productivity induced by the very much higher winter sunshine levels of the Huon coastal region.

The *Tridacna* probe has been tested further through growth band analysis of specimens collected from the Pleistocene raised reefs of Huon Peninsula (described by Chappell 1974; see also Chappell, chapter 4 this volume). Growth band structures, similar to those obtained from living specimens, (fig. 6) are well preserved. The isotopic variations within growth bands are similar to those found in living specimens. Figure 7b shows an example of 7 years of growth from a specimen that grew during the last interglacial interval, 133 000 years ago.

CONCLUDING DISCUSSION

During the past 25 years, analysis of oxygen isotopes in deep sea cores has provided one of the most valuable techniques contributing to Quaternary studies. Results have advanced considerably our understanding of the essential patterns of longer-term climatic and glacial changes. However, deep sea cores provide low temporal resolution. We believe

Figure 6   Transverse section through *Tridacna* shell from New Guinea. Specimen is 133 000 years old but indistinguishable from a modern example. The strong oblique pallial line separates the internal growth region (below) from the external region. Internal region grows about 2 cm yr^{-1}, while external region grows up to four times as fast. Major alternations of dark and light bands represent seasons; this specimen started life in the austral winter and died 6 years later in the summer

a. Modern Tridacna

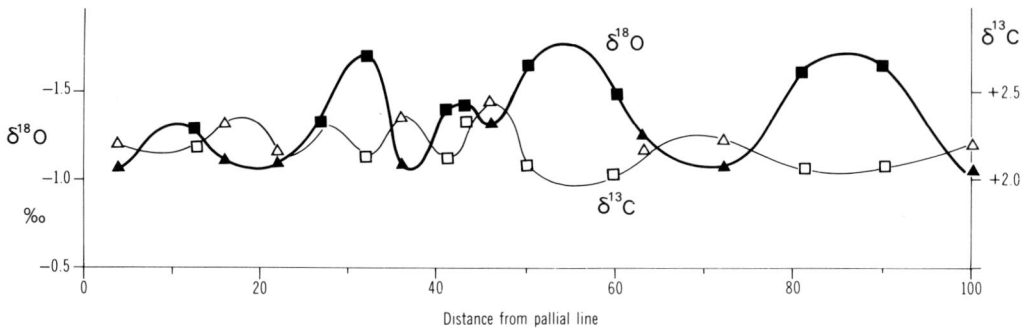

b. 133 000 B.P.

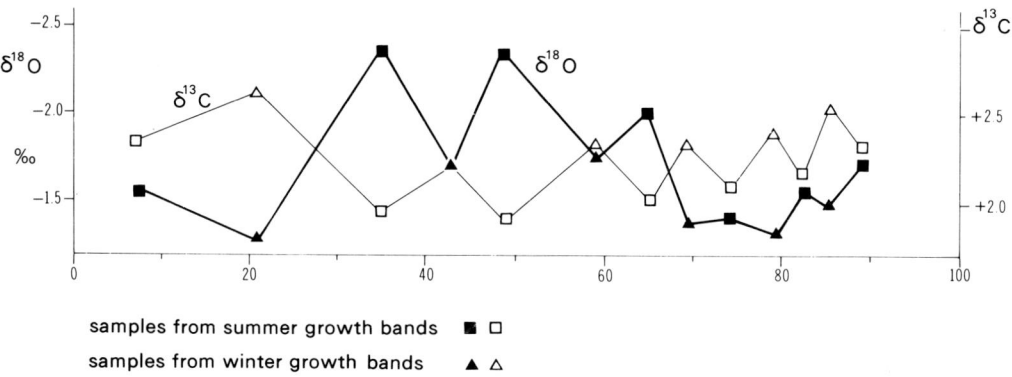

samples from summer growth bands   ■ □
samples from winter growth bands   ▲ △

Figure 7   Seasonal variations of $\delta^{18}O$ and $\delta^{13}C$, based on analyses from internal growth layers of
            Tridacna gigas.
            a. 5 years record from a modern sample.
            b. 7 years record from 133 000-year-old sample

that results which will emerge from extended analyses of high resolution records inherent
in annual layering will be just as stimulating. One prospect, particularly interesting in the
context of coral reefs, is reconstruction of detailed histories of environmental factors such
as temperature and salinity, and of community factors such as productivity.

Table 1 illustrates that the two geochemical parameters considered here, $\delta^{13}C$ and $\delta^{18}O$,
are influenced by many more than two factors. Hence, unambiguous interpretation
requires knowledge of how $\delta^{13}C$ and $\delta^{18}O$ vary in relation to the local system (illustrated by
the growth band study of *Tridacna,* above), as well as independent information about
certain of the external factors. For example, global ice volume, which affects all oceanic
$\delta^{18}O$ levels, can be allowed for through its $\delta^{18}O$ signal in deep sea cores, at relatively low
resolution, or allowed for indirectly at higher resolution through sea level studies
(discussed by Aharon *et al.* 1980). More generally, other geochemical parameters such as
trace element substitutions for calcium in the skeletal carbonate need further development. Strontium levels, for example, have been shown to be temperature-related in corals

(Smith *et al.* 1979). Finally, allowing for the species and site dependency of $\delta^{13}C$ and $\delta^{18}O$ variability, the corals have the potential to provide useful environmental information. Although the difficulties illustrated in figure 3 have appeared inimical to paleothermometry, they are vital data of a different kind (Land *et al.* 1975, 1977), the environmental meaning of which may well be resolved by parallel growth band studies of corals and *Tridacna* specimens from the same sites. In conclusion, the prospects are exciting for applications of these isotopic probes to various aspects of reef histories.

ACKNOWLEDGEMENTS

The authors thank Dr P. J. Cook, A.N.U., and Dr P. J. Davies, B.M.R., Canberra, for their stimulating support in writing this paper and Dr D. J. Barnes from A.I.M.S. for his constructive criticism and thorough review of the manuscript. We also thank Mr H. Polach, A.N.U., for supplying the *Tridacna* samples collected by members of the Great Barrier Reef Expedition, 1973.

LITERATURE CITED

Aharon, P., 1974. The geochemistry of selected elements in the "Solar Lake" Eilat Gulf. M.Sc. dissertation, The Hebrew University of Jerusalem, 106pp.

Aharon, P., 1980. Stable isotopes geochemistry of a late Quaternary reef environment, New Guinea: application of high resolution data to palaeoclimatology. Ph.D. dissertation, Australian National University, 259pp.

Aharon, P., J. Chappell and W. Compston, 1980. Stable isotope and sea-level data from New Guinea supports Antarctic ice-surge theory of ice ages. *Nature* 283,649-51.

Bowen, R., 1966. Paleotemperature Analysis. London: Elsevier Publishing.

Broecker, W. S., T. Takahashi, H. J. Simpson and T. H. Peng, 1979. Fate of fossil fuel carbon dioxide and the global carbon budget. *Science* 206, 409-18.

Buddemeier, R. W. 1974. Environmental controls over annual and lunar monthly cycles in hermatypic coral calcification. In, *Proceedings of the Second International Symposium on Coral Reefs*. Great Barrier Reef Committee, Brisbane, vol. 2, pp. 259-67.

Chappell, J., 1974. Geology of coral terraces, Huon Peninsula, New Guinea: a study of Quaternary tectonic movements and sea level changes. *Geological Society of America Bulletin* 85, 553-70.

Chappell, J., chapter 4 this volume. Sea level changes and coral reef growth.

Chappell, J. and H. A. Polach, 1972. Some effects of partial recrystallisation on ^{14}C dating late Pleistocene corals and molluscs. *Quaternary Research* 2, 244-52.

Craig, H. and L. I. Gordon, 1965. Isotopic oceanography: deuterium and oxygen 18 variations in the ocean and the marine atmosphere. In D. R. Schink and J. T. Corless (eds.), *Proceedings of the Symposium on Marine Geochemistry*. University of Rhode Island Occasional Publications, 3, pp. 277-374.

Dansgaard, W., 1964. Stable isotopes in precipitation. *Tellus* 16, 436-68.

Emiliani, C., 1955. Pleistocene temperatures. *Journal of Geology* 63, 538-78.

Emrich, K., D. H. Ehhalt and J. C. Vogel, 1970. Carbon isotope fractionation during the precipitation of calcium carbonate. *Earth and Planetary Science Letters 8*, 363-71.

Epstein, S. and T. Mayeda, 1953. Variations of ^{18}O content of waters from natural sources. *Geochimica et Cosmochimica Acta 4*, 213-24.

Epstein, S., R. Buchsbaum, H. Lowenstam and H. C. Urey, 1953. Revised carbonate-water isotopic temperature scale. *Geological Society of America Bulletin* 64, 1315-26.

Erez, J., 1978. Vital effect on stable-isotope composition seen in foraminifera and coral skeletons. *Nature* 273, 199-202.

Fairbanks, R. G. and R. E. Dodge, 1979. Annual periodicity of the $^{18}O/^{16}O$ and $^{13}C/^{12}C$ ratios in the coral *Montastrea annularis. Geochimica et Cosmochimica Acta* 43, 1009-20.

Friedman, I. and J. R. O'Neil, 1977. Compilation of stable isotope fractionation factors of geochemical interest. *United States Geological Survey Professional Paper* 440-KK.

Garlick, G. D., 1970. The stable isotopes of oxygen, carbon and hydrogen in the marine environment. In E. G. Goldberg (ed.), *The Sea, Ideas and Observation.* New York: John Wiley & Sons, vol. 5, pp. 393-425.

Goreau, T. J., 1977. Coral skeletal chemistry: physiological and environmental regulation of stable isotopes and trace metals in *Montastrea annularis. Proceedings of the Royal Society of London, Series B* 196, 291-315.

Hoefs, J., 1973. Stable Isotope Geochemistry. Berlin–New York: Springer-Verlag.

Killingley, J. S. and W. H. Berger, 1979. Stable isotopes in a mollusk shell: detection of upwelling events. *Science 205,* 186-8.

Land, L. S., J. C. Lang and D. J. Barnes, 1975. Extension rate: a primary control on the isotopic composition of West Indian (Jamaican) scleractinian reef coral skelton. *Marine Biology* 33, 221-33.

Land, L. S., J. C. Lang and D. J. Barnes, 1977. On the stable carbon and oxygen isotopic composition of some shallow water, ahermatypic, scleractinian coral skeletons. *Geochimica et Cosochimica Acta* 41, 169-72.

MacIntyre, F., 1979. Carbon-13 in tree rings indicates no record of sea-surface temperature. *Science* 205, 1127-9.

Mook, W. G., J. C., Bommerson and W. H. Staverman, 1974. Carbon isotopic fractionation between dissolved bicarbonate and gaseous dioxide. *Earth and Planetery Science Letters* 22, 169-76.

Pickard, G. L., J. R. Donguy, C. Henin and F. Rougerie, 1977, A review of the physical oceanography of the Great Barrier Reef and western Coral Sea. *Australian Institute of Marine Science Monograph Series.* Canberra: Australian Government Publishing Service, 2.

Sackett, W. M., W. R. Eckelmann, M. L. Bender and A. W. H. Be, 1965. Temperature dependence of carbon isotope composition in marine plankton and sediments. *Science* 148, 235-7.

Shackleton, N. J. 1977. Carbon-13 in *Uvigerina:* tropical rainforest history and the equatorial Pacific carbonate dissolution cycles. In N. R. Anderson and A. Malahoff (eds.), *The Fate of Fossil Fuel CO_2 in the Oceans.* New York–London: Plenum Press, pp. 401-27.

Shackleton, N. J. and N. J. Opdyke, 1973. Oxygen isotope and paleomagnetic stratigraphy of equatorial Pacific core V28-238; oxygen isotope temperatures and ice volumes in a 10^5 year and 10^6 year scale. *Quaternary Research* 3, 39-55.

Smith, S. V., R. W. Buddemeier, R. C. Redalje and J. E. Houck, 1979. Strontium-calcium thermometry in coral skeletons. *Science* 204, 404-7.

Tans, P. P., 1978. Carbon 13 and carbon 14 in trees and the atmospheric CO_2 increase. Ph.D. dissertation, University of Groningen, Holland.

Tarutani, T., R. N. Clayton and T. K. Mayeda, 1969. The effect of polymorphism and magnesium substitution on oxygen isotope fractionation between calcium carbonate and water. *Geochimica et Cosmochimica Acta* 33, 987-96.

Weber, J. N., 1977. Use of corals in determining glacial-interglacial changes in temperature and isotopic composition of seawater. *American Association of Petroleum Geologists. Studies in Geology* 4, 289-95.

Weber, J. N. and P. M. J. Woodhead, 1970. Carbon and oxygen isotope fractionation in the skeletal carbonate of reef-building corals. *Chemical Geology* 6, 93-117.

Weber, J. N. and P. M. J. Woodhead, 1972a. Temperature dependence of oxygen-18 concentration in reef coral carbonates. *Journal of Geophysical Research* 77, 463-73.

Weber, J. N. and P. M. J. Woodhead, 1972b. Stable isotopic ratio variations in non-scleractinian coelenterate carbonates as a function of temperature. *Marine Biology* 15, 293-7.

Wilson, A. T. and M. J. Grinsted, 1975. Paleotemperatures from tree rings and the D/H ratio of cellulose as a biochemical thermometer. *Nature* 257, 387-8.

Wilson, A. T. and M. J. Grinsted, 1977. $^{13}C/^{12}C$ in cellolose and lignin as paleothermometers. *Nature* 265, 133-5.

Yonge, C. M., 1936. Mode of life, feeding, digestion and symbiosis with zooxanthellae in the Tridacnidae. *Scientific Report of the Great Barrier Reef Expedition, 1928-29, British Museum (Natural History)* 1, 283-321.

Yonge, C. M., 1975. Giant clams *Scientific American* 232, 96-105.

York, D., 1969. Least squares fitting of a straight line with correlated errors. *Earth and Planetery Science Letters* 5, 320-24.

2 Calcium Carbonate Deposition by Reef Algae: Morphological and Physiological Aspects

Michael A. Borowitzka
School of Environmental and Life Sciences,
Murdoch University, W.A. 6150.

INTRODUCTION

The calcareous algae are an important biological and geological component of tropical coral reefs, contributing significantly to the total biogenically precipitated clacium carbonate (Chave *et al.* 1972; Stearn *et al.* 1977) as well as cementing carbonate deposits of non-algal origin. Calcareous algae are found in almost all reef habitats and most algal phyla have calcareous representatives (table 1; Borowitzka 1977). Algal calcium carbonate occurs as either the aragonitic (orthorhombic) or calcitic (hexagonal-rhombohedral) crystal isomorph of $CaCO_3$. This $CaCO_3$ may contain small amounts of other salts such as $MgCO_3$ and $SrCO_3$. Living plants deposit either one or the other crystal isomorph of $CaCO_3$. Mixtures of isomorphs may, however, be found in dead plants (cf. Alexandersson 1974). Algal calcification has recently been reviewed by Borowitzka (1977) and this paper will concentrate mainly on the major groups of calcareous algae observed in coral reefs. Littler (1976) has reviewed the ecological aspects of calcification in algae.

The calcareous reef algae may be separated into three groups on the basis of the location of their $CaCO_3$ deposits, the degree of organization of these deposits, and the nature of the crystal isomorph precipitated (table 2). There are also differences in the isotopic composition of algal carbonates; aragonitic algae such as *Halimeda* are enriched in ^{13}C and ^{18}O; the stable, naturally occurring isotopes of carbon and oxygen (cf. figure 1 in Borowitzka 1977 for detailed references). Calcite depositing algae have more ^{12}C and less ^{18}O than the aragonitic algae and are enriched with these isotopes with respect to seawater. These observations on the structure and chemistry of the algal $CaCO_3$ suggest that different mechanisms of calcification exist in the different groups outlined in table 2.

MAJOR GROUPS OF CALCAREOUS ALGAE

Coccolithophorids

The coccolithophorids (Pyrrophyta) are unicellular phytoplankters that invest themselves with delicately sculpted calcite plates. The shape and structure of these plates (known as coccoliths, cystoliths, placoliths, etc.) is species specific. In *Emiliania huxleyi* each plate is made up of highly oriented calcite crystals (Watabe 1967). Coccolithophores are

Taxa	Polymorph of $CaCO_3$	Site of deposition	Habitat
Cyanophyta	Aragonite (usually)	Extracellular in mucilage	Estuarine and intertidal
Rhodophyta Nemalionales	Aragonite	Extracellular in intercellular space	Shallow benthic
Cryptonemiales Peysonelliaceae Corallinaceae	Aragonite Calcite	Cell wall (?) Cell wall	Benthic
Pyrrhophyta Peridiniales	Calcite*	Cell wall of cyst	Benthic
Prymnesiophyta Prymnesiales	Calcite	Golgi	Planktonic
Isochrysidales	Calcite (?)	External mucilage	
Phaeophyta *Padina* (Dictyotales)	Aragonite	Extracellular	Benthic
Chlorophyta Dasycladales	Aragonite	Extracellular in intracellular space	Benthic
Caulerpales	Aragonite	Extracellular in intracellular space	Benthic

*Deposited only in the resting cysts

Table 1 The site of deposition of calcite and aragonite in calcareous reef algae (adapted from Borowitzka 1977)

predominantly mid-oceanic species and their quantitative contribution to reef carbonate sediments is very low. They are, however, important palaeoenvironmental indicators.

The calcite is deposited within a Golgi cisterna, usually upon a protein-polysaccharide base plate and within an organic vesicle (see Klaveness & Paasche 1979 for references). Organic material is also incorporated into the $CaCO_3$ crystal matrix (De Jong *et al.* 1976) The exact shape of the coccolith appears to be moulded by this organic material. Once formed the coccolith moves to the anterior end of the cell where it is egested and becomes lodged in the mucilage surrounding the cell. Considering the extreme organization of the $CaCO_3$ in the coccoliths, and the complex steps involved in their synthesis, it is not surprising that this process appears to be under close metabolic control. Not only must all the components be transported into the Golgi vesicle and synthesised but they must also

	Group A	Group B	Group C
Site of CaCO₃ deposit	Deposits formed on and within an organic matrix within Golgi cisternae	Deposits formed in organic matrix of cell wall or in mucilage	Deposits generally completely extra-cellular
Nature of CaCO₃ deposit	Calcite	Calcite	Aragonite
Organization	Extreme	Some	None
Stable isotope composition* $\delta^{13}C$	—	−0.9 (+3.2 to −3.6)	+3.6 (+0.1 to +5.9)
$\delta^{18}O$	—	−3.3 (−0.9 to −6.0)	−1.3 (+0.6 to −4.0)
Where found	Coccolithophores	Coralline red algae Some Cyanophyta	Green algae *Padina* Red algae (Nemalionales) (Peyssonneliaceae) Some Cyanophyta

* Results in ‰ relative to the PDB standard. Based on data from Craig (1953); Lowenstam & Epstein (1957); Gross (1964); Keith & Weber (1965); Gross & Tracey (1966): Devser & Hunt (1969); Milliman (1974). Seawater $\delta^{13}C$ values range from −1 to +2‰ and $\delta^{18}O$ values from 0 to +3‰

Table 2 The marine calcareous algae grouped according to the organization, isotopic composition and site of formation of the CaCO₃ deposits (modified from Borowitzka *et al.* 1974)

be assembled in a stepwise fashion. No model exists for the biochemical processes involved but detailed physiological and biochemical studies have shown that reducing power is required and is generally supplied by photosynthesis, although organic carbon compounds can be used as external energy sources in some conditions (Paasche 1968; Blankley 1971; Dorigan & Wilbur 1973). Production of coccoliths is not essential to the survival of these algae; they can be induced to lose all their coccoliths in acid media; at least one strain of *Emiliania huxleyi* has lost the ability to form coccoliths in culture. Further details on the physiology of coccolithophores may be found in the review of Klaveness & Paasche (1979).

Aragonite Depositing Algae

The calcareous green algae, including *Halimeda, Penicillus* and *Udotea,* the brown alga *Padina* and those red algae belonging to the Family Nemalionales, precipitate aragonite outside their cell walls (table 1). In *Padina* these deposits are wholly extracellular, while in the other genera they are intercellular. The arrangement and location of these deposits has been described by Marszalek (1971), Borowitzka *et al.* (1974), Flajs (1977) and Böhm *et al.* (1978a). Intracellular crystals of calcium oxalate have also been reported in *Acetabularia* (Vinogradov 1953), *Apjohnia* (Dawes 1969), *Penicillus, Rhipocephalus* and *Udotea* (Friedmann *et al.* 1972).

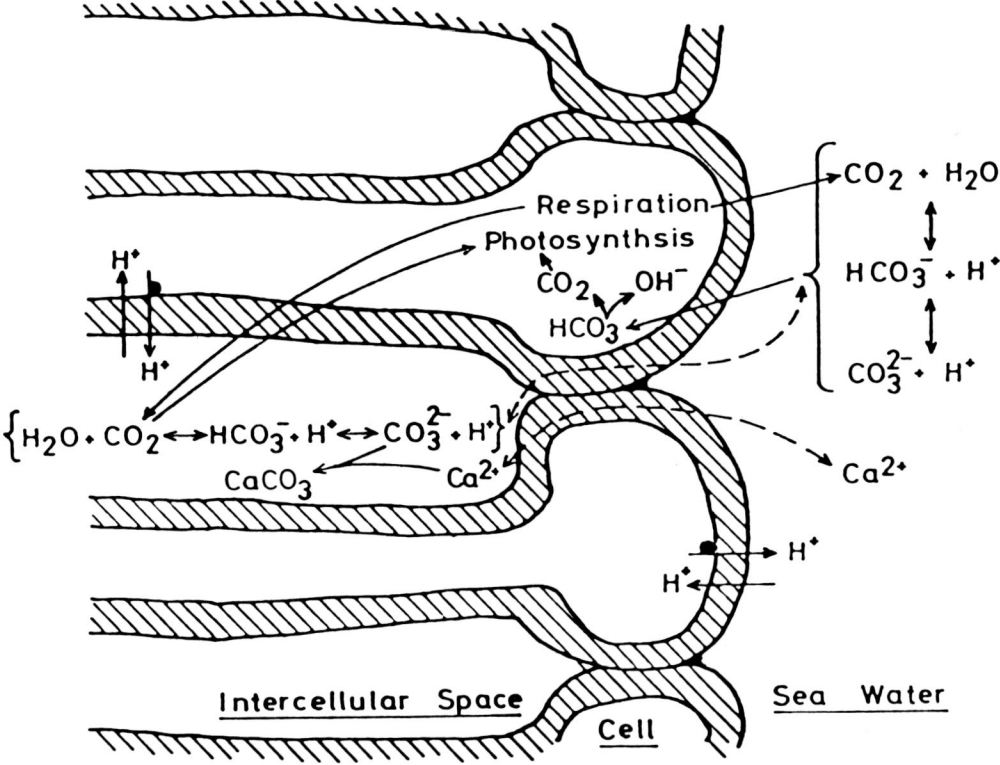

Figure 1   Schematic representation of the postulated ion fluxes affecting $CaCO_3$ precipitation in *Halimeda*. The black dot at the plasmalemma indicates that this flux is postulated to be active. Passage of ions from the sea water to the intercellular space is by diffusion across $20\mu m$ or more of cell wall of the appressed external cells (utricles); CO_2 for photosynthesis enters the cell by diffusion and CO_2 produced during respiration diffuses out of the cell; bicarbonate enters the cell probably by an energy-requiring process. Uptake of CO_2 from the intercellular space leads to a rise in pH and $[CO_3^{2-}]$ resulting in $CaCO_3$ (aragonite) precipitation; respiratory CO_2 evolution has the opposite effect. A light-stimulated proton efflux is also postulated (from Borowitzka & Larkum 1976b,c)

The aragonite needles of green algae, especially those of *Halimeda* species, make up a significant proportion of the fine carbonate sediment of reefs (Maxwell *et al.* 1964; Chave *et al.* 1972). *Halimeda* is the best studied of the calcareous algae. Table 3 lists measured calcification rates of several *Halimeda* species and some other aragonitic algae. There is a wide range of measured calcification rates reflecting the differences in physiology and environmental factors as well as differences in the measurement techniques. It must also be remembered that *Halimeda* tends to grow in spurts; long periods of apparent non-growth are interspersed with periods of rapid addition of new segments (cf. plate 8 in Milliman 1974).

The calcification mechanism of *Halimeda* has been studied by Wilbur *et al.* (1969), Stark *et al.* (1969), Böhm (1972 a,b), Böhm & Goreau (1973) and Borowitzka & Larkum (1972 a,b,c, 1977). Borowitzka & Larkum (1976b) presented a model of the calcification mechanism (fig. 1). This model proposes that the precipitation of the aragonite $CaCO_3$ in the intercellular spaces of the *Halimeda* thallus is dependent on the anatomy of the alga and the process of photosynthetic CO_2 uptake. The intercellular spaces where $CaCO_3$ precipitation occurs are separated from the external seawater by an outer layer of appressed utricles. Inorganic ions, such as HCO_3^-, CO_3^{2-} and Ca^{2+}, and CO_2 must either pass through the cells or diffuse through the cell walls of the outer utricles; a distance of at least 200 μm. The total diffusion path is probably significantly longer due to unstirred layers on the outer surface of the thallus. The intercellular space is therefore a semi-isolated compartment. Removal of CO_2 from this compartment during photosynthesis raises the intercellular pH and CO_3^{2-} concentration (see appendix in Borowitzka & Larkum 1976b), resulting in precipitation of $CaCO_3$ from seawater as aragonite. This model accounts for all current physiological data on *Halimeda* calcification and can also be used to explain aragonite deposition by other marine algae — with the exception of *Padina*. Aragonite deposition always occurs in a semi-isolated part of the thallus outside the cell wall. In *Rhipocephalus phoenix, Penicillus pyriformis, Udotea cyanthiformis* and *U. conglutinata,* deposition occurs in a layer of organic material surrounding the internal cell filaments (Böhm *et al.* 1978a). It appears that this layer acts further to reduce diffusion, probably enhancing the semi-isolation of the interior of the algal thallus and enabling the formation of a region sufficiently alkaline for $CaCO_3$ precipitation. The chemical composition of this organic layer needs to be determined before any further possible function may be assigned to it.

Calcification in *Padina* occurs in bands on the outer surface of the fan-like thallus, (Borowitzka *et al.* 1974) and it appears likely that the mechanism may be similar to that of the freshwater alga *Chara* (Lucas 1979) where OH ions are excreted from the cell in certain regions of the thallus producing a localized pH shift which results in $CaCO_3$ precipitation.

The main problem with the *Halimeda* calcification model is the fact that other algae with similar anatomy do not calcify. It is possible that these non-calcified algae excrete organic compounds such as phosphates or organic acids which inhibit nucleation of $CaCO_3$. Many such 'crystal poisons' have been reported (Simkiss 1964; Degens 1976). Confirmation of this hypothesis requires more detailed comparative examination of the excretory products of calcareous and non-calcareous algae.

Calcite Depositing Algae

The coralline red algae (Corallinaceae, Rhodophyta), especially the crustose species, are probably the most abundant calcareous reef algae. They precipitate calcite, which often contains a high proportion of magnesium (Kolesar 1978), in their cell walls (Borowitzka *et al.* 1974; Flajs 1977). The calcite crystals are intimately associated with the organic cell wall material (Borowitzka & Vesk 1978, 1979). Near the cell, they are oriented at right angles to the cell (Baas-Becking & Galliher 1931; Borowitzka *et al.* 1974), but they become more disoriented and larger further away from the cell.

Coralline algal $CaCO_3$ is enriched in ^{12}C and ^{16}O compared to seawater, suggesting that at least some of the carbonate is derived from metabolic CO_2 rather than seawater inorganic carbon. The isotopic evidence is supported by data of Okazaki *et al.* (1970) who showed that ^{14}C from respired ^{14}C-pyruvate was incorporated into the $CaCO_3$. Pulse-chase experiments with ^{14}C provide additional support for incorporation of metabolically derived CO_2 in $CaCO_3$ (Borowitzka, unpublished results). An attempt at a stable isotope balance sheet, similar to that prepared by Goreau (1977) for the coral *Montastrea annularis*, could provide valuable insight into the coralline algal calcification mechanisms.

Detailed studies of the exchange of calcium between alga and seawater (Böhm 1978; Borowitzka 1979) have shown that there appear to be at least two kinetically distinguishable organic Ca^{2+}-exchanging compartments in coralline algae. Borowitzka (1979) has suggested that these might represent different anion exchange sites on the cellulose and non-cellulose (pectic) cell wall components (cf. Wuytack & Gillett 1978). The role of this organic matrix in calcification remains unknown. The organic material may act as a template for epitaxial nucleation (Degens 1976), or may influence calcification in some other way. Whatever the influence of the organic material, it is likely that it is the main factor causing the precipitation of calcite rather than the aragonite that would be expected from inorganic precipitation of $CaCO_3$ from seawater (Kinsman & Holland 1969; Kitano *et al.* 1969).

Calcification rates of only a few coralline algae have been measured (table 4). As in other calcareous algae, the rate is highest in the light (Goreau 1963; Okazaki *et al.* 1970; Borowitzka 1979) and in the younger parts of the algal thallus (Pearse 1969; Pentecost 1978; Borowitzka 1979). The calcification rate shows a linear relationship to the photosynthetic rate (Pentecost 1978; Borowitzka, unpublished results) and may be inhibited by a range of photosynthetic or metabolic inhibitors (Ikemori 1970). Studies on the effects of seawater pH and inorganic carbon concentration on photosynthesis and calcification in *Bossiella orbigniana* and *Amphiroa* species indicate that there are at least two mechanisms of calcification in these algae (Smith & Roth 1979; Borowitzka 1980): 1) precipitation due to localized changes in carbonate concentration as a result of CO_2 uptake in the light and 2) an active, metabolically controlled carbonate deposition mechanism. The nature of this metabolically controlled mechanism has not yet been determined. It has been suggested that the enzyme carbonic anhydrase, which catalyses the reversible hydration of CO_2:

$$CO_2 + H_2O \leftrightarrow HCO_3^- + H^+$$

stimulates calcification by increasing the supply of inorganic carbon at the calcification site. This seems unlikely since there is no significant difference in carbonic anhydrase activity between calcareous and non-calcareous algae (Okazaki 1972; Graham & Smillie 1976). *Serraticardia maxima* and other Corallinaceae have, however, been shown to have a Ca^{2+}-dependent ATPase, apparently localized on the plazma membrane (Okazaki 1977). This enzyme could not be detected in other calcareous and non-calcareous algae. It is possible that this ATPase uses ATP formed during photosynthesis for active transport of Ca^{2+} to the calcification site. As yet, however, there exist too few data to allow the construction of a testable model for the mechanism of calcification in coralline red algae.

MEASUREMENT OF CALCIFICATION RATES

It is essential to biochemical, physiological or ecological studies of algal calcification to be able to measure the calcification rate accurately and reproducibly. As can be seen from tables 3 and 4, the measured calcification rates show great inter- and intraspecific variation. This variation can be attributed to variations in the environmental conditions during measurement, differences in the physiological state of the algae and differences in the techniques used.

Species	Locality	Calcification rate (nmol g dry weight⁻¹ min⁻¹) Light	Dark	Temp (°C)	Method	Labelling time	Washing time	Refs.
Amphiroa fragillissima	Bermuda	207.8 ± 124.7	—	28	titration	—	—	(2)
		332.5	—	28	^{45}Ca time course	—	—	(2)
A. foliacea	G.B.R.	14–68	6–27	—	^{45}Ca time course	—	—	(4) *
A. foliacca	G.B.R.	20–122	21–120	—	^{14}C time course	—	—	(4) *
Corallina officinalis	Wales	31.9–63.9	—	20	^{14}C	240	"brief"	(3) **
C. officinalis	Wales	20.8–34.7	—	10	^{14}C	240	"brief"	(3) ***
Serraticardia maxima	Japan	213.1	55.6	18–19	^{45}Ca	?	?	(1) ****
Serraticardia maxima	Japan	185.7	90.0	18–19	^{14}C	?	?	(1) ****
Phymatolithon calcareum	Baltic Sea	2.3	—	10	^{45}Ca time course	—	—	(5)

*Range of rates is taken over all parts of thallus

**In the laboratory

***In the field

****Recalculated using %N content figures given by Okazaki (1972)

References: (1) Okazaki et al. (1970); (2) Böhm (1978); (3) Pentecost (1978); (4) Borowitzka (1979); (5) Böhm et al. (1978b)

G.B.R. = Great Barrier Reef

Table 3 Calcification rates of representative calcite-depositing coralline red algae measured by various methods

Species	Locality	Calcification rate (nmol g dry weight⁻¹ min⁻¹) Light	Dark	Temp (°C)	Method	Labelling time (h)	Washing time (min)	Refs.
Halimeda discoidea	Jamaica	55.8 ± 37.1 (4)	57.1 ± 11.3 (4)	29	^{45}Ca	1.5-2.5	120	(1) *
H. discoidea	Puerto Rico	178	145	28	^{45}Ca	12	"brief"	(2) **
H. incrassata	Bermuda	415.6 ± 249.4	—	28	titration	—	—	(5)
H. incrassata	Bermuda	789.7	—	28	^{45}Ca time course	—	—	(5)
H. opuntia	Jamaica	166.2 ± 41.6	—	28	titration	—	—	(3) ***
H. opuntia	Bermuda	207.8	—	28	^{45}Ca time course	—	—	(5)
H. opuntia	Jamaica	111.3 ± 17.8 (5)	157.6 ± 171.5 (8)	29	^{45}Ca	1.5-2.5	120	(1) *
H. opuntia	Puerto Rico	347	145	28	^{45}Ca	12	"brief"	(2) **
H. tuna	Jamaica	94.5 ± 71.4 (10)	109.4 ± 142.6 (10)	29	^{45}Ca	1.5-2.5	120	(1) *
H. tuna	G.B.R.	137.6 (3)	—	28	alkalinity	2	1-2	(4)
H. tuna (mean) (range)	G.B.R.	149.8 ± 26.8 (143) (76.9 - 208.4)	67.7 ± 7.3 (121) (21.5 - 93.9)	28	^{14}C	2	1-2	(4)
H. tuna (mean) (range)	G.B.R.	459.9 ± 26.7 (59) (336.4 - 735.5)	241.9 ± 9.6 (48) (240.2 - 336.4)	26	^{45}Ca	2	1-2	(4)
Cymopolia barbata	Bermuda	332.5 ± 166.2	—	28	^{45}Ca time course	—	—	(5)
		294.4	—	28	titration	—	—	(5)
Penicillus flabelliforms	Bermuda	166.2 ± 124.7	—	28	^{45}Ca time course	—	—	(5)
		706.6	—	28	titration	—	—	(5)
Udotea flabellum	Bermuda	457.2 ± 166.2	—	28	^{45}Ca time course	—	—	(5)
		498.8	—	28	titration	—	—	(5)
Padina sanctae-crucis	Bermuda	498.8 ± 249.4	—	28	^{45}Ca time course	—	—	(5)
		290.9	—	28	titration	—	—	(5)

*Data of Goreau (1963) were recalculated assuming the same dry weight:protein relationship as for *H. tuna* (Borowitzka & Larkum 1974)

**Data taken from the published graphs

***Caculated from Böhm & Goreau (1973) by Böhm (1978)

References: (1) Goreau (1963); (2) Stark *et al.* (1969); (3) Böhm & Goreau (1973); (4) Borowitzka & Larkum (1974); (5) Böhm (1978)

G.B.R. = Great Barrier Reef

Table 4 Calcification rates of representative aragonite-depositing green and brown marine algae measured by various methods

Goreau (1963) pioneered work in this area by introducing the use of radioactive isotopes of calcium (^{45}Ca) and carbon (^{14}C). His method, like that of many subsequent workers, consisted of placing the alga in seawater with radioisotope for a period, then washing the thallus in unlabelled seawater to remove contaminating labelled seawater. The thallus was then dried, the label extracted, and the radioactivity incorporated into the inorganic constituents of the alga determined. Because of technical difficulties with ^{14}C isotope extraction, most workers have used ^{45}Ca. These radioisotope methods assume a constant uptake of radioisotope over time, and that the radioisotope is only incorporated into the $CaCO_3$ by net accretion. However, radioisotope uptake curves for *Halimeda* (Borowitzka & Larkem 1976a), the coralline algae *Serraticardia* (Okazaki *et al.* 1970), *Bossiella* (Pearse 1972) and other algae (Böhm 1978; Borowitzka 1979) are non-linear. Detailed studies on calcium exchange kinetics of *Halimeda opuntia* (Böhm & Goreau 1973), *Halimeda tuna* (Borowitzka & Larkum 1976a) and *Amphiroa foliacea* (Borowitzka 1979), as well as other algae (Böhm 1978), showed that radioisotope was incorporated into the $CaCO_3$ but that, additionally, radioisotope exchanged onto the $CaCO_3$ crystals and exchanged with calcium bound to organic cell wall constituents.

This single time-point method using radioisotopes is therefore unsatisfactory because of isotopic exchange on the $CaCO_3$ crystal surface and other exchange phenomena which lead to an overestimation of the true accretion rate. Recently Böhm (1978) and Borowitzka (1979) developed a kinetic method for measuring algal calcification rates using either ^{45}Ca or ^{14}C. In this method, the algae are first allowed to reach a steady state between uptake and efflux. Once all compartments have reached isotopic equilibrium, it is assumed that net uptake of isotope can only occur by accretion of $CaCO_3$. The method consists of measuring the time-course of isotope uptake and then calculating the 'true' calcification rate from the slope of the linear portion of the uptake graph. This method is quite sensitive although it requires more plant material (Böhm 1978; Borowitzka 1979).

Radioisotope methods are by far the most sensitive in studies of calcification and permit measurements on small parts of the plant, an important factor in biochemical and physiological experiments.

However, other methods are also available. Most of these require larger quantities of alga. Böhm (1978) has measured calcification by determining the amount of calcium remaining in the medium after incubation. This method is tedious at present, but with the advent of more sensitive calcium-ion-sensitive electrodes, which are less influenced by high concentrations of Na^+ or other ions, it may prove to be the easiest and most accurate. Another chemical method has been used by Borowitzka & Larkum (1976a). This method measures the change in total alkalinity resulting from CO_3^{2-} removal from the seawater during $CaCO_3$ precipitation. The method, based on that of Kinsey (1972) and Smith (1973), makes a number of assumptions, the major one being that there is little or no delay between uptake of CO_2 species and release of balancing ions. Furthermore, it is assumed that changes in pH and total alkalinity caused by nutrient fluxes (Brewer & Goldman 1976) are negligible. Calculations by Kinsey (1978) suggest this to be the case.

Other gross methods of measuring calcification (growth) rates are also available. Some idea of calcification rate can be gained from measurements of increases in thallus size (Johansen & Austin 1970), area (Adey 1970; Adey & Vassar 1975), thickness (Digby 1977) or weight (Bak 1976; Stearn *et al.* 1977). Gross measurements of accretion rate should be possible using the dyes, alizarin red and blue which have been used to measure coral growth rates (Barnes 1973).

In recent years there have been great advances in our understanding of algal calcification and, with the development of accurate methods for measuring algal calcification rates, it will be possible to examine further the algal calcification mechanisms and to understand better the role of the calcareous algae in the reef system.

LITERATURE CITED

Adey, W. H., 1970. The effects of light and temperature on growth rates in boreal-subarctic crustose corallines. *Journal of Phycology* 6, 269–76.

Adey, W. H. and J. M. Vassar, 1975. Colonization, succession and growth rates of tropical crustose coralline algae (Rhodophyta, Cryptonemiales). *Phycologia* 14, 55–69.

Alexandersson, T., 1974. Carbonate cementation in coralline algal nodules in the Skagerrak, North Sea: biochemical precipitation in undersaturated waters. *Journal of Sedimentary Petrology* 44, 7–26.

Baas-Becking, L. G. M. and E. W. Galliher, 1931. Wall structure and mineralization in coralline algae. *Journal of Physical Chemistry* 35, 467–79.

Bak, R. P. M., 1976. The growth of coral colonies and the importance of crustose coralline algae and burrowing sponges in relation with carbonate accumulation. *Netherlands Journal of Sea Research* 10, 285–337.

Barnes, D. J., 1973. Growth in colonial scleractinians. *Bulletin of Marine Science* 23, 280–98.

Blankley, W. F., 1971. Autotrophic and heterotrophic growth and calcification in coccolithophorids. Ph.D. dissertation, University of California, San Diego.

Böhm, E. L., 1972a. Cation-anion balance in some calcium carbonate depositing algae and the detection of organic calcium fractions in the calcareous alga *Halimeda opuntia* (Chlorophyta, Udoteaceae). *Internationale Revue der Gesellschaft für Hydrobiologie* 57, 685–93.

Böhm, E. L., 1972b. Composition and binding properties of some water soluble polysaccharides in the calcareous alga *Halimeda opuntia* (L.) (Chlorophyceae, Udoteaceae). *Internationale Revue der Gesellschaft für Hydrobiologie* 58, 117–26.

Böhm, L., 1978. Application of the ^{45}Ca tracer method for determination of calcification rates in calcareous algae: effects of calcium exchange and differential saturation of algal calcium pools. *Marine Biology* 47, 9–14.

Böhm, E. L. and T. F. Goreau, 1973. Rates of turnover and net accretion of calcium and the role of calcium binding polysaccharides during calcification in the calcareous alga *Halimeda opuntia (L.)*. *Internationale Revue der Gesellschaft für Hydrobiologie* 58, 723–40.

Böhm, L., D. Fütterer and E. Kaminski, 1978a. Algal calcification is some Codiaceae (Chlorophyta): ultrastructure and location of the skeletal deposits. *Journal of Phycology* 14, 486–93.

Böhm, L., W. Schramm and V. Rabsch, 1978b. Ecological and physiological aspects of some coralline algae from the Western Baltic. Calcium uptake and skeleton formation in *Phymatolithon calcareum*. *Kielermeeresforschungen* 4, 282–8.

Borowitzka, M. A., 1977. Algal Calcification. *Oceanography and Marine Biology an Annual Review* 15, 189–223.

Borowitzka, M. A., 1979. Calcium exchange and the measurement of calcification rates in the calcareous coralline red alga *Amphiroa foliacea*. *Marine Biology* 50, 339–47.

Borowitzka, M. A. and A. W. D. Larkum, 1976a. Calcification in the green alga *Halimeda*. II. The exchange of Ca^{2+} and the occurrence of age gradients in calcification and photosynthesis. *Journal of Experimental Botany* 27, 864–78.

Borowitzka, M. A. and A. W. D. Larkum, 1976b. Calcification in the green alga *Halimeda*. III. The source of inorganic carbon for photosynthesis and calcification and a model of the mechanism of calcification. *Journal of Experimental Botany* 27, 879–93.

Borowitzka, M. A. and A. W. D. Larkum, 1976c. Calcification in the green alga *Halimeda*. IV. The action of metabolic inhibitors on photosynthesis and calcification. *Journal of Experimental Botany* 27, 894–907.

Borowitzka, M. A. and A. W. D. Larkum, 1977. Calcification in the green alga *Halimeda*. I. An ultrastructure study of thallus development. *Journal of Phycology* 13, 6-16.

Borowitzka, M. A., A. W. D. Larkum and L. J. Borowitzka, 1978. A preliminary study of algal turf communities of a shallow coral reef lagoon using an artificial substratum. *Aquatic Botany* 5, 365-81.

Borowitzka, M. A., A. W. D. Larkum and C. E. Nockolds, 1974. A scanning electron microscope study of the structure and organization of the calcium carbonate deposits of algae. *Phycologia* 13, 195-203.

Borowitzka, M. A. and M. Vesk, 1978. Ultrastructure of the Corallinaceae. I. *The Vegetative cells of Corallina officinalis* and *C. cuvierii*. *Marine Biology* 46, 295-304.

Borowitzka, M. A. and M. Vesk, 1979. Ultrastructure of the Corallinaceae (Rhodophyta). II. Vegetative cells of *Lithothrix aspergillum*. *Journal of Phycology* 15, 146-53.

Brewer, P. G. and J. C. Goldman, 1976. Alkalinity changes generated by phytoplankton growth. *Limnology and Oceanography* 21, 108-17.

Chave, K. E., S. V. Smith and K. J. Roy, 1972. Carbonate production by coral reefs. *Marine Geology* 12, 123-40.

Craig, H., 1953. The geochemistry of stable carbon isotopes. *Geochimica et Cosmochimica Acta* 3, 53-92.

Dawes, C. J., 1969. A study of the ultrastructure of a green alga *Apjohnia laetivirens* Harvey with emphasis on cell wall structure. *Phycologia* 8, 77-84.

Degens, E. T., 1976. Molecular mechanisms on carbonate, phosphate and silica deposition in the living cell. *Topics in Current Chemistry* 64, 1-112.

De Jong, E. W., B. Lendert and P. Westbrock, 1976. Isolation and characterization of a Ca^{2+}-binding polysaccharide associated with coccoliths of *Emiliania huxleyi* (Lohmann) Kamptner. *European Journal of Biochemistry* 70, 611-21.

Devser, W. G. and J. M. Hunt, 1969. Stable isotope ratios of dissolved inorganic carbon in the Atlantic. *Deep Sea Research* 16, 221-5.

Digby, P. S. B., 1977. Growth and calcification in the coralline algae, *Clathromorphum circumscriptum* and *Corallina officinalis*, and the significance of pH in relation to precipitation. *Journal of the Marine Biological Association of the United Kingdom* 57, 1095-109.

Dorigan, J. L. and K. M. Wilbur, 1973. Calcification and its inhibition in coccolithophorids. *Journal of Phycology* 9, 450-6.

Flajs, G., 1977. Die Ultrastrukturen des Kalkalgenskeletts. *Palaeontographica* 160B, 69-128.

Friedmann, E. I., W. C. Roth, J. B. Turner and R. S. McEwen, 1972. Calcium oxalate crystals in the aragonite-producing green alga *Penicillus* and related genera. *Science* 177, 891-3.

Goreau, T. F., 1963. Calcium carbonate deposition by coralline algae and corals in relation to their roles as reef-builders. *Annals of the New York Academy of Sciences* 109, 127-67.

Goreau, T. J., 1977. Coral skeletal chemistry: physiological and environmental regulation of stable isotopes and trace metals in *Montastrea annularis*. *Proceedings of the Royal Society of London, Series B* 196, 291-315.

Graham, D. and R. M. Smillie, 1976. Carbonate dehydratase in marine organisms of the Great Barrier Reef. *Australian Journal of Plant Physiology* 3, 113-19.

Gross, M. G., 1964. Variations in the O^{18}/O^{16} and C^{13}/C^{12} ratios in diagenetically altered limestones in the Bermuda Islands. *Journal of Geology* 72, 170-94.

Gross, M. G. and I. J. Tracey, 1966. Oxygen and carbon isotopic composition of limestones and dolomites, Bikini and Eniwetok Atolls. *Science* 151, 1082-4.

Ikemori, M., 1970. Relation of calcium uptake to photosynthetic activity as a factor controlling calcification in marine algae. *Botanical Magazine* 83, 152-62.

Johansen, H. W. and L. F. Austin, 1970. Growth rates in the articulated coralline *Calliarthron* (Rhodophyta). *Canadian Journal of Botany* 48, 125-32.

Keith, M. L. and J. N. Weber, 1965. Systematic relationships between carbon and oxygen isotopes in carbonates deposited by modern corals and algae. *Science* 150, 498–501.

Kinsey, D. W., 1972. Preliminary observations on community metabolism and primary productivity of the pseudo-atoll reef at One Tree Island, Great Barrier Reef. In C. Mukundan and C. Gopinadha Pillai (eds.), *Proceedings of the Symposium on Corals and Coral Reefs.* Cochin: The Marine Biological Association of India, pp. 13–32.

Kinsey, D. W., 1978. Alkalinity changes and coral reef calcification. *Limnology and Oceanography* 23, 989–91.

Kinsman, D. J. J. and H. D. Holland, 1969. The co-precipitation of cations with $CaCO_3$. IV. The co-precipitation of Sr^{2+} with aragonite between 16^0 and 96^0 C. *Geochimica et Cosmochimica Acta* 33, 1–17.

Katano, Y., N. Kanamori and A. Tokuyama, 1969. Effects of organic matter on solubilities and crystal form of carbonates. *American Zoologist* 9, 681–8.

Klaveness, D. and E. Paasche, 1979. Physiology of coccolithophorids. In M. Levandowsky and S. H. Hutner (eds.), *Biochemistry and Physiology of Protozoa.* New York: Academic Press, vol 1, 2nd ed., pp. 191–213.

Kolesar, P. T., 1978. Magnesium in calcite from a coralline alga. *Journal of Sedimentary Petrology* 48, 815–20.

Littler, M. M., 1976. Calcification and its role among the macroalgae. *Micronesica* 12, 27–41.

Lowenstam, H. A. and S. Epstein, 1957. On the origin of sedimentary aragonite needles of the Great Bahama Bank. *Journal of Geology* 65, 364–75.

Lucas, W. J., 1979. Alkaline band formation in *Chara corallina.* due to OH^- efflux of $H+$ influx. *Plant Physiology* 63, 247–54.

Marszalek, D. S., 1971. Skeletal ultrastructure of sediment producing green algae. In O. Johari and T. Corvin (eds.), *Scanning Electron Microscopy, Part 1.* Chicago: Scanning Electron Microscopy, Inc., pp. 273–280.

Maxwell, W. G. H., J. S. Jell and R. G. McKellar, 1964. Differentiation of carbonate sediments on the Heron Island reef. *Journal of Sedimentary Petrology* 34, 194–308.

Milliman, J. D., 1974. Marine Carbonates. Springer-Verlag, Berlin, 375pp.

Okazaki, M., 1972a . Comparison of SO_3 and ash contents in calcareous and non-calcareous algae. *Bulletin Tokyo Gakugei University, Series IV, Mathematical and Natural Sciences* 24, 220–5.

Okazaki, M., 1972b. Carbonic anhydrase in the calcareous red alga, *Serraticardia maxima.* *Botanica Marina* 15, 133–38.

Okazaki, M., T. Ikawa, K. Furuya, K. Nisizawa and T. Miwa, 1970. Studies on calcium carbonate deposition of a calcareous red alga *Serraticardia maxima.* *Botanical Magazine* 83, 193–201.

Paasche, E., 1964. A tracer study of the inorganic carbon uptake during coccolith formation and photosynthesis in the coccolithophorid *Coccolithus huxleyi.* *Physiologia Plantarum,* 17, Supplement 3, 5–82.

Pearse, V. B., 1972. Radioisotopic study of calcification in the articulated corralline alga *Bossiella orbigniana.* *Journal of Phycology* 8, 88–97.

Pentecost, A., 1978. Calcification and photosynthesis in *Corallina officinalis* L. using the $^{14}CO_2$ method. *British Phycological Journal* 13, 383–90.

Simkiss, K., 1964. Phosphates as crystal poisons of calcification. *Biological Reviews of the Cambridge Philosophical Society* 39, 487–505.

Smith, A. D. and A. A. Roth, 1979. Effect of carbon dioxide concentration on calcification in the red coralline alga *Bossiella orbigniana.* *Marine Biology* 52, 217–25.

Smith, S. V., 1973. Carbon dioxide dynamics: a record of organic carbon production, respiration, and calcification in the Eniwetok reef flat community. *Limnology and Oceanography* 18, 106–20.

Stark, L. M., L. Almodovar and R. W. Krauss, 1969. Factors affecting the rate of calcification in *Halimeda opuntia* (L.) Lamouroux and *Halimeda discoidea* Decaisne. *Journal of Phycology* 5, 305-12.

Stearn, C. W., T. P. Scoffin and W. Martindale, 1977. Calcium carbonate budget of a fringing reef on the west coast of Barbados. Part I-Zonation and productivity. *Bulletin of Marine Science* 27, 479-510.

Vinogradov, A. P., 1953. The elementary chemical composition of marine organisms. *Sears Foundation for Marine Research Memoir* 2, 647.

Watabe, N., 1967. Crystallographic analysis of the coccolith of *Coccolithus huxleyi*. *Calcified Tissue Research* 1, 114-21.

Wilbur, K. M., Colinvaux, L. H. and N. Watabe, 1969. Electron microscope study of calcification in the alga *Halimeda* (Order Siphonales). *Phycologia* 8, 27-35.

Wuytack, R. and C. Gillet, 1978. Nature des liaisons de l'ion calcium dans la paroi de *Nitella flexilis*. *Canadian Journal of Botany* 56, 1439-43.

3 Calcification by Corals and Other Animals on the Reef

Bruce E. Chalker *
Department of Zoology-Entomology, Auburn University,
Auburn, Alabama, 36830, U.S.A.
* Present address: Australian Institute of Marine Science, P.M.B. No. 3, M.S.O.,
Townsville, Qld. 4810.

INTRODUCTION

The physiology of growth and skeletal deposition by scleractinian corals has been the subject of experimentation and speculation for more than one hundred years. Investigators in the late nineteenth and early twentieth centuries noted that almost all reef-building (hermatypic) corals contain large populations of an endosymbiotic alga variously identified as the dinoflagellate *Symbiodinium microadriaticum* (Freudenthal 1962), *Gymnodinium microadriaticum* (Taylor 1971a) and *Zooxanthella microadriatica* (Loeblich & Sherley 1979). They speculated that the alga might contribute to the metabolism and growth of coral. These efforts were expanded and reviewed by C. M. Yonge following his participation in the Great Barrier Reef Expedition of 1928-1929 (Yonge 1930, 1931, 1940, 1944, 1957, 1958, 1963, 1968, 1973; Yonge & Nicholls 1931a,b; Yonge *et al.* 1932) and by the Japanese at Palau. It is now known that approximately one half of all the carbon fixed by the endosymbiotic algae is translocated to the animal tissues where it is metabolized. For reviews of this topic see Smith *et al.* (1969), Goreau *et al.* (1971), Muscatine (1971, 1973, 1974), Taylor (1971b, 1973a, b), Porter (1976), and Muscantine & Porter (1977).

It is also known that corals which contain endosymbiotic algae calcify many times faster in the light than in the dark. This was first demonstrated chemically by Kawaguti & Sakumoto (1948) and later substantiated in studies with radioisotopes (Goreau 1959; Goreau & Goreau 1959; Vandermeulen *et al.* 1972. More than twenty years after the pioneering radioisotopic investigations of Goreau (1959; Goreau & Goreau 1959, 1960a,b) the biochemical mechanisms of calcification and of light-enhanced calcification are still unknown.

This article reviews current knowledge about coral calcification. Of necessity it is restricted in scope, dealing only with the central biochemical questions and with short-term variations in rates. Several important areas have been omitted entirely. Ecological omissions include studies of the general metabolic significance of organic carbon excreted by endosymbiotic algae and all studies of growth and form. Geological omissions include the mineralogy, chemical composition, and stable isotopic composition of coral skeletons and the use of coral skeletons as paleoenvironmental indicators. Some of these topics are discussed elsewhere in this volume.

MEASUREMENT OF CALCIFICATION

Calcification by scleractinian corals has been measured by a variety of methods including long-term growth studies (Shinn 1966; Lewis *et al.* 1968; Barnes 1972; Bak 1973), X-ray densitometry (Buddemeier *et al.* 1974; Dodge & Thompson 1974; Baker & Weber 1975; Weber *et al.* 1975; Hudson *et al.* 1976 Highsmith 1979), pH and alkalinity techniques (Smith 1973; Smith & Key, 1975), and radioisotopes (Goreau 1959; Goreau & Goreau 1959; Clausen & Roth 1975; Barnes & Crossland 1977; Crossland & Barnes 1977). The application of many of these methods has been reviewed by Buddemeier & Kinzie (1976) and by Stoddart & Johannes (1978). These reviews are comprehensive, and the traditional methods for the measurement of coral calcification will not be discussed further.

With the exception of many long-term growth studies, which are not suitable for biochemical investigations, these methods are destructive of either the medium or of the corals themselves. The calcium ion-selective electrode would appear to be an ideal alternative method to be used for the continuous, non-destructive measurement of calcium in seawater. A liquid membrane calcium ion-selective electrode has been developed by Ross (1967) and has been used to measure calcium activity in seawater (Thompson & Ross 1966). It is now manufactured commercially by Orion. Unfortunately liquid membrane ion-selective electrodes are inherently subject to drift and noise. In seawater, they have a limited useful life due to ion exchange with sodium ions. Preliminary calculations indicate that a successful electrode system must be sufficiently stable to detect changes in calcium ion activity of approximately 0.5% h^{-1}. Commercially available electrodes are barely adequate. Fortunately, electrode technology is advancing rapidly. Recent developments include improved ion-exchange resins (Ruzicka *et al.* 1973; Brown *et al.* 1976), matrix membrane ion-selective electrodes in which the ion-exchanger is cast in collodion (Schultz *et al.* 1968) or in PVC (Moody *et al.* 1970; Craggs *et al.*1974), coated -wire electrodes (Cattrall & Freiser 1971), the 'Selectrode' (Ruzicka *et al.* 1973, which is now manufactured commercially by Radiometer, and the use of synthetic electrically neutral ion carriers in electrode construction (Ammann *et al* 1976; Morf & Simon 1978). In the near future calcium ion-selective electrodes may well replace radioisotopes in many laboratory and field situations.

All of the methods mentioned above have specific advantages and specific limitations. All will continue to be used successfully in specific situations.

CALCIFICATION MODELS

Several models have been proposed to explain the biochemical mechanism of calcification in general and of light-enhanced calcification in particular. For reviews, see Muscatine (1971) and Vandermeulen & Muscatine (1974). These hypotheses are divided into two groups: those that emphasize the removal of unwanted substances from the sites skeletogenesis and those that emphasize the translocation of reduced carbon compounds from the algae to the sites of skeletogenesis. In the removal models, the interfering substances are usually the end products of proposed physical-chemical precipitation reactions. These end products are assimilated by actively photosynthesizing endosymbiotic algae. Substances that may be removed according to these hypotheses include:

1. carbon dioxide (Goreau 1961a,b, 1963),
2. general metabolic wastes (Yonge 1940, 1968),
3. phosphate which may act as a crystal poison (Simkiss 1964a,b,c) and
4. ammonium ion (Campbell & Speeg 1968, 1969; Crossland & Barnes 1974).

In the translocation models, reduced organic carbon produced by algal photosynthesis is used as substrate for the formation of an organic matrix upon which crystal growth occurs (Goreau 1961a; Wainwright 1963; Pearse & Muscatine 1971; Degens 1976). The incorporation of organic material into coral skeletons has been most extensively investigated by Young (1971, 1973; Young *et al.* 1971). Johnston (1977, 1979) has demonstrated that this organic material has an organized structure that is consistent with its function as a site for the initiation of the crystalization of aragonite.

The issue is complicated by the fact that the best evidence for an organized matrix comes from studies of the Pacific finger coral *Pocillopora damicornis*. In this species, 85% of all the amino compounds in the matrix is glucosamine from chitin (Young 1971). Many coral species that are known to possess endosymbiotic algae and show light-enhanced calcification do not have chitin (Young 1971). It is unclear to what extent generalizations based upon a regular chitinous structure in the skeletons of *Pocillopora damicornis* can be extrapolated to proteins found in the skeletons of other species.

Vandermeulen & Muscatine (1974) argue that the matrix is extremely important in light-enhanced calcification. Barnes (1971) concludes the opposite, that the organic material recovered from coral skeletons is trapped during skeletogenesis and is not necessary for aragonite formation. Towe (1972) emphasizes that the skeleton itself is the best template for the initiation of further crystal growth during skeletogenesis, and that an organic matrix is not necessary for continuing skeletogenesis. Investigations of matrix formation and of the functional significance of the organic materials found in coral skeletons will surely continue. Priority must be given to the question: 'is matrix formation rate limiting in light-enhanced calcification?' It is also possible that algal photosynthate may be metabolized by the coral and used not only as a carbon source but also as an energy source either for matrix formation or for the active transport of calcium ions during skeletogenesis. This possibility is suggested by Chapman (1974) and by Chalker & Taylor (1975).

A closely related topic is the mechanism by which skeletal components move through the cytoplasm to the sites of skeletogenesis. In a classic review article, Lehninger (1970) noted that mitochondria have an exceptionally high affinity for calcium and speculated that they may participate in calcification. He further speculated that calcium crystals may be organized in the mitochondria and then transported through the cytoplasm to the skeleton. Lehninger's article has stimulated a successful search for intracellular crystals in many invertebrates (Simkiss 1976a). Crystalline areas within the secretory cells located in the calicoblastic epidermis (i.e., coral epithelium located next to the skeleton) of scleractinians are well known. These structures have occasionally been described as aragonitic calcium carbonate crystals (Kawaguti & Sato 1968; Goreau & Hays 1977; Hays & Goreau 1977). In every case where X-ray crystallography has been attempted, it is demonstrated that these structures are not crystals. In every case where electron microprobe analysis has been attempted, it is demonstrated that these structures are relatively rich in calcium, but that the concentration of calcium is far less than that of a carbonate. Therefore, I must conclude that there is, at present, no evidence for the intracellular formation of aragonitic crystals in scleractinians.

If calcium is not transported as a crystal, four alternative pathways are possible (Simkiss 1974, 1976b):
1. a polarized pumping membrane,
2. pinocytotic vesicles,
3. a calcium binding protein (CaBP) as a calcium carrying system, and
4. intercellular calcium transport.

The suggestion by Weber (1973) that calcium and strontium diffuse through living cells to the skeleton can be rejected for both biological and kinetic reasons (Chalker 1976). The exact free (ionized) intracellular calcium concentration in scleractinians is unknown. The experiments of Goreau & Bowen (1955) demonstrate that it is less than seawater, which is approximately 10 millimolar (Goldberg 1963) of which 91% exists in the ionized form (Garrels & Thompson 1962). The true intracellular calcium ion concentration for scleractinians is probably very similar to that of other marine invertebrates or approximately 0.3 micromolar (Brinley 1973). If this is true, corals are capable of transporting massive amounts of calcium for skeleton formation and are capable of simultaneously maintaining an intracellular calcium concentration that is 30 000 times lower than that of seawater. Scleractinians should be excellent test organisms for the study of not only skeletogenesis but also calcium ion regulation.

Recent mechanistic studies of scleractinian skeletogenesis have concentrated on the transport of divalent cations. Aragonite, obviously, is composed not only of calcium, strontium, and a little magnesium, but also of carbonate ions. Goreau (1961a) suggested that the supply of carbonate ions may be rate-limiting in calcification. Pearse (1970, 1971) demonstrated that some metabolic carbon dioxide is incorporated into coral skeleton. The remaining carbonate ions are presumed to come from seawater, but the mechanism of transport is totally unknown. A study of bicarbonate transport is scleractinians is very much needed.

LIGHT-SATURATION CURVES

Since light is a dominant environmental factor affecting the rate of coral calcification, quantitative models of coral growth must begin with a simulation of light-saturation curves for photosynthesis. Light-saturation curves plot the rate of photosynthesis *v.* irradiance. All such curves have a characteristic shape. Initially the rate of gross photosynthesis (P) is directly proportional to irradiance (I). Thereafter, the rate of photosynthesis approaches a horizontal asymptote which is the photosynthetic capacity (Pm). The irradiance at which the initial slope of the line intercepts the horizontal asymptote is I_k (Talling 1957). I_k is interpreted as a measure of the adaptation of a plant to its light regime (Steemann Nielsen 1975). Gross photosynthesis (P) is customarily defined as the sum of net photosynthesis (Pn) and the dark rate of respiration (R). Several functions have been used to simulate photosynthesis curves (Jassby & Platt 1976). Three have been applied to photosynthesis by corals or by their isolated endosymbiotic algae. These are the right rectangular hyperbola:

$$P = Pm(I/I_k) / (1 + I/I_k) \qquad (1)$$

(Wethy & Porter 1976a, b), a simple exponential function,

$$P = Pm(1 - \exp(-I/I_k)) \qquad (2)$$

(Graus & Macintyre 1976; Graus 1977), and the hyperbolic tangent function,

$$P = Pm \tanh(I/I_k) \qquad (3)$$

(Chalker & Taylor 1978). For light-saturation curves constructed of measurements taken simultaneously, equation (3) yields the best fit to the data (Chalker 1980).

Functions similar to equations (1)-(3) can be applied directly to simulate light-saturation curves for calcification:

$$C = Cm(I/I_k)/(1 + I/I_k) \qquad (4)$$
$$C = Cm(1 - \exp(-I/I_k)) \qquad (5)$$
$$C = Cm \tanh(I/I_k) \qquad (6)$$

where C is the rate of light-enhanced calcification, and Cm is the maximal possible rate of calcification. A form similar to equation (5) has been used by Graus & Macintyre (1976) and by Graus (1977). The comparative utility of equations (4)-(6) has not yet been evaluated. Since light-enhanced calcification appears to be directly proportional to photosynthesis, it is likely that equation (6) will prove to be the most useful.

All of the equations discussed thus far ignore the possibility of photoinhibition. Photoinhibition of photosynthesis and calcification is commonly observed in transplant experiments in which deep-water corals are transported to shallower depths (Barnes & Taylor 1973). A similar phenomenon could occur when corals that normally reside in turbid water experience a sudden influx of clear water. Figure 1 is a light-saturation curve for calcification by the Atlantic staghorn coral *Acropora cervicornis* collected from a depth of 17 m. At local solar noon, the ambient irradiance at 17 m depth is 390 μ Einsteins m^{-2} sec^{-1}. Photoinhibition is evident after three hours only at an irradiance of 1950 μEinsteins m^{-2} sec^{-1}. This is five times greater than the coral has ever experienced in the field. Thus, it seems likely that corals that grow at moderate depths in clear waters are never photoinhibited. The possibility of photoinhibition in corals which grow in very shallow depths and in waters of variable turbidity has not yet been adequately investigated.

DAILY VARIATION

Several investigators have studied daily changes in photosynthesis by corals *in situ* on the reef and in flowing sea water in aquaria under solar irradiance. These studies have utilized both polarographic oxygen electrodes (Wells *et al.* 1973; Wells 1977) and alkalinity techniques (Smith & Kinsey 1978). In all cases the patterns are similar. Photosynthesis increases with increasing solar irradiance until a saturating irradiance is reached. For four or more hours about local solar noon, photosynthesis is nearly constant. Since calcification appears to be directly proportional to photosynthesis, it is expected that daily patterns of calcification will be similar. Figure 2 shows daily variations in the rate of light-enhanced calcification by *Acropora cervicornis*. The coral was collected from a depth of 17 m and incubated at the surface with ambient irradiance reduced to that recorded at 17 m by inserting neutral density filters. Calcification was measured by the incorporation of the radioisotope calcium-45. The vertical bars are the means for five individuals. The curve is a mathematical simulation of these data. It illustrates the general trend which is commonly observed in this type of measurement. Barnes & Crossland (1978) examined daily variations in carbon fixation and in the incorporation of ^{14}C-bicarbonate into the skeletons of the Pacific staghorn coral *Acropora acuminata*. In general they observe a pattern similar to that described above except that the rates of skeletogenesis are somewhat higher in the afternoon than in the morning.

It is now known that these daily variations are due not only to daily changes in solar irradiance but also to daily changes in the efficiency of algal photosynthesis. Isolated, axenic cultures of *Gymnodinium microadriaticum* show an endogenous circadian rhythm in photosynthetic efficiency at both saturating and sub-saturating irradiances. When the

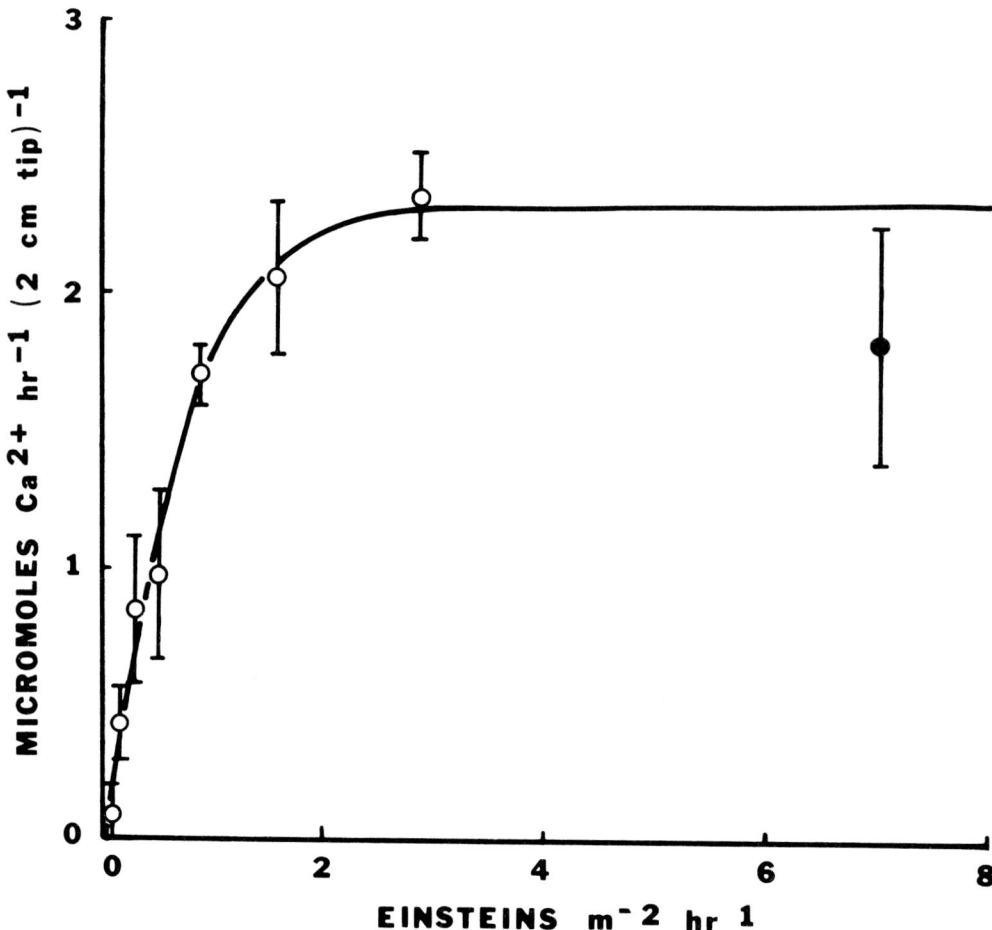

Figure 1   Light-saturation curve for light-enhanced calcification by 2 cm tips of *Acropora cervicornis*, incubated with solar irradiance for two hours about local solar noon. Mean ± one standard error for four individuals. The data are fitted to the function, $C = Cm \tanh (I/I_k)$. Cm equals 2.34 μmoles Ca^{2+} h^{-1}. 2 cm tip $^{-1}$. I_k equals 1.01 Einsteins m^{-2} h^{-1}. The filled circle Φ represents corals that are presumed to be photoinhibited. These data were not used in curve fitting

algae are grown on alternating light and dark periods twelve hours long, the greatest potential efficiency occurs approximately one hour before the start of the light period and one hour after the end of the light period. A similar rhythm is observed in photosynthesis by the intact coral branches and in calcification (Chalker 1977; Chalker & Taylor, 1978). With solar irradiance, corals are relatively efficient at photosynthesis and calcification during the early morning hours and late in the afternoon. On the reef these rhythms appear to function as an effective homeostatic mechanism providing a relatively constant flow of photosynthetic carbon to the coral during most of the day. Concomitantly, light-enhanced calcification remains at a relatively constant level through much of the day.

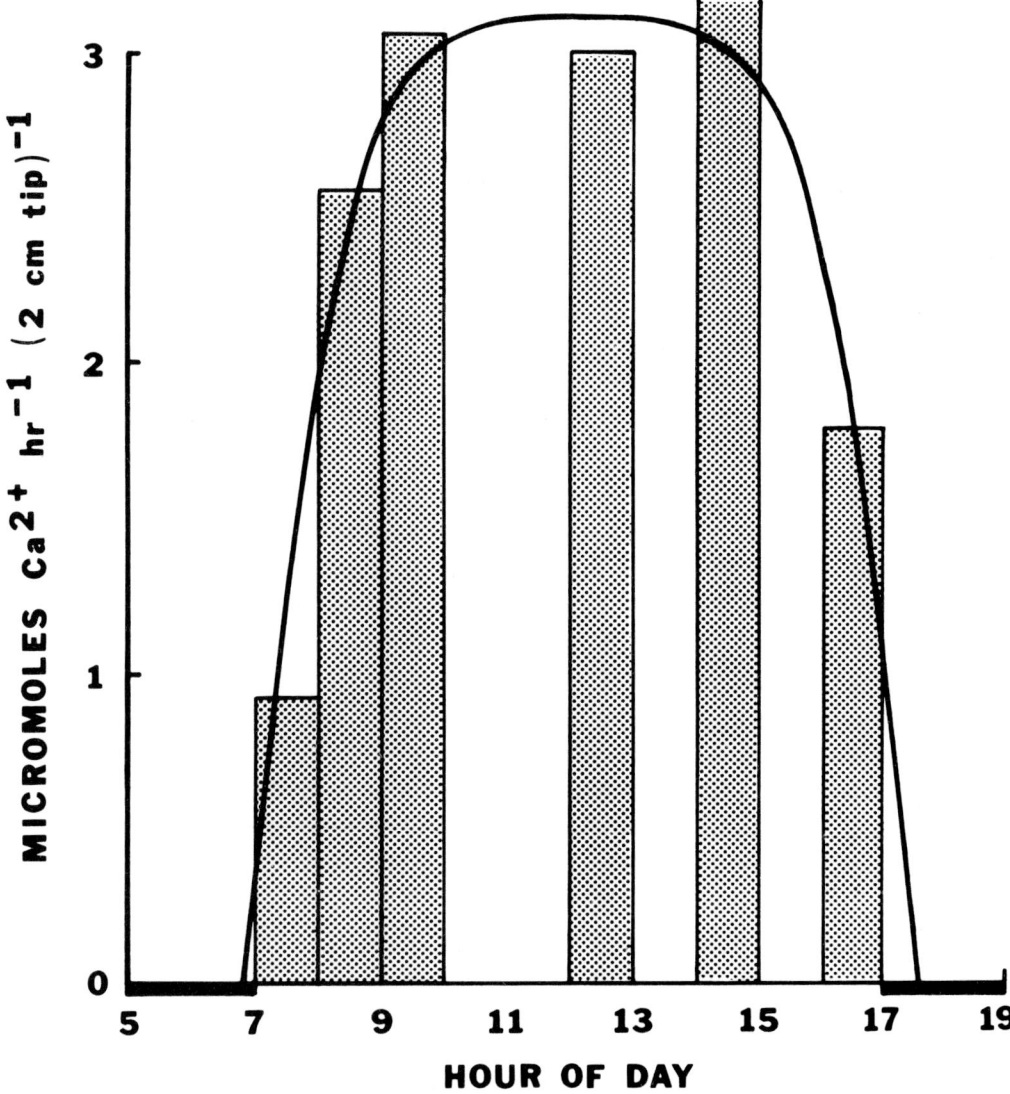

Figure 2   Daily variation in light-enhanced calcification by 2 cm tips of *Acropora cervicornis*, incubated with solar irradiance reduced to the level of 17 m with neutral density screens (bar graphs). The standard error is ± 9% of the mean for five individuals. The curve which is superimposed is an estimate of the instantaneous rate of calcification. The maximum value of Cm observed during this day is 4.5 μmoles Ca^{2+} h^{-1} 2 cm tip^{-1}, and the minimum value of Cm is 3.4 μmoles Ca^{2+} h^{-1} 2 cm tip^{-1}. I_k is 1.19 Einsteins m^{-2} h^{-1}. The irradiance at local solar noon is 1.86 Einsteins m^{-2} h^{-1} at a depth of 17 m. The length of the day is 10.5 h

DEPTH VARIATIONS

Several investigators have attempted to correlate calcification rates and depth. The conclusions which are reached depend in part upon the units of rate measurement. These are usually rates of linear increase, rates of mass deposition, and rates of radioisotope incorporation. In the first 5 m, rates of linear increase are subject to extreme local variability. In particular, corals growing in back-reef environments may have less than optimum conditions for growth. Below 5 m, rates of linear growth decline rapidly particularly in the vertical direction (Buddemeier *et al.* 1974; Weber & White 1977; Baker & Weber 1975; Dustan 1975). For example, at Enewetak Atoll *Porites lutea* at 4 m depth grows approximately 2.5 times faster than the same species at 30 m depth (Buddemeier *et al.* 1974). Studies of mass deposition and radioisotopic incorporation show similar trends, although the changes are often much less dramatic (Barnes & Taylor 1973; Drew 1973; Gladfelter & Monahan 1977). This change may be in part due to altered skeletal density with depth and to altered growth morphology. Several physical factors could be responsible for changes in coral growth rates with depth. These include light, temperature, salinity, suspended inorganic particles, and water movement. By far the greatest attention is focused on light quantity and quality.

It is generally recognized that the decrease in calcification rates with depth is less than the decrease in solar irradiance. In fact, on the reefs of Belize, *Montastrea cavernosa* grows at approximately the same rate from the surface to a depth of 18 m (Weber & White 1977), and *Montastrea annularis* grows at approximately the same rate from depths of 4.5 to 18 m (Baker & Weber 1975). Clearly biological adaptations must be occurring. Biological adaptations to depth may include changes in algal density (Drew 1972), photosynthetic capacity and I_k (Wethy & Porter 1976a,b), respiration (Spencer Davies 1977), and polyp size, distribution, and behavior (Lasker 1976, 1977, 1979). Many coral species also adapt a flattened shape when growing in deep or turbid waters (Barnes 1973; Dustan 1975; Graus & Macintyre 1976, Graus 1977). It is not known exactly how these changes individually affect calcification rates. There is now a considerable amount of interest in the genetic variability of local coral populations. One major question is: 'what are the genetic limitations on the adaptations of coral planulae to the environmental conditions found at different depths?' This question can, in part, be investigated by transplanation experiments. Controlled environment studies, however, will probably provide the best correlations.

OTHER ANIMALS

In addition to algae and coelenterates, several groups of animals make major contributions to the calcareous sediments of coral reefs. These include foraminifera (Muller 1974; Ross 1977), sponges (Hartman 1977), bryozoans (Cuffey 1977), molluscs and echinoderms (Chave *et al.* 1972; Maxwell 1973; Orme 1977). In most cases far more is known about the sedimentology of skeletal parts than about the biochemistry of skeletogenesis. However, some general reviews of invertebrate skeletogenesis are available including Istin (1975), Degens (1976), Simkiss (1976b), and Krampitz & Witt (1970). Much of the biochemical and ultrastructural work on invertebrate skeletogenesis has focused on the foraminifera and to a much greater extent on the molluscs.

Foraminifera

Many foraminifera contain endosymbiotic algae. Unlike the symbiosis found in scleractinians, foraminiferan endosymbionts include representatives from several algal classes including Chlorophyceae (Lee *et al.* 1974), Bacillariophyceae (Schmaljohann & Röttger (1976), and Dinophyceae (Müller-Merz & Lee 1976). One foram, *Heterostigina depressa,* can grow in the light (45 to 20 000 lux) without added food and is thus capable ofa complete autotrophic mode of existence (Röttger 1972). Light-enhanced calcification has been demonstrated for *Archais angulatus* during laboratory experiments (Lee & Zucker 1969) but not in the field (Lee & Bock 1976).

The most extensive series of experiments on foraminiferan calcification has been conducted by Duguay & Taylor (1978) on *Archais angulatus* which contains the unicellular green alga *Chlamydomonas hedleyi* (Lee *et al.* 1974). *A. angulatus* Shows light-enhanced calcification which can be completely inhibited by the addition of the photosynthetic inhibitor 3-(3,4-dichlorophenyl)-1,1-dimethylurea (DCMU). Light-saturation curves for photosynthesis and calcification are identical in form to those of scleractinians. In the dark, *A. angulatus* shows a relatively high initial rate of radioisotope incorporation. This is attributed primarily to isotopic exchange. Thus the kinetics of calcification in foraminifera appear to be nearly identical to those of scleractinians, despite the obvious ultrastructural differences. As yet almost nothing is known of the biochemical mechanisms of calcification in these animals.

Molluscs

More is known about calcification in molluscs than in any other group of invertebrates. This information has been frequently reviewed. The physical structure of the calcium carbonate shell and associated organic materials have been reviewed by Grégoire (1972). Both the physical structure of the shell and the biochemical mechanisms of calcification have been discussed by Wilbur and co-workers in a comprehensive series of reviews (Wilbur 1960, 1964, 1972, 1976; Wilbur & Simkiss 1968; Wilbur & Watabe 1963, 1967). As summarized by Wilbur (1976), the following is a series of events that occur when a shell consisting of multiple layers is deposited.

1. A specialized area of the mantle secretes a sheet of protein (the periostracum) which becomes tanned and is the substratum on which the outer crystalline shell layer is deposited.

2. Calcium, bicarbonate, and CO_2 pass through the mantle epithelium to the extrapallial fluid between the mantle and the inner shell surface. Crystals of calcium carbonate are formed.

3. The mantle secretes organic material, largely protein, into the extrapallial fluid between the mantle and the shell or directly onto the inner shell surface. This provides the organic matrix of the shell.

4. Crystal nucleation occurs on the organic matrix or on the surface of the crystals previously deposited.

5. Oriented crystal growth takes place on and within the organic matrix. Lateral crystal growth displaces the surrounding matrix which becomes sandwiched between crystals as they grow toward each other and form a single layer.

Much of the biochemical effort in molluscan calcification has focused on the nature of the organic matrix. The matrix consists primarily of protein and glycoprotein with both soluble and insoluble fractions (Grégoire 1967; Wilbur & Simkiss 1968). The amino acid composition of these proteins may vary widely between species (Wilbur & Simkiss 1968). The soluble glycoproteins often have a high affinity for calcium ions (Crenshaw 1972), and may serve as calcium binding proteins during skeletogenesis.

The exact functional role of the matrix is the subject of some dispute. Two hypotheses have been advanced. The compartment hypothesis proposes that the matrix is merely an enclosed space in which ions can be concentrated, and where calcification can occur. The epitaxic hypothesis emphasizes the role of the organic matrix as a template for the initiation and control of crystallization. The epitaxic hypothesis appears to be ascendant (Wilbur 1976), but it is not universally supported (Towe 1972).

CONCLUSIONS

The 'problem' of calcification and biomineralization by invertebrates has occupied a number of competent investigators for many years. It is a general hope that progress made in understanding biomineralization in one group of organisms may increase the understanding of events occurring in invertebrates and even in vertebrates. The process is more complex than coral reef biologists might have hoped. Future work will probably concentrate on the areas of regulation and control. On the biochemical level, considerable information is needed on the mechanisms for the regulation of calcium ion concentration, and on the transport of calcium and carbonate. Increased attention must focus on the biochemical adaptations of calcareous marine organisms to their environment. In addition most investigators still hope to identify a single rate-limiting factor in light-enhanced calcification. On a larger scale, increasing attention will probably focus on environmental factors which control the rates of calcification and on the inherent variability of reef ecosystems. Meaningful correlations will require observations over several years. Environmentally, there are many interesting and important problems, but no easy solutions.

ACKNOWLEDGEMENTS

This work was supported by U.S. National Science Foundation Grant DES 75-17256 to Professor D. L. Taylor, Rosenstiel School of Marine and Atmospheric Science, University of Miami. Ship operations were supported by the U.S. National Science Foundation. I acknowledge the assistance and co-operation of the master and mate of *R/V Calanus*. I thank the government of the Bahamas for continuing permission to conduct research in their waters. The figures were drafted by Ms Susan Markley. The manuscript was reviewed by Dr D. J. Barnes, Australian Institute of Marine Science.

LITERATURE CITED

Ammann, D., R. Bissig, Z. Cimerman, U. Fiedler, M. Güggi, W. E. Morf, M. Oehme, H. Osswald, E. Pretsch and W. Simon, 1976. Synthetic neutral carriers for cations. In M. Kessler, L. C. Clark Jr., D. W. Lübbers, I. A. Silver and W. Simon (eds), *Ion and Enzyme Electrodes in Biology and Medicine*. Baltimore University Park Press, pp. 22-37.

Bak, R. P. M., 1973. Coral weight increment *in situ*. A new method to determine coral growth. *Marine Biology* 20, 45-49.

Baker, P. A. and J. N. Weber, 1975. Coral growth rate: Variation with depth. *Earth and Planetary Science Letters* 27, 57-61.

Barnes, D. J., 1971. A study of growth, structure, and form in modern coral skeletons. Ph.D. dissertation, University of Newcastle-upon-Tyne, 180 pp.

Barnes, D. J., 1972. The structure and formation of growth-ridges in scleractinian coral skeletons. *Proceedings of the Royal Society of London, Series B* 182, 331-50.

Barnes, D. J., 1973. Coral reef project — papers in memory of Dr. Thomas F. Goreau. 12. Growth in colonial scleractinians. *Bulletin of Marine Science* 23, 280-98.

Barnes, D. J. and C. J. Crossland, 1977. Coral calcification: Sources of error in radioisotopic techniques. *Marine Biology* 42, 119–29.

Barnes, D. J. and C. J. Crossland, 1978. Diurnal productivity and apparent ^{14}C-calcification in the staghorn coral, *Acropora acuminata*. *Comparitive Biochemistry and Physiology* 59A, 133–8.

Barnes, D. J. and D. L. Taylor, 1973. *In situ* studies of calcification and photosynthetic carbon fixation in the coral, *Montastrea annularis*. *Helgoländer wissenschaftliche Meeresuntersuchungen* 24, 284–91.

Brinley, F. J., Jr., 1973. Calcium and magnesium transport in living cells. *Federation Proceedings, Federation of American Societies for Experimental Biology* 32, 1735–9.

Brown, H. M., J. P. Pemberton and J. D. Owen, 1976. A calcium-sensitive microelectrode suitable for intracellular measurement of calcium (II) activity. *Analytica Chimica Acta* 85, 261–76.

Buddemeier, R. W. and R. A. Kinzie III, 1976. Coral growth. *Oceanography and Marine Biology an Annual Review* 14, 183–225.

Buddemeier, R. W., J. E. Maragos and D. K. Knutson, 1974. Radiographic studies of reef coral exoskeletons: Rates and patterns of coral growth. *Journal of Experimental Marine Biology and Ecology* 14, 179–200.

Campbell, J. W. and K. V. Speeg Jr., 1968. Theoretical considerations of the possible role of ammonia in the biological deposition of calcium carbonate. *American Zoologist* 8, 770–1.

Campbell, J. W. and K. V. Speeg Jr., 1969. Ammonia and biological deposition of calcium carbonate. *Nature* 224, 725–6.

Cattrall, R. W. and H. Freiser, 1971. Coated wire ion selective electrodes. *Analytical Chemistry* 43, 1905–6.

Chalker, B. E., 1976. Calcium transport during skeletogenesis in hermatypic corals. *Comparative Biochemistry and Physiology* 54A, 455–9.

Chalker, B. E., 1977. Daily variation in the calcification capacity of *Acropora cervicornis*. In D. L. Taylor (ed.), *Proceedings: Third International Coral Reef Symposium*. Miami: University of Miami, vol. 2, pp. 417–23.

Chalker, B. E., 1980. Modelling light saturation curves for photosynthesis: an exponential function. *Journal of Theoretical Biology* 84, 205–15.

Chalker, B. E. and D. L. Taylor, 1975. Light-enhanced calcification and the role of oxidative phosphorylation in calcification of the coral *Acropora cervicornis*. *Proceedings of the Royal Society of London, Series B* 190, 323–31.

Chalker, B. E. and D. L. Taylor, 1978. Rhythmic variations in calcification and photosynthesis associated with the coral *Acropora cervicornis* (Lamarck). *Proceedings of the Royal Society of London, Series B* 201, 179–89.

Chapman, G., 1974. The skeletal system. In L. Muscatine and H. M. Lenhoff (eds.), *Coelenterate Biology: Reviews and New Perspective*. New York: Academic Press, pp. 93–128.

Chave, K. E., S. V. Smith and K. J. Roy, 1972. Carbonate production by coral reefs. *Marine Geology* 12, 123–40.

Clausen, C. D. and A. A. Roth, 1975. Estimation of coral growth-rates from laboratory 45-Ca incorporation rates. *Marine Biology* 33, 85–91.

Craggs, A., G. J. Moody and J. D. R. Thomas, 1974. PVC matrix membrane ion-selective electrodes: Construction and laboratory experiments. *Journal of Chemical Education* 51, 541–4.

Crenshaw, M. A., 1972. The soluble matrix from *Mercenaria mercenaria* shell. *Biomineralization* 6, 6–11.

Crossland, C. J. and D. J. Barnes, 1974. The role of metabolic nitrogen in coral calcification. *Marine Biology* 28, 325–32.

Crossland, C. J. and D. J. Barnes, 1977. Coral calcification: Variations in apparent skeletal incorporation of radioisotopes due to different methods of processing. *Marine Biology* 43, 57–62.

Cuffey, R. J., 1977. Bryozoan contributions to reefs and bioherms through geologic time. *American Association of Petroleum Geologists. Studies in Geology* 4, 181-94.

Degens, E. T., 1976. Molecular mechanisms on carbonate, phosphate, and silica deposition in the living cell. *Topics in Current Chemistry* 64, 1-112.

Dodge, R. E. and J. Thompson, 1974. The natural radiochemical and growth records in contemporary hermatypic corals from the Atlantic and Caribbean. *Earth and Planetary Science Letters* 23, 313-22.

Drew, E. A., 1972. The biology and physiology of algal-invertebrate symbiosis. II. The density of symbiotic algal cells in a number of hermatypic hard corals and alcyonarians from various depths. *Journal of Experimental Marine Biology and Ecology* 9, 71-5.

Drew, E. A., 1973. The biology and physiology of algal-invertebrate symbiosis. III. *In situ* measurements of photosynthesis and calcification in some hermatypic corals. *Journal of Experimental Marine Biology and Ecology* 13, 165-79.

Duguay, L. E. and D. L. Taylor, 1978. Primary production and calcification by the soritid foraminiferan *Archais angulatus* (Fichtel & Moll). *Journal of Protozoology* 25, 356-61.

Dustan, P., 1975. Growth and form in the reef-building coral *Montastrea annularis. Marine Biology* 33, 101-7.

Freudenthal, H. D., 1962. *Symbiodinium gen. nov.* and *Symbiodinium microadriaticum sp. nov.,* a zooxanthella: taxonomy, life cycle and morphology. *Journal of Protozoology* 9, 45-52.

Garrles, R. M. and M. E. Thompson, 1962. A chemical model for sea water at 25°C and one atmosphere total pressure. *American Journal of Science* 260, 57-66.

Gladfelter, E. H. and R. K. Monahan, 1977. Primary production and calcium carbonate deposition rates in *Acropora palmata* from different positions in the reef. In D. L. Taylor (ed.), *Proceedings: Third International Coral Reef Symposium*. Miami: University of Miami, vol. 1, pp. 389-94.

Goldberg, E., 1963. The oceans as a chemical system. In M. N. Hill (ed.), *The Sea*. New York: Interscience, vol. 2, pp. 2-25.

Goreau, N. I. and R. L. Hays, 1977. Nucleation catalysis in coral skeletogenesis. In D. L. Taylor (ed.), *Proceedings: Third International Coral Reef Symposium*. Miami: University of Miami, vol. 2, pp. 439-45.

Goreau, T. F., 1959. The physiology of skeleton formation in corals. I. A method for measuring the rate of calcium deposition by corals under different conditions. *Biological Bulletin* 116, 59-75.

Goreau, T. F., 1961a. Problems of growth and calcium deposition in reef corals. *Endeavour* 20, 32-9.

Goreau, T. F., 1961b. On the relation of calcification to primary productivity in reef building organisms. In H. M. Lenhoff and W. F. Loomis (eds.), *The Biology of Hydra and Some Other Coelenterates*. Miami: University of Miami Press, pp. 269-85.

Goreau, T. F., 1963. Calcium carbonate deposition by coralline algae and corals in relation to their roles as reef-builders. *Annals of the New York Academy of Sciences* 109, 127-67.

Goreau, T. F. and V. T. Bowen, 1955. Calcium uptake by a coral. *Science* 122, 1188-9.

Goreau, T. F. and N. I. Goreau, 1959. The physiology of skeleton formation in corals. II. Calcium deposition by hermatypic corals under various conditions in the reef. *Biological Bulletin* 117, 239-50.

Goreau, T. F. and N. I. Goreau, 1960a. The physiology of skeleton formation in corals. III. Calcification rate as a function of colony weight and total nitrogen content in the reef coral *Manicina areolate* (Linnaeus). *Biological Bulletin* 118, 419-29.

Goreau, T. F. and N. I. Goreau, 1960b. The physiology of skeleton formation in corals. IV. On isotopic equilibrium exchanges of calcium between corallum and environment in living and dead reef building corals. *Biological Bulletin* 119, 416-27.

Goreau, T. F. , N. I. Goreau and C. M. Yonge, 1971. Reef corals: Autotrophs or heterotrophs? *Biological Bulletin* 141, 247-60.

Graus, R. R., 1977. Investigation of coral growth adaptations using computer modelling. In D. L. Taylor (ed.), *Proceedings: Third International Coral Reef Symposium*. Miami: University of Miami, vol. 2, pp. 463–9.

Graus, R. R. and I. G. Macintyre, 1976. Light control of growth form in colonial corals: Computer simulation. *Science* 193, 895–7.

Grégoire, C., 1967. Sur la structure des matrices organiques de coquilles de mollusques. *Biological Reviews of the Cambridge Philosophical Society* 42, 653–88.

Grégoire, C., 1972. Structure of the molluscan shell. In M. Florkin and B. T. Sheer (eds.), *Chemical Zoology*. New York: Academic Press, vol. 7, pp. 45–102.

Hartman, W. D., 1977. Sponges as reef builders and shapers. *American Association of Petroleum Geologists. Studies in Geology* 4, 127–34.

Hays, R. L. and N. I. Goreau, 1977. Intracellular crystal-bearing vesicles in the epidermis of scleractinian corals, *Astrangia danae* (Agassiz) and *porites porites* (Pallas). *Biological Bulletin* 152, 26–40.

Highsmith, R. C., 1979. Coral growth rates and environmental control of density banding. *Journal of Experimental Marine Biology and Ecology* 37, 105–25.

Hudson, J. H., E. A. Shinn, R. B. Halley and B. Lidz, 1976. Sclerochronology: A tool for interpreting past environments. *Geology* 4, 361–4.

Istin, M., 1975. The structure and formation of calcified tissue. In D. C. Malins and J. R. Sargent (eds.), *Biochemical and Biophysical Perspectives in Marine Biology*. New York: Academic Press, vol 2, pp. 1–68.

Jassby, A. D. and T. Platt, 1976. Mathematical formulation of the relationship between photosynthesis and light for phytoplankton. *Limnology and Oceanography* 21, 540–7.

Johnston, I. S., 1977. Aspects of the structure of the skeletal organic matrix, and the process of skeletogenesis in the reef-coral *Pocillopora damicornis*. In D. L. Taylor (ed.), *Proceedings: Third International Coral Reef Symposium*. Miami: University of Miami, vol. 2, pp. 448–53.

Johnston, I. S., 1979. The organization of a structural organic matrix within the skeleton of a reef-building coral. In O. Johari and R. P. Becker (eds.), *Scanning Electron Microscopy/1979*. Chicago: Scanning Electron Microscopy, Inc., Part II, pp. 421–31.

Kawaguti, S. and D. Sakumoto, 1948. The effect of light on the calcium deposition of corals. *Bulletin of the Oceanographic Institute of Taiwan* 4, 65–70.

Kawaguti, S. and K. Sato. 1968. Electron microscopy on the polyp of staghorn corals with specific reference to its skeleton formation. *Biological Journal of Okayama University* 14, 87–98.

Krampitz, G. and W. Witt, 1979. Biochemical aspects of biomineralization. *Topics in Current Chemistry* 78, 57–144.

Lasker, H. R., 1976. Intraspecific variability of zooplankton feeding in the hermatypic coral *Montastrea cavernosa*. in G. O. Mackie (ed.), *Coelenterate Ecology and Behavior*. New York: Plenum Publishing Corporation, pp. 101–9.

Lasker, H. R., 1977. Patterns of zooxanthellae distribution and polyp expansion in the reef coral *Montastrea cavernosa*. in D. L. Taylor (ed.), *Proceedings: Third International Coral Reef Symposium*. Miami: University of Miami, vol. 1, pp. 607–13.

Lasker, H. R., 1979. Light dependent activity patterns among reef corals: *Montastrea cavernosa*. *Biological Bulletin* 156, 196–211.

Lee, J. J. and W. D. Bock, 1976. The importance of feeding in two species of soritid foraminifer with algal symbionts. *Bulletin of Marine Science* 26, 530–7.

Lee, J. J. and W. Zucker, 1969. Algal flagellate symbiosis in the foraminifer *Archaias*. *Journal of Protozoology* 16, 71–81.

Lee, J. J., L. J. Crockett, J. Hagen and R. Stone, 1974. The taxonomic identity and physiological ecology of *Chlamydomonas hedleyi sp. nov.*: algal flagellate symbiont from the foraminifer *Archaias angulatus*. *British Phycological Journal* 9, 407–22.

Lehninger, A. L., 1970. Mitochondria and calcium ion transport. *Biochemical Journal* 119, 129–38.

Lewis, J. B., F. Axelson, I. Goodbody, C. Page and G. Chislett, 1968. Comparative growth rates of some reef corals in the Caribbean. *Marine Science Manuscript Report, McGill University* 10, 1-27.

Loeblich, A. R., III and J. L. Sherley, 1979. Observations on the theca of the motile phase of free-living and symbiotic isolates of *Zooxanthella microadriatic* (Freudenthal) *comb. nov.* Journal of the Marine Biological Association of the United Kingdom 59, 195-205.

Maxwell, W. G. H., 1973. Sediments of the Great Barrier Reef Province. In O. A. Jones and R. Endean (eds.), *Biology and Geology of Coral Reef.* New York: Academic Press, Volume I, Geology 1, pp. 299-345.

Moody, G. J., R. B. Oke and J. D. R. Thomas, 1970. A calcium-sensitive electrode based on a liquid ion exchanger in a poly(vinyl chloride) matrix. *Analyst* 95, 910-18.

Morf, W. E. and W. Simon, 1978. Recent developments in the field of neutral carrier based ion-selective electrodes. In E. Pungor (ed.), *Ion-Selective Electrodes.* Amsterdam: Elsevier Scientific, pp. 149-59.

Muller, P. H., 1974. Sediment production and population biology of the benthic foraminifer *Amphistegina madagascariensis. Limnology and Oceanography* 19, 802-9.

Müller-Merz, E. and J. J. Lee, 1976. Symbiosis in the larger foraminiferan *Sorites marginalis* (with notes on *Archaias spp.). Journal of Protozoology* 23, 390-6.

Muscatine, L., 1971. Endosymbiosis of algae and coelenterates. In H. M. Lenhoff, L. Muscatine and L. V. Davis (eds)., *Experimental Coelenterate Biology.* Honolulu: University of Hawaii Press, pp. 179-91.

Muscatine, L., 1973. Nutrition of corals. In O. A. Jones and R. Endean (eds.), *Biology and Geology of Coral Reefs.* Academic Press, New York, vol. 2, Biology 1, pp. 77-115.

Muscatine, L., 1974. Endosymbiosis of cnidarians and algae. In L. Muscatine and H. M. Lenhoff (eds.), *Coelenterate Biology: Reviews and New Perspectives.* New York: Academic Press, pp. 359-95.

Muscatine, L. and J. W. Porter, 1977. Reef corals: Mutualistic symbiosis adapted to nutrient-poor environments. *Bioscience* 27, 454-60.

Orme, G. R., 1977. Aspects of sedimentation in the coral reef environment. In O. A. Jones and R. Endean (eds.), *Biology and Geology of Coral Reefs.* New York: Academic Press, vol. 4, Geology 2, pp. 129-82.

Pearse, V. B., 1970. Incorporation of metabolic CO_2 into coral skeleton. *Nature* 228, 383.

Pearse, V. B., 1971. Sources of carbon in the skeleton of the coral *Fungia scutaria.* In H. M. Lenhoff, L. Muscatine and L. V. Davis (eds.), *Experimental Coelenterate Biology.* Honolulu: University of Hawaii Press, pp. 239-45.

Pearse, V. B. and L. Muscatine, 1971. Role of symbiotic algae (zooxanthellae) in coral calcification. *Biological Bulletin* 141, 350-63.

Porter, J. W., 1976. Autotrophy, heterotrophy, and resource partitioning in Caribbean reef-building corals. *American Naturalist* 110, 731-42.

Ross, C. A., 1977. Calcium carbonate fixation by large reef-dwelling foraminifera. *American Association of Petroleum Geologists. Studies in Geology* 4, 219-30.

Ross, J. W., Jr., 1967. Calcium-selective electrode with liquid ion exchanger. *Science* 156, 1378-9.

Röttger, R., 1972. Die Kultur von *Heterostegina depressa* (Foraminifera: Nummulitidae). *Marine Biology* 15, 150-9.

Ruzicka, J., E. H. Hansen and J. C. Tjell, 1973. Selectrode — the universal ion selective electrode. Part VI. The calcium (II) selectrode employing a new ion exchanger in a nonporous membrane and a solid-state reference system. *Analytica Chimica Acta* 67, 155-78.

Schmaljohann, R. and R. Röttger, 1976. Die Symbionten der Grossforaminifere *Heterostegina depressa* sind Diatomeen. *Naturwissenschaften* 63, 486-7.

Schultz, F. A., A. J. Peterson, C. A. Mask and R. P. Buck, 1968. Solid ion-exchange electrode selective for calcium ion. *Science* 162, 267-8.

Shinn, E. A., 1966. Coral growth rate, and environmental indicator. *Journal of Paleontology* 40, 233–40.

Simkiss, K., 1964a. Phosphates as crystal poisons of calcification. *Biological Reviews of the Cambridge Philosophical Society* 39, 487–505.

Simkiss, K., 1964b. The inhibitory effects of some metabolites on the precipitation of calcium carbonate from artificial and natural sea water. *Journal du Conseil, Conseil Permanent. International pour l'Exploration de la Mer* 29, 6–18.

Simkiss, K., 1964c. Possible effects of zooxanthellae on coral growth. *Experentia* 20, 140.

Simkiss, K., 1974. Calcium translocation by cells. *Endeavour* 33, 119–23.

Simkiss, K., 1976a. Intracellular and extracellular routes in biomineralization. *Symposia of the Society for Experimental Biology* 30, 423–44.

Simkiss, K., 1976b. Cellular aspects of calcification. In N. Watabe and K. M. Wilbur (eds.), *The Mechanisms of Mineralization in the Invertebrates and Plants.* Columbia, South Carolina: University of South Carolina Press, pp. 1–31.

Smith, D., L. Muscatine and D. Lewis, 1969. Carbohydrate movements from autotrophs to heterotrophs in parasitic and mutualistic symbiosis. *Biological Reviews of the Cambridge Philosophical Society* 44, 17–90.

Smith, S. V., 1973. Carbon dioxide dynamics: a record of organic carbon production, respiration, and calcification in the Eniwetok reef flat community. *Limnology and Oceanography* 18, 106–20.

Smith, S. V. and G. S. Key, 1975. Carbon dioxide and metabolism in marine environments. *Limnology and Oceanography* 20, 493–5.

Smith, S. V. and D. W. Kinsey, 1978. Calcification and organic carbon metabolism as indicated by carbon dioxide. In D. R. Stoddart and R. E. Johannes (eds.), *Coral Reefs: Research Methods.* UNESCO, Paris, Monographs on Oceanographic Methodology, No. 5, pp. 469–84.

Spencer Davies, P., 1977. Carbon budgets and vertical zonation of Atlantic reef corals. In D. L. Taylor (ed.), *Proceedings: Third International Coral Reef Symposium.* University of Miami, Miami, vol. 1, pp. 391–6.

Steemann Nielsen, E., 1975. Marine Photosynthesis with Special Emphasis on the Ecological Aspects. Amsterdam Elsevier Scientific Publishing.

Stoddart, D. R. and R. E. Johannes (eds.), 1978 *Coral Reefs: Research Methods.* UNESCO, Paris, Monographs on Oceanographic Methodology, Number 5, 581 pp.

Talling, J. F., 1957. Photosynthetic characteristics of some freshwater diatoms in relation to underwater radiation. *New Phytologist* 56, 29–50.

Taylor, D. L., 1971a. Ultrastructure of the "zooxanthella" *Endodinium chatonii in situ. Journal of the Marine Biological Association of the United Kingdom* 51, 227–34.

Taylor, D. L., 1971b. Patterns of carbon translocation in algal-invertebrate symbiosis. In K. Nisizawa (ed.), *Proceedings of the Seventh International Seaweed Symposium.* Tokyo: University of Tokyo Press, pp. 590–7.

Taylor, D. L., 1973a. Symbiotic pathways of carbon in coral reef ecosystems. Present status and future prospectives. *Helgoländer wissenschaftliche Meeresuntersuchungen* 24, 276–83.

Taylor, D. L., 1973b. Cellular interactions of algal-invertebrate symbiosis. *Advances in Marine Biology* 11, 1–56.

Thompson, M. E. and J. W. Ross Jr., 1966. Calcium in sea water by electrode measurement. *Science* 154, 1643–4.

Towe, K. M., 1972. Invertebrate shell structure and the organic matrix concept. *Biomineralization Research Reports* 4, 1–14.

Vandermuelen, J. H. and L. Muscatine, 1974. Influence of symbiotic algae on calcification in reef corals: Critique and progress report. In W. B. Vernberg (ed.), *Symbiosis in the Sea.* Columbia, South Carolina: University of South Caroline Press, pp: 1–20.

Vandermuelen, J. H., N. Davis and L. Muscatine, 1972. The effect of inhibitors of photosynthesis on zooxanthellae from corals and other invertebrates. *Marine Biology* 16, 185–91.

Wainwright, S. A., 1963. Skeletal organization in the coral *Pocillopora damicornis. Quarterly Journal of Microscopical Science* 104, 169-83.

Weber, J. N., 1973. Incorporation of strontium into reef coral skeletal carbonate. *Geochimica et Cosmochimica Acta* 37, 2173-90.

Weber, J. N. and E. W. White, 1977. Caribbean reef corals *Montastrea annularis* and *Montastrea cavernosa* — long-term growth data as determined by skeletal x-radiography. *American Association of Petroleum Geologists. Studies in Geology* 4, 171-9.

Weber. J. N., E. W. White and P. H. Weber, 1975. Correlation of density banding in reef corals with environmental parameters: The basis for interpretation of chronological records preserved in the coralla of corals. *Paleobiology,* 137-49.

Wells, J. M., 1977. A comparative study of the metabolism of tropical benthic communities. In D. L. Taylor (ed.), *Proceedings: Third International Coral Reef Symposium.* Miami: University of Miami, vol 1, pp. 545-9.

Wells, J. M., A. H. Wells and J. G. VanDerwalker, 1973. *In situ* studies of metabolism in benthic reef communities. *Helgoländer wissenschaftliche Meeresuntersuchungen* 24. 78-81.

Wethey, D. S. and J. W. Porter, 1976a. Sun and shade differences in productivity of reef corals. *Nature* 262, 281-2.

Wethey, D. S., and J. W. Porter, 1976b Habitat-related patterns of productivity of the foliaceous reef coral, *Pavona praetorta* Dana. In G. O. Mackie (ed.), *Coelenterate Ecology and Behavior.* New York: Plenum Publishing, pp. 59-66.

Wilbur, K. M., 1960. Shell structure and mineralization in molluscs. In R. G. Sognnaes (ed.), *Calcification in Biological Systems.* Washington: American Association for the Advancement of Science, pp. 15-40.

Wilbur, K. M., 1964. Shell formation and regulation. In K. M. Wilbur and C. M. Yonge (eds.), *Physiology of Mollusca.* New York: Academic Press, vol. 1, pp. 243-82.

Wilbur, K. M., 1972. Shell formation in molluscs. In M. Florkin and B. T. Scheer (eds.). *Chemical Zoology.* New York: Academic Press, vol. 7, pp. 103-45.

Wilbur, K. M., 1976. Recent studies of invertebrate mineralization. In N. Watabe and K. M. Wilbur (eds.), *The Mechanisms of Mineralization in the Invertebrates and Plants.* Columbia, South Carolina: University of South Carolina Press, pp. 79-108.

Wilbur, K. M. and K. Simkiss, 1968. Calcified shells. In M. Florkin and E. H. Stotz (eds.), *Comprehensive Biochemistry.* Amsterdam: Elsevier Publishing Company, vol, 26, part A, pp. 229-95.

Wilbur, K. M. and N. Watabe, 1963. Experimental studies on calcification in molluscs and the alga *Coccolithus huxleyi. Annals of the New York Academy of Sciences* 109, 82-112.

Wilbur, K. M. and N. Watabe, 1967. Mechanism of calcium carbonate deposition in coccolithophorids and molluscs. *Studies in Tropical Oceanography* 5, 133-54.

Yonge, C. M., 1930. Studies on the physiology of corals. I. Feeding mechanisms and food. *Scientific Report of the Great Barrier Reef Expedition, 1928-29, British Museum (Natural History)* 1, 13-57.

Yonge, C. M. 1931. The significance of the relationship between corals and zooxanthellae. *Nature* 128, 309-11.

Yonge, C. M., 1940. The biology of reef-building corals. *Scientific Report of the Great Barrier Reef Expedition, 1928-29, British Museum (Natural History)* 1, 353-91.

Yonge, C. M., 1944. Experimental analysis of the association between invertebrates and unicellular algae. *Biological Reviews of the Cambridge Philosophical Society* 19, 68-80.

Yonge, C. M., 1957. Symbiosis. *Memoirs of the Geological Society of America* 67, 429-42.

Yonge, C. M., 1958. Ecology and physiology of reef-building corals. In A. A. Buzzati-Traverso (ed.), *Perspectives in Marine Biology.* Berkley, University of California Press, pp. 117-35.

Yonge, C. M., 1963. The biology of coral reefs. *Advances in Marine Biology* 1, 209-60.

Yonge, C. M., 1968. Living corals. *Proceedings of the Royal Society of London, Series B* 169, 329-44.

Yonge, C. M., 1973. Coral reef project — papers in memory of Dr Thomas F. Goreau. 1. The nature of reef-building (hermatypic) corals. *Bulletin of Marine Science* 23, 1-15.

Yonge, C. M. and A. G. Nicholls, 1931a. Studies on the physiology of corals. IV. The structure, distribution, and physiology of the zooxanthellae. *Scientific Report of the Great Barrier Reef Expedition, 1928-29, British Museum (Natural History)* 1, 135-76.

Yonge, C. M. and A. G. Nicholls, 1931b. Studies on the physiology of corals. V. The effects of starvation, in light and in darkness, on the relationship between corals and zooxanthellae. *Scientific Report of the Great Barrier Reef Expedition, 1928-29, British Museum (Natural History)* 1, 177-211.

Yonge, C. M., M. J. Yonge and A. G. Nicholls, 1932. Studies on the physiology of corals. VI. The relationship between respiration in corals and the production of oxygen by their zooxanthellae. *Scientific Report of the Great Barrier Reef Expedition, 1928-29, British Museum (Natural History)* 1, 213-51.

Young, S. D., 1971. Organic material from scleractinian coral skeletons. I. Variation in composition between several species. *Comparative Biochemistry and Physiology* 40B, 113-20.

Young, S. D., 1973. Calcification and synthesis of skeletal organic material in the coral, *Pocillopora damicornis* (L.) (Astrocoeniidae, Scleractinia). *Comparative Biochemistry and Physiology* 44A, 669-72.

Young, S. D., J. D. O'Connor and L. Muscatine, 1971. Organic material from scleractinian coral skeletons. II. Incorporation of ^{14}C into protein, chitin and lipid. *Comparative Biochemistry and Physiology* 40B, 945-58.

4 Sea Level Changes and Coral Reef Growth

John Chappell
Research School of Pacific Studies, Australian National University, P.O. Box 4, Canberra, A.C.T. 2601

INTRODUCTION

There are two positions from which the matter of sea level change and reef growth can be viewed. Geologists concerned with crustal movements (tectonic and isostatic), and with 'absolute' sea level (eustatic) changes, view reefs as indicators of relative movement of the land-sea juxtaposition. This is because reef crest and reef flat materials, preserved within ancient reef structures, accurately express the position of the low tide datum at their time of formation. An alternative viewpoint is represented by those who investigate the effects of sea level change on reef growth itself. The two viewpoints are implicit within Darwin's (1842) theory of atoll formation, where atolls are explained by crustal subsidence beneath the island on which the embryonic reef developed, with submergence being matched by continued upward coral growth, constrained to the low tide level. Barriers impounding deep lagoons are argued as forming in the same way, and Darwin's *(op. cit.)* world map of atolls, barrier, and fringing reefs stands as the first global picture of vertical crustal movements.

Darwin's theory is seminal in several aspects. Firstly, it contains the central concept that the well known arrangement of reef flat, reef crest, and fore-reef zones persists throughout the course of relative sea level rise. The extent to which this may be true is examined later. Secondly, it requires that the potential for upward growth on the oceanic side of a barrier or atoll reef exceeds that on the lagoonal side. If this were not so, submergence of a conical island or sloping shelf would lead to development of an ever-widening fringing reef — a difficulty which has led to the idea that barriers develop on pre-existing basement prominences, following their submergence (Purdy 1974). Finally, the basement subsidence concept was brought into focus by Daly (1915), who pointed out that the lagoons of large atolls are of similar depths, a coincidence which he explained as caused by global sea level rise following melting of the great Pleistocene icecaps of the northern continents.

The fact is that vertical crustal movements, downwards in some parts of the world and upwards in some others, have proceeded for the last several million years. Superimposed on these have been major oscillations of sea level associated with repeated growth and decay of the Pleistocene icecaps. Although these glacio-eustatic changes have not been equal throughout the world, due to the effects of changing water-ice distributions on the figure of the earth itself (Clark *et al.* 1978), roughly speaking the sea level has oscillated through a range of 100 m, with rates of change often exceeding 10 m per 1000 years,

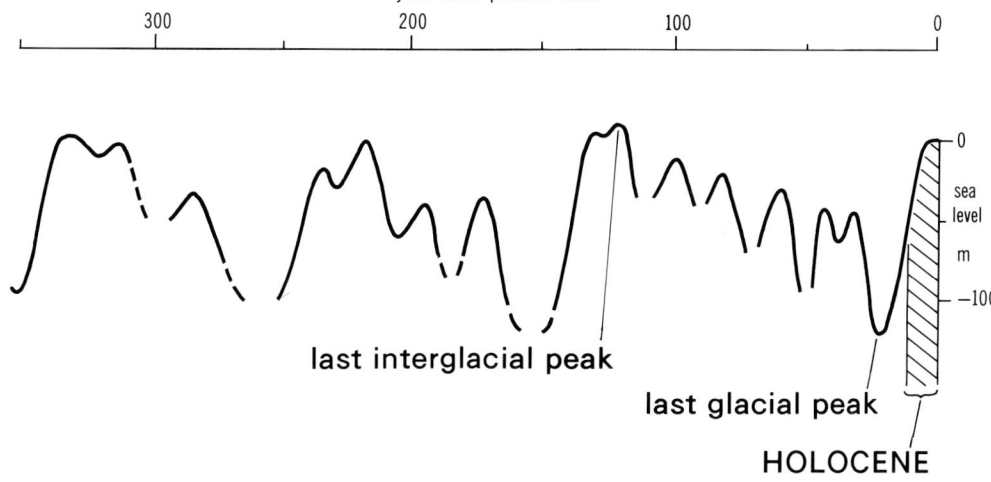

last interglacial peak

last glacial peak

HOLOCENE

Figure 1   Sea level changes for last 350 000 years, based on Chappell (1974), Chappell & Veeh (1978). Note that this is a global average curve the amplitude of which is expected to vary from one part of the world to another (cf. Clarke *et al.* 1978)

upwards and downwards. As these glacio-eustatic rates well exceed the known rates of vertical crustal movement outside of glacio-isostatically rising areas, all coral reefs which are today at or near sea level are growing on substrates that stood above sea level many times throughout Pleistocene times. Figure 1 shows the general pattern of sea level changes for the last 350 000 years; similar but less well known variations extend back at least to the beginning of Pleistocene times, nearly 2 million years ago. The context of reef development is set by these oscillations, conditioned by repeated episodes of subaerial landform development. Within this frame, the processes of reef building, by growth of primary coral framework and by sedimentation of a host of materials, have acted to produce the reefs of today.

Variation of reef form in a province such as the Great Barrier Reef of Australia, ranging from the ribbon-like outer barriers in the north through large ovoid patch or platform reefs in mid-shelf to the inner fringing reefs, indicates a history of formation more complex than can be embraced by Darwin's theory of atoll and barrier development. Inheritance of subaerial landforms from glacial low sea level times is expressed dramatically by the 'blue holes' of some reefs, formed by coral growth upon drowned karstic dolines (Backshall *et al.* 1979) and is an important factor in general (Purdy 1974; Stoddart *et al.* 1978). In order to place the effects of previous topography in perspective, however, we commence by harking back to Darwin and examining the reef structures that develop when a simple sloping substrate is drowned. This is best seen in Pleistocene reefs that have been raised well above see level by tectonic uplift.

RAISED REEFS ON A RECTILINEAR SUBSTRATE

Flights of raised coral reef terraces occur on many tropical islands where convergence of great crustal plates causes continuing uplift. These are particularly well known from Huon Peninsula, New Guinea (Chappell 1974), Barbados in the Caribbean (Mesolella *et al.* 1970), and Atauro Island north of Timor (Chappell & Veeh 1978a). These reefs have been dated radiometrically and are interpreted as having formed during successive glacio-eustatic sea level oscillations, superimposed on the rising islands. Deep ravines cutting through the terrace flights at Huon Peninsula and at Atauro expose internal structures of the reefs, as well as their substrate geometries, sufficiently well for their growth patterns to be analysed.

Figure 2 shows sections through reefs that formed before and during the last interglacial episode, 135 000 to 120 000 years ago, at Huon Peninsula and at Atauro Island (simplified from Chappell 1974; Chappell & Veeh 1978a). All formed on a sloping rectilinear substrate, as shown. Figure 2 also indicates the course of sea level changes during the period when these reefs formed, and the tectonic uplift rates for each section. Before proceeding to examine the growth forms of these reefs, some explanation is needed of how the sea level change and uplift factors are known.

Sea level change relative to a given reef is interpreted from the internal age structure of its shallowest-water coral and sediment assemblages (shallow water facies). To illustrate, figure 3 shows a cliff section cut into a young raised reef at Huon Peninsula; at the base of the photograph the reef is dated by radiocarbon as 7500 years old, and becomes progressively younger upwards, showing 6500 years at the top. As the facies, dominated by *Acropora humilis,* is characteristic of shallow water reef crests at Huon Peninsula, the section indicates that sea level was rising while the reef grew upwards, 7500 to 6500 years ago (a lengthier account of this simple argument is given by Chappell & Polach 1976). By extension to the larger sections shown in figure 2, using $^{230}Th/^{234}U$ dating, the course of sea level changes relative to each section has been determined.

Interpretation of uplift requires knowledge of the 'absolute' position of sea level for at least one moment in the past, plus assumption of the constancy or otherwise of the uplift vector. The paleo-datum generally used is the last interglacial sea level of 120 000 years ago, taken as 4 m to 8 m above present sea level on the basis of raised reefs on islands and coasts in several parts of the world that are remote from tectonically active areas (reviewed by Chappell & Veeh 1978a). Although uplift rates may not be constant, the simple approximation of uniform uplift fits all known data from Huon Peninsula (Bloom *et al.* 1974) and from Atauro (Chappell & Veeh 1978a) sufficiently well for it to be adopted for our purposes. Hence, on the basis of the present elevations of the 120 000 yr reef at each section, the uplift vectors shown in figure 2 are estimated.

The three reefs shown in figure 2 developed at the same time under qualitatively similar conditions. Each formed on a sloping rectilinear surface during a period of sea level rise (transgression) which led to an interglacial episode of only minor sea level change, followed by sea level fall (regression). Boundary condition differences amount merely to differences of substrate gradient and of uplift rate. Yet, there are significant differences of reef form, and of growth history as it is known from the published radiometric dates. The Huon Peninsula reefs can be said to be transgression-limited, in that growth of the main structure (reef VIIa, fig. 2) ceased when sea level stopped rising, 135 000 years ago. The small cap of reef VIIb was added during the small transgression prior to 120 000 years ago. The Atauro reefs, on the other hand, continued to grow after the transgression ceased and well into the regression. In this case, development is growth-limited, as growth lags well behind sea level change except at the narrow shoreline fringe. Specifically, the Huon Reefs grew 50 m upwards in about 10 000 or fewer years during the transgression, that is, about 0.5 cm yr^{-1}, while the Atauro reefs (about 40 m thick) grew through the entire interval shown, 145–115 thousand years, that is, about

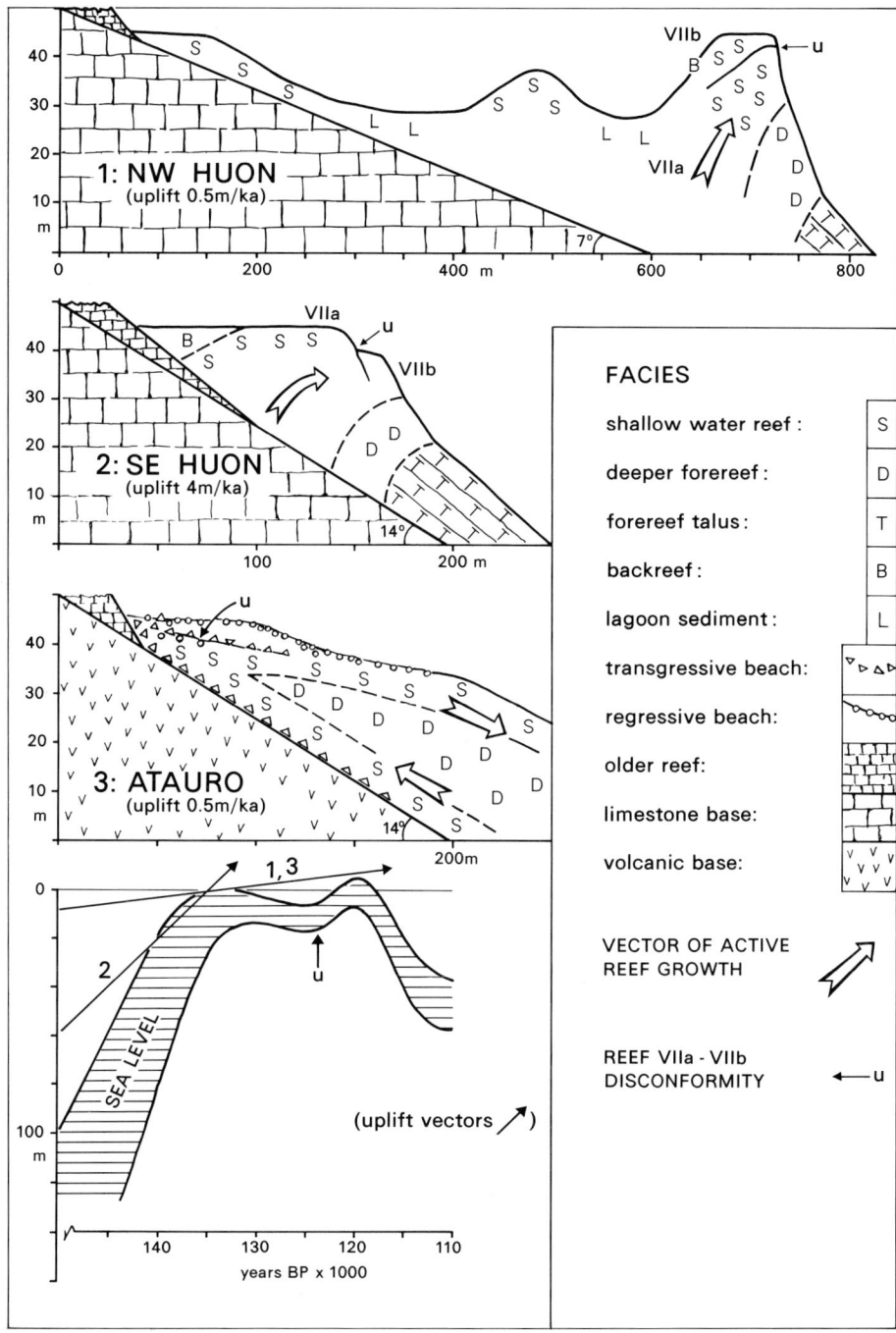

Figure 2 Cross sections of late Pleistocene reefs from Huon Peninsula, New Guinea (Chappell 1974) and Atauro Island (Chappell & Veeh 1978). These reefs grow during the sea level changes shown at bottom; symbol 'u' indicates growth hiatus (disconformity) related to minor dip in sea level about 125 000 years ago. Uplift vectors are explained in the text

Figure 3   Typical section of shallow water (upper fore-reef) growth at Huon Peninsula showing the sort of evidence on which figure 2 is based. In this case, base of section is radiocarbon dated as 7500 years old; top is 6500 years old

0.13 cm yr^{-1} (the low growth rates at Atauro probably are due to high turbidity during the wet season, when muddy clastic sediment flux from the land is high; this is not a factor at the Huon reefs).

Some important points emerge. Firstly, it is clear that barrier-lagoon structure can develop through submergence of a simple sloping substrate. It is stressed that the example shown in figure 2 is representative of a large number of occurrences of this structure at Huon Peninsula. Secondly, when reef growth rate is much lower than the sea level rise rate (Atauro case), a shelving reef without a barrier or even a reef flat develops. Finally, the S.E. Huon example (fig. 2) is particularly interesting in that here a fringing reef only has formed, due to interplay of two factors. The substrate slope is twice as steep as at N.W. Huon, giving less physical space for a lagoon to form behind the growing reef. Additionally, the relative rate of sea level rise for this section is substantially less than at N.W. Huon, due to the high uplift rate of 4 m per 1000 years. Hence, the S.E. Huon case is transgression-limited to a greater degree than the N.W. Huon case. In fact, the mean transgression rate for S.W. Huon is about 6 m per 1000 years (cf. about 10 m per 1000 years at the N.W. section), which is less than the highest growth rates identified by Chappell & Polach (1976) in their dating study of young (Holocene) reef growth. The conclusion from the S.E. section is that when potential reef growth rate exceeds transgression rate, a fringing reef will form.

COMPLICATIONS: RATE VARIATIONS, AND IRREGULAR SUBSTRATES

The foregoing sections make it clear that major growth of modern reefs occurred during glacio-eustatic transgressions, where the rate of sea level rise at any locality represents that due to ice melting, modified by local tectonic and possible isostatic factors. It was shown above that reef development on a simple sloping substrate can lead to different forms, according to rate relationships between sea level rise and potential reef growth. Both rates were treated as though they are constant at a given site throughout a transgression, which is not the case. Opinions differ about the exact course of sea level rise during the last (post-Glacial) transgression, with some (Fairbridge 1961; Morner 1971) suggesting that oscillations of about 10 m amplitude and 1000 years duration were superimposed on the more general rise of over 100 m which commenced about 17 000 years ago and terminated about 6000 years ago. Others (e.g. Bloom 1971) argue that sea level rose monotonically, without oscillations, increasing in rate to about 11 000 years ago and decreasing thereafter (cf. fig. 1). Scrutiny of published data does not seem to confirm the secondary oscillations claimed by Fairbridge and others, at least not as global events (Chappell & Thom 1978). At present Bloom's appears the more realistic interpretation, allowing that transgression rate may have altered episodically. However, reef development on a simple substrate is not only a function of relative transgression and growth rates, but is also affected by vagaries of reef establishment and disturbances to reef growth. These factors alone could generate reef growth at more than one place on a sloping surface, during transgression.

To illustrate, figure 4 shows a double barrier reef structure formed at Huon Peninsula during the last (post-Glacial) transgression. Two episodes of reef initiation clearly are indicated, but whether these reflect still-stands during the sea level rise, or adventitious successes of two embryonic reefs, is not known. Indeed, in this case it is not known whether substrate irregularities played a part in reef initiation, although any such features would have been no more than narrow terraces if they existed at all. Hence, the

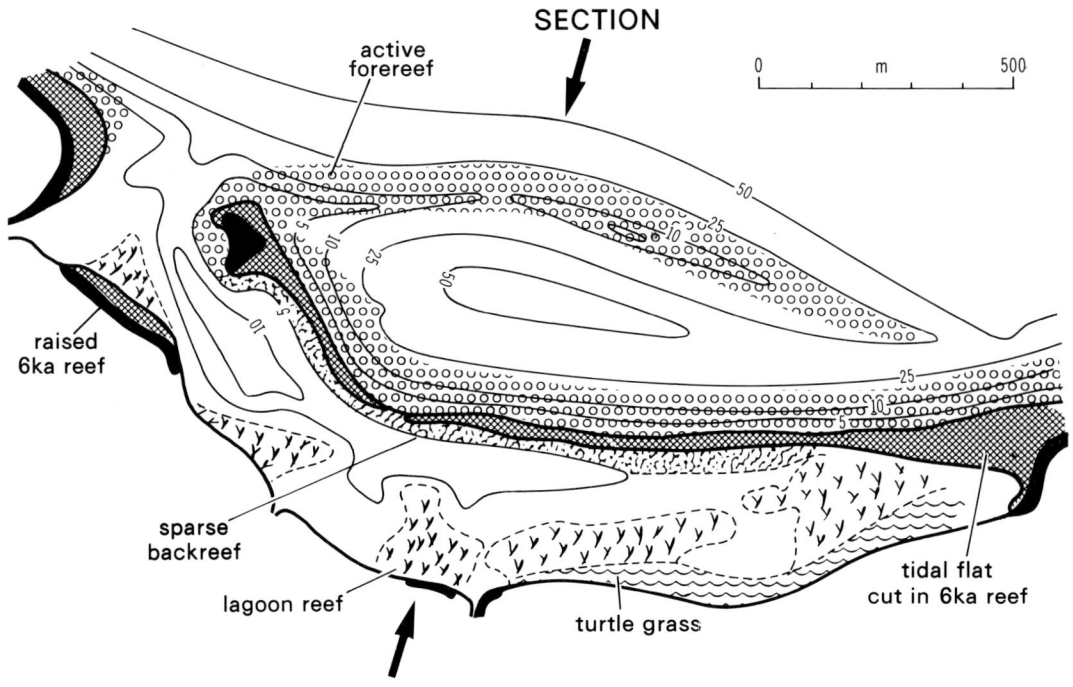

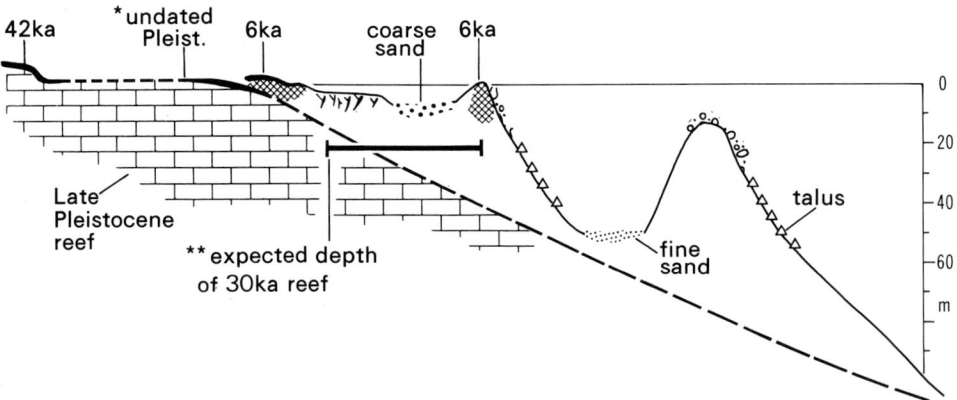

Figure 4   Double barrier-lagoon structure at Gitua, Huon Peninsula, developed during the Holocene (post-Glacial) transgression. Radiometric ages (^{14}C and ^{230}Th/^{234}U) were obtained at positions shown (note: ka = 1000 yrs). The surface of the Late Pleistocene reef is unknown but suggested to be as shown by analogy with the older reefs shown in figure 2. This pre-Holocene surface may have been terraced by a 29 ka reef (known elsewhere at Huon Peninsula where uplift rate is higher; Chappell & Veeh 1979), but is believed not to have shown any pre-Holocene barriers or major ramparts

example again affirms that barriers can develop on sloping surfaces when growth and transgression rate relationships are suitable, and also shows that a single transgression can lead to composite reef structures.

Further conclusions can be drawn from figure 4. Clearly, both barriers have amplified their initial form during transgression (i.e. vertical growth exaggerated their initial topography), implying more rapid growth on the crest than on the flanks. Higher levels of intercepted light at the crest will induce higher coral calcification rates (Chalker, chapter 3 this volume), and self-amplifying growth of the outer barrier (still up to 15 m below sea level) may be due to this factor alone. However, the inner barrier is transgression-limited in that growth at the crest ceased 6000 years ago (shown by radiocarbon dating) implying that its growth virtually matched sea level rise. It seems likely, therefore, that limitation of growth on its inner flank (necessary for self-amplification) must be caused by some factor other than light diminution alone. It is suggested that back-reef coral growth is inhibited by sediment driven inwards from the active crest. Observation of the modern reef supports this. Hence, a distinction is made between light-limited self-amplification for the outer reef, and sediment-limited amplification for the inner one. Finally, as shown below, self-amplification is significant only when sea level rise rates fall within a certain range.

OTHER DIVERSE EFFECTS OF SEA LEVEL CHANGES

We have shown that reef development is influenced strongly by general relationships between rates of sea level rise and potential growth rate of the coral framework. It is acknowledged that substrate topography strongly influences the forms of reefs at a regional level and that it affects the growth structure of any given reef during a single transgression (see Davies, chapter 6 this volume). Figure 4 also makes it clear that reef growth can be self-amplifying in a vertical sense during sea level rise, and it is the conjugation of this tendency during transgressions plus the down-cutting effects of fluvial streams during times of low sea levels that causes maintenance of a regional pattern of reef islands throughout a long succession of transgressions and regressions.

The important self-amplifying tendency is not, however, independent of transgression rates and growth rates. Again we consider the case with a simple sloping substrate. Firstly, when coral growth rate is much lower than transgression rate, a barrier does not form at all and hence there is no significant self-amplification (Atauro case, fig 2). Alternatively, when sea level rise is slow compared with potential coral growth rate, transgression leads to an ever-widening fringing reef. Intermediate between these is the general case where the self-amplifying tendency of a barrier is offset by infilling of its lagoon, through deposition of materials eroded from the reef itself as well as through organisms growing within the lagoon. Huon Peninsula examples in figures 2 and 4 show the infilling effect to different degrees, with the outer lagoonal trough seen in figure 4 being least infilled, essentially because its own seaward barrier lies between 5 m and 10 m depth, which is generally below the depth of effective wave erosion in the region. For barriers that continue to approach sea level during transgression, the infilling tendency is the sum of two factors. First, as potential lagoon area increases with decreasing substrate slope (or some other topographic shape factor) the infilling tendency decreases because the length of sediment-yielding barrier increases only as the approximate square root of the lagoon area. Second, as lagoon depth increases the growth potential of its infilling benthos (dominantly light-limited corals) decreases. Hence, lagoon depth behind a

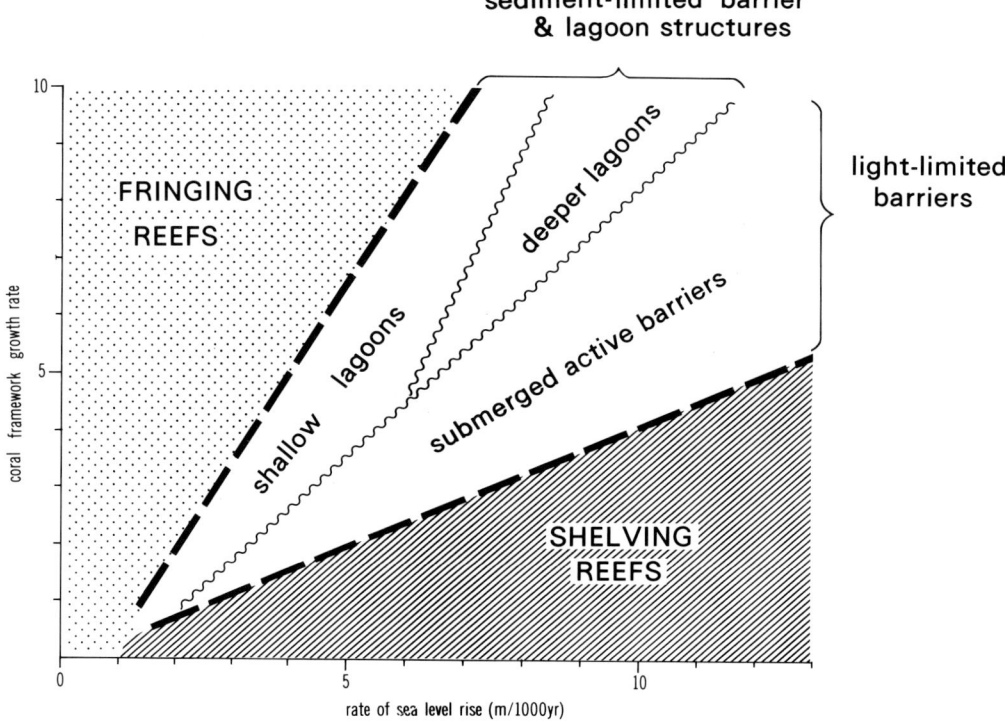

Figure 5   Suggested dependency of reef form on the relationship between rate of sea level rise and potential rate of growth of coral framework (note; Shelving reefs are represented by the Atauro example, figure 2)

transgression-limited barrier is related to transgression rate as much as it is to lagoon size. The conclusions of this and of the preceding discussion are summarized in figure 5.

Finally, we have so far considered effects of major sea level changes, in particular transgressions. The effects of smaller changes may be more subtle, but significant. Thom & Chappell (1978) have attributed some of the differences between reefs in the northern Great Barrier Reef to differentials between their respective sea level histories over the last 8000 years or so, due partly to isostatic movements across the shelf. They suggest that the post-Glacial transgression terminated nearly 6000 years ago on the inner shelf areas, whereas it continued several thousand years longer at the outer margins of the shelf, especially where it is broadest and deepest. Their argument is not reviewed here, but examination of the effects of slow or minor sea level changes raises a different issue, fundamental to the geological development of any large feature of the earth, viz., the extent to which the uniformitarian or gradualist argument is applicable. It is a question of scale. As regards the geologic development of large and complex reef provinces, the major glacio-eustatic oscillations in figure 1 constitute catastrophies — the gross effects of which must be identified. At the scale of a large transgression over 10 000 years or so the evidence reviewed suggests that, in general, reef growth is most affected by mean rates of sea level change and coral growth, although the double lagoon in figure 4 indicates that rate changes or an adventitious event can initiate substantial complications in reef structure. When rates are slow, however, (i.e. lie near the origin in figure 5) effects of disturbances such as minor sea level variations, or the occurrence of a cluster of cyclones, may be more consequential than the effects of a slow trend of sea level change.

LITERATURE CITED

Backshall, D. G., J. Barnett, P. J. Davies, D. C. Duncan, N. Harvey, D. Hopley, P. J. Isdale, J. N. Jennings and R. Moss, 1979. Drowned dolines — the blue holes of the Pompey Reefs, Great Barrier Reef. *BMR Journal of Australian Geology and Geophysics* 4, 99-109.

Bloom, A. L., 1971. Glacio-Eustatic and isostatic controls of sea level since the last glaciation. In K. Turekian (ed.), *Late Cenozoic Glacial Ages*. Yale University Press, pp. 355-79.

Bloom, A. L., W. S. Broecker, J. Chappell, R. S. Mathews and K. J. Messlella, 1974. Quaternary sea level fluctuations on a tectonic coast: new ^{230}Th/^{234}U dates from the Huon Peninsula, New Guinea. *Quaternary Research* 4, 185-205.

Chalker, B. E., chapter 3 this volume. Calcification by corals and other animals on the reef.

Chappell, J., 1974. Geology of coral terraces, Huon Peninsula, New Guinea: a study of Quaternary tectonic movements and sea level changes. *Geological Society of America Bulletin* 85, 553-70.

Chappell, J. and H. A. Polach, 1976. Holocene sea-level change and coral-reef growth at Huon peninsula, Papua New Guinea. *Geological Society of America Bulletin* 87, 235-40.

Chappell, J. and B. G. Thom, 1978. Sea levels and coasts. In J. Allen and J. Golson (eds.), *Sunda and Sahul*. Canberra: Australian National University Press, pp. 275-91.

Chappell, J. and H. H. Veeh, 1978a. Late Quaternary tectonic movements and sea-level changes at Timor and Atauro Island. *Geological Society of America Bulletin* 89, 356-68.

Chappell, J. and H. H. Veeh, 1978b. ^{230}Th/^{234}U age support of an interstadial sea level of -40 m at 30,000yr BP. *Nature* 276, 602-3.

Clark, J. A., W. E. Farrell and W. R. Peltier, 1978. Global changes in post glacial sea level: A numerical calculation. *Quaternary Research* 9, 265-87.

Daly, R. A., 1915. The glacial-control theory of coral reefs. *Proceedings American Academy of Arts and Sciences* 51, 155-251.

Darwin, C., 1842. Coral Reefs. Republished by University of California Press, Berkeley (1962), 214 pp.

Davies, P. J., chapter 6 this volume. Reef growth.

Fairbridge, R., 1962. World sea level and climatic changes. *Quaternaria* 6, 111-33.

Mesolella, K. J., H. A. Sealy and R. K. Matthews, 1970. Facies geometries within Pleistocene reefs of Barbados, West Indies. *American Association of Petroleum Geologists Bulletin* 54, 1899-917.

Morner, N. A., 1971. Late Quaternary isostatic, eustatic, and climatic changes. *Quaternaria* 14, 65-83.

Purdy, E. G., 1974. Reef configurations: cause and effect. *Society of Economic Paleontologists and Mineralogists, Special Publication* 18, 9-76.

Stoddart, D. R., R. J. McLean, T. P. Scoffin, B. G. Thom and D. Hopley, 1978. Evolution of reefs and islands, northern Great Barrier Reef, synthesis, and interpretation. *Philosophical Transactions of the Royal Society of London, Series B* 248, 149-61.

Thom, B. G. and J. Chappell, 1978. Holocene sea level change: an interpretation. *Philosophical Transactions of the Royal Society of London, Series A* 291, 187-94.

5 Dissolved Nutrients in Coral Reef Waters

C. J. Crossland

C.S.I.R.O., Division of Fisheries and Oceanography, P.O. Box 20, North Beach, W.A. 6020.

INTRODUCTION

Coral reef environments are regarded as having low dissolved nutrient concentrations (table 1). Despite this, an apparently high gross primary productivity is sustained at least in the algal-dominated benthic communities (see Kinsey 1978).

An understanding of the interaction of biomass production and maintenance in relation to nutrient availability raises several fundamental questions about coral reef ecosystems. If low concentration of dissolved nutrients is a common factor to coral reefs, how is the seemingly high primary production and biotic growth sustained? Is there a seasonality of nutrient input, e.g. rainfall, groundwater, which is then stored by some reef organisms or the reef ecosystem and subsequently released to the biota? Is tight nutrients cycling between individual organisms or biotic zones of more importance than continual nutrient inputs? Is our knowledge of nutrient chemistry and inorganic v. organic interactions in seawater sufficient to allow interpretation of growth-nutrient processes in coral reef environment?

WHAT NUTRIENTS ARE DETERMINED?

The low, dissolved nutrient status ascribed to coral reefs is based primarily on analyses of inorganic nutrients in waters surrounding oceanic reef areas. Few data are available on the relative concentration of organic nutrients.

Nutrients usually analysed in reef waters are shown in table 2, along with further nutrient components that have occasionally been measured. Values for nitrogen and phosphorus concentrations have received most attention as supply of both elements is fundamental for tissue growth. Trace element and vitamin distribution in reef water have received little attention. Indeed, low concentrations of these compounds could contribute to the low standing-crop of phytoplankton characteristic of reef waters (Jeffrey 1968).

Dissolved organic compounds have been investigated to a lesser extent. Essentially, this has been a reflection of equipment and/or sample storage facilities, manpower and logistics limitations in earlier coral reef studies. The role of organic fractions in nutrient budgets of reef communities was highlighted particularly by the SYMBIOS team in their nitrogen study of Enewetak Atoll (Webb et al. 1975). Considerable variation has existed in the methods applied to organic nutrient analyses, and this has made comparison between reef studies difficult. However, some degree of methodological uniformity has evolved recently (see Monographs in Oceanographic Methodology, 5. Coral Reefs: Research Methods. Editors D. R. Stoddart and R. E. Johannes, UNESCO 1978).

Nutrient	Temperate	Tropical
Nitrate[b] ($NO_3^- -N$)	2.0–5.0	0.1–0.3
Ammonia ($NH_4^+ -N$)	<1.0	0.2–0.5
Phosphate ($PO_4^{3-} -P$)	0.5–2.0	<0.3
Silicate[c] ($SiO_4^{2-} -Si$)	<0.2(<50)	<2.0
Dissolved Organic Phosphate (DOP–P)	<1.3	<0.15
Dissolved Organic Nitrogen (DON–N)	5.0–20.0	3.0–13.8
Dissolved Organic Carbon (DOC–C)	1500–2000	500–800

[a] From reviews by Spencer (1975) and Williams (1975) (especially for temperate water) and citations in tables 4 and 5

[b] ($NO_3^- + NO_2^-$) often determined, especially where automated methods are used, but NO_2^- is generally a small component of the total value (<10%)

[c] High values in temperate waters reflect upwelling areas and fresh water intrusions

Table 1 Generalized dissolved nutrient concentrations in surface sea waters of temperate and tropical reef regions[a]. (All values expressed as μg atom l^{-1} except DOC which is μg C l^{-1})

Inorganic:	Nitrate[a]	Silicate[b]	
	Nitrite[a]	Sulphate[c]	
	Ammonia[a]	Soluble iron[c]	
	Total nitrogen[b], including organics	Trace elements[c]	
	Phosphate — Orthophosphate[a], inorganic soluble reactive (ISR)[d], polyphosphate (PP)[c,d]		
Organic:	Dissolved organic carbon (DOC)[b,e] Dissolved organic nitrogen (DON)[b,f]		
	Vitamins[c]		
	Dissolved organic phosphorus – organic soluble reactive (OSR), [c,d] organic soluble unreactive (OSU)[c,d], polyphosphate (PP), [c,d] enzyme hydrolysable (EHP), [c,d] total phosphate (EHP + OSU + PP), [b,d] soluble reactive (OSR + 1SR)[a,d]		

a-c Determined often (a); less frequently (b); rarely (c)
d Phosphate species are discussed by Strickland and Parsons (1972)
e Carbohydrates, lipids and small molecular weight compounds have been determined infrequently (D'Elia 1978)
f Urea, peptides, amino acids have been determined infrequently (Webb 1978)

Table 2 Dissolved nutrients species determined in reef water studies

NUTRIENT PROCESSES

Particulate and dissolved organic compounds released to seawater by benthic algae and autotrophic associations provide nutrients to downstream animals and microorganisms. Biocycling of organic compounds and mineralized deposits, especially by planktonic and interstitial micro-organisms, undoubtedly increase the molecular species of nutrient elements (fig. 1). These processes may contribute to the presence or absence of specific biota and thus to the diversity of downstream communities within the reef system. It is possible that periodic nutrient inputs can be trapped in the reef ecosystem by organisms

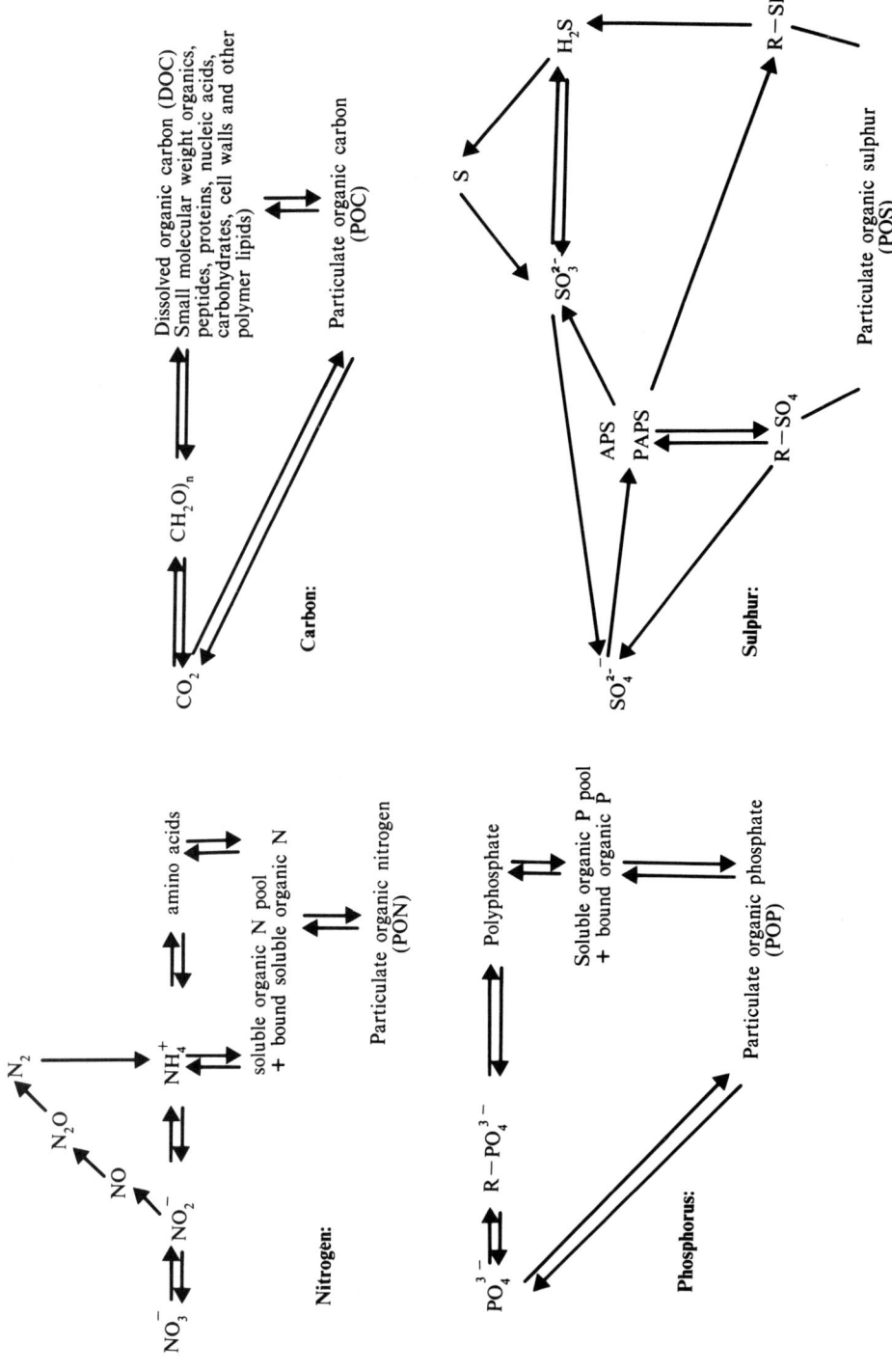

Figure 1   Nutrient relationships and potential interconversions in coral reef waters and sediments   (see Wiebe, 1979)

showing seasonal growth patterns (Tsuda 1974). They also can be stored in mineralized deposits and later released. This could provide an important temporal or spatial buffer of nutrient. Pomeroy (1970) has suggested there is comparatively little substrate storage of nutrients in reef communities, unlike shallow coastal communities with considerable sediments. However, sulphate reduction in sediments has been strongly implicated in the cycling and availability of other nutrients, e.g., phosphorus (Patrick & Khalid 1974). The cycling of sulphate, especially in lagoonal sediments (Skyring & Chalmers 1976), may be an essential mechanism in coral reef nutrient processes.

Rates and sources of nutrient production, degradation and cycling within coral ecosystems require considerable research. Little of this sequencing has been character- ized in tropical waters compared with temperate coastal and estuarine systems. However, hypotheses have been advanced relating to phosphorus (Pilson & Betzer 1973; Pomeroy *et al.* 1974) and nitrogen (Webb *et al.* 1975) fluxes in coral reef benthic communities.

APPROACHES TO NUTRIENT STUDIES

In most nutrient studies of coral reefs, the nutrient 'standing-stock' in seawater has been evaluated, that is, the nutrient concentrations present in a volume of the water column at a single point in time. Such studies provide little information of either nutrient changes in the water column or nutrient flux between the benthos, sediments and seawater.

Nutrient flux studies require a detailed understanding of hydrological characteristics of the particular reef system, that is, the history of the water before it impinges on the reef, water residency and velocity. Most nutrient flux studies have not dealt adequately with these measurements.

The nutrients contained and stored by the biota and sediments can be considered a measure of the nutritional status of reefs in much the same way such analyses are used in terrestrial studies. However, little attention has been given to such analyses. This is not surprising in light of the physical difficulties in sampling and the diversity of organisms on coral reefs. As a consequence, the study of dissolved nutrients has been adopted as a primary approach for understanding nutritional interaction in reefs, in a way analogous to the use the medical profession makes of the human blood system.

Standing-Stock Nutrients

Standing-stock nutrient concentrations in coral reef waters are summarized in tables 3 and 4. These data provide some clue of the nutrient status of a system. However, they can provide little information as to the nutrient dynamics of reef systems. Indeed, the values could be regarded as residues or thresholds. In fact, we do not know precisely how to interpret standing-stock nutrient data.

The grouping of reefs into four categories depending on gross geographical description (tables 3 and 4) has been made more in anticipation of future information than for reasons of difference in nutrient data. Low oceanic reefs and those associated with high islands may require separate consideration due to possible nutrient input variables such as ground water (Marsh 1977) and surface run-off (Wade 1976), trace elements from igneous rocks and water circulation patterns.

Lizard Island in the Great Barrier Reef (GBR) has been referred to as a moated, high island. Although there is limited information on the hydrology of the GBR (Pickard *et al.* 1977), the outer GBR formations could be considered as forming an inshore 'pond' of seawater. Limited exchange between oceanic waters and the inshore Coral Sea would

Island Type & Location	Nitrate	Nitrite	Ammonia	Phosphate	Silicate	Reference
Oceanic						
Canton Atoll, Lagoon	0.02–2.40		0.09–1.30	0.03–0.45	2.10–3.20	Smith & Henderson (1973); Smith & Jokiel (1975).
Enewetak Atoll, offshore	0.02					1. Odum & Odum (1955).
(Marshall Is.), reef1	0.06–1.00		0.30	0.26–0.64		2. Webb et al. (1975).
, reef2	0.08–0.30		0.20–0.29			3. Johannes et al. (1972).
, reef3	0.11–0.17		0.24–0.29	0.17		
, lagoon2	0.07		0.28			
Fanning Atoll	0.48–1.98		0.36	0.05–0.38		Krasnick (1973); Johannes et al. (1979).
Tarawa Atoll, general (Gilbert Is.), offshore	0.05–2.60		0.27	0.01		
, reef	0.04–0.68		0.31–0.54	0.03–0.16		
, lagoon	0.03–0.40		0.34–0.36			
Takapoto Atoll, offshore (Tuamotu Is.), lagoon	0.36		0.10	0.26	1.01	Sournia & Ricard (1976).
	0.22		0.10	0.12	0.28	
High						
Kaneohe Bay, general (Hawaii), outside bay (Ulupau)		0.30–0.94	1.60–2.40	0.16–0.36		Henderson, Smith & Evans (1976).
, inside bay (HIMB)		0.62–2.64	1.90–2.40	0.19–0.22		
Palau, general	0.30–1.20	0.05–0.85	0.40–2.40	0.28–1.01		Birkeland et al. (1976); Randall et al. (1978).
, reef	0.21			0.09–0.32		
, lagoon	0.09–0.32			0.14		
US Virgin Is., general	0.12	0.00		0.10–0.18		Dong et al. (1972).
				0.06		
, modified	1.00–2.50	0.20–0.50		0.05–0.25		
Kingston Harbour, outer (Jamaica), inner (modified)		0.10–3.75		0.02–0.70	1.00–11.15	Wade (1976).
		0.15–4.75		0.35–5.50	0.99–117.55	
High, Moated						
Lizard Is., offshore (Gt. Barrier Reef), windward reef	0.54	0.14	0.32	0.25	1.32	Crossland & Barnes (in prep.).
	0.22–1.02	0.11–0.17	0.22–0.26	0.22–0.30	1.05–1.54	
, lagoon	0.59–0.82	0.17	0.25–0.34	0.18–0.24	1.19–1.41	
, leeward reef	0.54–0.58	0.07–0.14	0.23–0.38	0.15–0.23	1.01–1.14	
High Latitude						
Abrolhos Is., general (Western Australia)	0.79–5.17	0.01–0.50	0.07–11.00	0.16–2.92	1.29–6.73	Crossland (unpublished data).

Table 3   Standing-stock inorganic nutrient concentration in coral reef waters (Values are µg atoms l^{-1})

Island Type & Location		DON	Urea	Total N	DOP	DOC	TOC	Reference
Oceanic:								
Enewetak Atoll (Marshall Is.)	offshore reef	3.0 / 1.7–2.3		2.9	0.15	1.03 / 1.08–1.30		Johannes *et al.* (1972); Webb *et al.* (1975); Marshall *et al.* (1975).
	lagoon	2.3				1.20		
Tarawa Atoll		3.8–5.6		6.1–9.4				Johannes *et al.* (1979);
Grand Cayman	offshore					0.15–0.80	0.30–0.96	Westrum & Meyers (1978);
Andros Atoll (Caicos Is.)	lagoon					1.07 / 0.38–0.59		Marshall *et al.* (1975).
High								
Kaneohe Bay (Hawaii)		3.4–7.5	0.4–2.0		0.37	1.52–1.77		Smith, Henderson & Evans (1976); Marshall *et al.* (1975).
High, Moated								
Lizard Is. (Gt. Barrier Reef)	offshore	5.0						Crossland & Barnes (in prep.).
	windward reef	3.0–4.4						
	lagoon	4.2–4.6						
	leeward reef	3.9–4.8						
High Latitude								
Abrolhos Is. (Western Australia)	general	0.1–7.4			0.02–4.89			Crossland (unpublished data).

Table 4   Standing-stock dissolved organic nutrient concentrations in coral reef waters (Values are μg atom l^{-1} except DOC and TOC; mg C l^{-1})

contribute to retention of nutrients derived from continental and biotic sources. Indeed, the nutrient levels of waters surrounding Lizard Island are generally greater than those associated with other low latitude reefs.

Initial data (Crossland, unpublished) from a high latitude coral reef system (Houtman Abrolhos Group, Western Australia) suggest nutrient values comparable to temperate coastal waters. Sites of temperate-tropical community interface such as the Abrolhos Group may provide valuable study areas for elucidating factors that determine gross biotic assemblages and processes.

Studies have been made of a number of coral reefs subject to nutrient enrichment, such as pollution by man-made effluents (Dong *et al.* 1972; Johannes 1973; Banner 1974; Hansen & Gunderson 1976; Wade 1976; Smith 1977; Johannes *et al.* 1979). Investigations of the effects of sewage outfall in Kaneohe Bay, Hawaii indicate little variation in dissolved nutrient concentration but considerable changes in the benthos (see Smith 1977). Interpretation of effluent effects on reef processes is hampered by the lack of a basic understanding of how coral reef ecosystems function. The availability of tank facilities that can be used for small community assemblages, e.g., the microcosm facilities at Hawaii Institute of Marine Biology (HIMB) and Ulupau, Hawaii (Henderson *et al.* 1976; Henderson & Smith 1978) will aid expansion of our knowledge of key reef processes. Also, field experiments such as those directed to the effects of high N and P fertilizer on reef communities (Kinsey & Domm 1974; Kinsey & Davies 1979) provide a beginning for controlled studies in this sphere.

Nutrient Fluxes

An understanding of the dynamics of dissolved nutrients in coral reefs requires study of seasonal and diurnal changes in nutrients as a defined water mass passes across the reef communities. A unidirectional flow of water is ideal for a study area. However, this is not a requirement as long as predictable water flow patterns can be ascertained. An ecological description of the study area is needed and should include features such as zones and types of community assemblages, relative areas of benthic organisms and sediment-water interfaces, and pelagic fish communities.

Several studies have been made of nutrient fluxes in reef ecosystems; at Enewetak Atoll (SYMBIOS Team, see Johannes *et al.* 1972) and Lizard Island (Crossland & Barnes, in prep.). Both studies were carried out by large interdisciplinary teams in a concentrated study period.

Enewetak Atoll windward reef study areas showed a major reefal export of nitrogen to the lagoon, especialy dissolved organic nitrogen (DON) (table 5). An open nitrogen cycle was operating which particularly involved nitrogen fixation by benthic blue-green algae (Wiebe *et al.* 1975). While dissolved organic phosphate was exported to the lagoon from the algal community, a closed phosphorus cycle operated in coral communities (Pilson & Betzer 1973; Pomeroy *et al.* 1974).

The Lizard Island study was carried out across the whole reef system with water flowing across a windward reef to the lagoon, then crossing a leeward reef (see Moriarty 1979). Community assemblages differed greatly. Generally, there was net export of dissolved nutrients by the windward reef, similar to that for Enewetak Atoll (table 6). However, the leeward reef communities further affected the dissolved nutrient content of the water and hence the total system import-export balance. Study of different community assemblages on both reefs showed variable nutrient exchange (both in sign and magnitude) occurred as waters moved downstream from one biotic zone to the next

Transect II

Community Structure: (windward to leeward)	algae		corals
Nutrient Uptake	Nitrate		
Nutrient Output	Ammonia		Nitrate, Ammonia, DON
Net Export		DON, Ammonia, Phosphate loss neglible	

Transect III

Community Structure:		algae	
Nutrient Uptake		Nitrate, Ammonia, inorganic Phosphate	
Net Export		DON, Ammonia, DOP	

Table 5 Gross nitrogen and phosphorus flux across the windward reef of Enewetak Atoll (summarized from Webb *et al.* 1975)

Nutrient	Windward Reef	Leeward Reef	Net
Ammonia	Uptake (day) Nil change (night)	Uptake (day) Nil change (night)	Consumption
Nitrate	Output (day) Nil change (night)	Uptake (day) Output (night)	Nil change
Nitrite	Output	Uptake	Consumption
DON	Output	Output	Production
Inorganic Phosphate	Uptake	Output	Nil
Silicate	Uptake (day) Output (night)	Uptake (day) —	Consumption

Table 6 Gross nutrient flux across the reef system of Lizard Island, Great Barrier Reef

(table 7). While considerable dissolved nutrient flux occurred, the concentration and species of dissolved nutrient in waters downstream of the total reef system was similar to that of inflowing seawater near the windward reef; that is, while individual reef communities caused measurable changes the reef system as a whole caused little net change to the nutrient chemistry of the seawater.

Windward Reef Transect
Benthic Communities: (windward to lagoon)

Nutrients	Coral	Coral algae	Coral rubble filamentous algae	Coral patches, rubble filamentous algae	Coral patches sand	Net change across total transect
Ammonia	C	—	P	P	P	C
Nitrate	P	P	P	P	C	P
DON	C	P	P	P	C	P (low)
inorg. Phosphate	P	P	C	C	P	C (low)
Silicate	C	—	—	—	—	C (low)

Leeward Reef Transect
Benthic Communities: (lagoon to leeward)

Nutrients	Coral	Coral (hard & soft) algae	Coral–Halimeda patches filamentous algae, sand	Coral	Net change across total transect
Ammonia	P	C		C	C
Nitrate	C	C	C	P	C (low)
DON	P	C	P	P	P
inorg. Phosphate	P	P		P	P (low)
Silicate	C	C	—	—	C (low)

Table 7   Gross nutrient flux across benthic communities of windward and leeward reef transects of Lizard Island, Great Barrier Reef. Production (P) and consumption (C) of nutrients were calculated from diurnal rates of nutrient change. No detectable nutrient flux is indicated as (—).

Although both studies provide information on the diurnal dynamics of dissolved nutrient flux in coral reef ecosystems, the effects of seasonal variations in physical, chemical and biological factors is unknown. This point needs emphasizing. Coral reefs do show seasonal changes and cannot be considered invariant communities.

Dissolved nutrient flux studies have also been carried out with individual plant and animal species to determine not only the physiological characteristics of dominant reef organisms but also their role as possible sources of sinks for nutrients. Much of this work has been related to benthic macro-organisms (see Muscatine 1973; D'Elia 1977; D'Elia & Webb 1977; Webb *et al.* 1977; Muscatine & D'Elia 1978), with limited studies on benthic and water column micro-organisms and sediments (Sorokin 1973, 1978; Di Salvo 1974; LIMER 1975 Expedition team 1976; Skyring & Chalmers 1976; Weibe 1978; Moriarty 1979). In view of the major role micro-organisms have in nutrient cycling, much greater effort should be made to examine their roles in coral reef systems.

CONCLUSIONS

The concentration of dissolved nutrients in seawater provides a tool for use in furthering our understanding of coral reefs at all levels of spatial and temporal organization. The identity and concentration of individual dissolved nutrients can be determined with a reasonable degree of certainty. However, these data are of little value unless a greater knowledge is gained of nutrient cycling in coral reef environments. Nutrient concentrations alone provide minimal information, and future study should be aimed towards appraising the dynamics of nutrient fluxes. However, dissolved nutrients can not be studied in isolation but should be integrated with companion studies of particulate nutrients, parameters of the physical environment, description of ecological assemblages, and hydrological and geological features of study sites. The requirement for a multidisciplinary team-approach is obvious.

Comparative studies on a global basis are required. Here, determination of diurnal and seasonal flux dynamics of dissolved nutrients and extension of these studies to identifying real sources and sinks for nutrients would make a large contribution towards understanding processes in reef ecosystems. The information gained could benefit interpretation of coral community structure and provide a basis for comparative study between oceanic and coastal reefs, and reefs at different latitudes. In addition, the ability to monitor nutrient variations *and understand the processes they reflect* in coral reef ecosystems may prove to be of practical value to environmental management agencies. Attainment of coherent study objectives, by this or a similar approach, would provide valuable insight into a major global ecosystem (Smith 1978).

ACKNOWLEDGEMENTS

I wish to thank Drs R. E. Johannes, S. V. Smith and W. J. Wiebe for helpful discussion and comment.

LITERATURE CITED

Birkeland, C., R. T. Tsuda, R. H. Randall, S. S. Amesbury and F. Cushing, 1976. Limited current and underwater biological surveys of a proposed sewer outfall site on Malakal Island, Palau. *University of Guam Marine Laboratory Technical Report* no. 25.

Banner, A. H., 1974. Kaneohe Bay, Hawaii: urban pollution and a coral reef ecosystem. In, *Proceedings of the Second International Symposium on Coral Reefs.* Brisbane; Great Barrier Reef Committee, vol. 2, pp. 685–702.

D'Elia, C. F., 1977. The uptake and release of dissolved phosphorus by reef corals. *Limnology and Oceanography* 22, 301–15.

D'Elia, C. F., 1978. Dissolved nitrogen, phosphorus and organic carbon. In D. R. Stoddart and R. E. Johannes (eds.), *Coral Reefs: Research Methods*. UNESCO, Paris, Monographs on Oceanographic Methodology, no 5, pp. 485–97.

D'Elia, C. F. and K. L. Webb, 1977. The dissolved nitrogen flux of reef corals. In D. L. Taylor (ed.), *Proceedings: Third International Coral Reef Symposium*. Miami: University of Miami, vol, 1, pp. 325–30.

DiSalvo, L. H., 1974. Soluble phosphorus and amino nitrogen released to seawater during recoveries of coral reef regenerative sediments. In, *Proceedings of the Second International Symposium on Coral Reefs*. Brisbane: Great Barrier Reef Committee, vol. 1, pp. 11–19.

Dong, M., J. Rosenfeld, G. Redman, K. Ronnholm, D. Kenisberg, M. Elliot, P. Novak, J. Balazy, C. Cunningham, B. Poole and C. Karnow, 1972. The role of man-induced stresses in the ecology of Long Reef and Christiansted Harbour, St. Croix, U.S. Virgin Islands. *Report, Virgin Islands Project for Environmental Research (VIPER)* U.S. Virgin Islands: Fairleigh Dickinson University.

Hansen, R. B. and K. Gundersen, 1976. Influence of sewage discharge on nitrogen fixation and nitrogen flux from coral reefs in Kaneohe Bay, Hawaii. *Applied and Environmental Microbiology* 31, 942–8.

Henderson, R. S. and S. V. Smith, 1978. Flow-through microcosms for simulation of marine ecosystems: Changes in biota and oxygen production of semi-tropical benthic communities in response to nutrient enrichments. *Naval Ocean Systems Centre Technical Report 310 NOSC-TR-310*. San Diego, California: Naval Ocean Systems Centre.

Henderson, R. S., S. V. Smith and E. C. Evans, III, 1976. Flow-through microcosms for simulation of marine ecosystems: Development and intercomparison of open coast and bay facilities. *Report, Naval Undersea Research and Development Centre, San Diego, NUC-TP-519. United States Department of the Navy.*

Jeffrey, S. W., 1968. Photosynthetic pigments of the phytoplankton of some coral reef waters. *Limnology and Oceanography* 13, 350–5.

Johannes, R. E., 1973. Coral reefs and pollution. In M. Ruivo (ed.), *Marine Pollution and Sealife*. New York: Unipub. Inc., pp. 364–75.

Johannes, R. E., W. Kimmerer, R. Kinzie, E. Shiroma and T. Walsh, 1979. The impacts of human activities on Tarawa Lagoon. *Report to Government of Gilbert Islands.*

Johannes, R. E., J. Alberts, C. D'Elia, R. A. Kinzie, L. R. Pomeroy, W. Sottile, W. Wiebe, J. A. Marsh Jr., P. Helfrich, J. Maragos, J. Meyer, S. Smith, D. Crabtree, A. Roth, L. R. McCloskey, S. Betzer, N. Marshall, M. E. Q. Pilson, G. Telek, R. I. Clutter, W. D. DuPaul, K. L. Webb and J. M. Wells Jr., 1972. The metabolism of some coral reef communities: a team study of nutrient and energy flux at Eniwetok. *Bioscience* 22, 541–3.

Kinsey, D. W., 1978. Productivity and calcification estimates using slack-water periods and field enclosures. In D. R. Stoddart and R. E. Johannes (eds.), *Coral Reefs: Research Methods*. Paris: UNESCO, Monographs on Oceanographic Methodology, no. 5, pp. 439–68.

Kinsey, D. W. and P. J. Davies, 1979. Effects of elevated nitrogen and phosphorus on coral reef growth. *Limnology and Oceanography* 24, 935–40.

Kinsey, D. W. and A. Domm, 1974. Effects of fertilization on a coral reef environment — primary production studies. In, *Proceedings of the Second International Symposium on Coral Reefs*. Brisbane: Great Barrier Reef Committee, vol. 1, pp. 49–66.

Krasnick, G. 1973. Phytoplankton pigments and nutrient concentrations in Fanning Lagoon. *Final Report, Fanning Island Expedition*. Hawaii Institute of Geophysics, HIG-73/13, pp. 51–60.

Limer 1975 Expedition Team, 1976. Metabolic processes of coral reef communities at Lizard Island, Queensland. *Search* 7, 463–8.

Marsh, J. A., Jr., 1977. Terrestrial inputs of nitrogen and phosphorus on fringing reefs of Guam. In D. L. Taylor (ed.), *Proceedings: Third International Coral Reef Symposium*. Miami: University of Miami, vol. 1, pp. 331-6.

Marshall, N., A. G. Durbin, R. Gerber and G. Telek, 1975. Observations on particulate and dissolved organic matter in coral reef areas. *Internationale Revue der gesamten Hydrobiologie* 60, 719-36.

Moriarty, D., 1979. Biomass of suspended bacteria over coral reefs. *Marine Biology* 53, 193-200.

Muscatine, L., 1973. Nutrition of corals. In O. A. Jones and R. Endean (eds.), *Biology and Geology of Coral Reefs*. Academic Press, New York, vol. I Biology 1, pp. 77-115.

Muscatine, L. and C. F. D'Elia, 1978. The uptake, retention, and release of ammonium by reef corals. *Limnology and Oceanography* 23, 725-34.

Odum, H. T. and E. P. Odum, 1955. Trophic structure and productivity of a windward coral reef community on Eniwetok Atoll. *Ecological Monographs* 25, 291-320.

Patrick, W. H., Jr. and R. A. Khalid, 1974. Phosphate release and absorption by soils and sediments: Effect of aerobic and anaerobic conditions. *Science* 186, 53-5.

Pickard, G. L., J. R. Donguy, C. Henin and F. Rougerie, 1977. A review of the physical oceanography of the Great Barrier Reef and western Coral Sea. *Australian Institute of Marine Science Monograph Series*. Canberra: Australian Government Publishing Service, 2.

Pilson, M. E. Q. and S. B. Betzer, 1973. Phosphorus flux across a coral reef. *Ecology* 54, 581-8.

Pomeroy, L. R., 1970. The strategy of mineral cycling. *Annual Review of Ecology and Systematics* 1, 171-90.

Pomeroy, L. R., M. E. Q. Pilson and W. J. Wiebe, 1974. Tracer studies of the exchange of phosphorus between reef water and organisms on the windward reef of Eniwetok atoll. In, *Proceedings of the Second International Symposium on Coral Reefs*. Brisbane: Great Barrier Reef Committee, vol. 1, pp. 87-96.

Qasim, S. Z. and V. N. Sankaranarayanan, 1970. Production of particulate organic matter by the reef on Kavaratti Atoll (Laccadires). *Limnology and Oceanography* 15, 574-8.

Randall, R. H., C. Birkeland, S. S. Amesbury, D. Lassay and J. R. Eads, 1978. Marine survey of a proposed resort site at Arakabesan Island, Palau. *University of Guam Marine Laboratory Technical Report* 44.

Skyring, G. W. and L. A. Chambers, 1976. Biological sulphate reduction in carbonate sediments of a coral reef. *Australian Journal of Marine and Freshwater Research* 27, 595-602.

Smith, S. V., 1977. Kaneohe Bay: a preliminary report on the responses of a coral reef/estuary ecosystem to relaxation of sewage stress. In D. L. Taylor (ed.), *Proceedings: Third International Coral Reef Symposium*. Miami: University of Miami, vol. 2, pp. 577-83.

Smith, S. V., 1978. Coral-reef area and the contributions of reefs to processes and resources of the world's oceans. *Nature* 273, 225-6.

Smith, S. V. and R. S. Henderson, 1976. An environmental survey of Canton Atoll lagoon, 1973. *Report, Naval Undersea Research and Development Centre, San Diego. NUC-TP-395*. United States Department of the Navy.

Smith, S. V. and P. L. Jokiel, 1975. Water composition and biogeochemical gradients in the Canton Atoll lagoon: 2. Budgets of phosphorus, nitrogen, carbon dioxide, and particulate materials. *Marine Science Communications* 1, 165-207.

Sorokin, Yu. I., 1973. Microbiological aspects of the productivity of coral reefs. In O. A. Jones and R. Endean (eds.), *Biology and Geology of Coral Reefs*. New York: Academic Press, vol. 2. Biology 1, pp. 17-45.

Sorokin, Yu. I., 1978. Microbial production in coral-reef community. *Archiv fr Hydrobiologie* 83, 281-323.

Sournia, A. and M. Ricard, 1976. Données sur l'hydrologie et la productivité du lagon d'un atoll fermé (Takapoto, Iles Tuamotu). *Vie et Milieu* 26, 243-79.

Spencer, C. P., 1975. The micronutrient elements. In J. P. Riley and G. Skirrow (eds.), *Chemical Oceanography*. London: Academic Press, vol. 2, pp. 245-99.

Strickland, J. D. H. and T. R. Parsons, 1972. A practical handbook of seawater analysis. *Fisheries Research Board of Canada, Bulletin* 167, 2nd ed., Ottawa, 310 pp.

Tsuda, R. T., 1974. Seasonal aspects of the Guam Phaeophyta (brown algae). In, *Proceedings of the Second International Symposium on Coral Reefs*. Brisbane: Great Barrier Reef Committee, vol. 1, pp. 43-7.

Wade, B. A., 1976. The pollution ecology of Kingston Harbour, Jamaica. Scientific Report of the U.W.I.-O.D.M. Kingston Harbour Research Project, 1972-1975. *Research Report from the Zoology Department, University of the West Indies, No. 5*. Mona: University of the West Indies.

Webb, K. L., 1978. Nitrogen determination. In D. R. Stoddart and R. E. Johannes (eds.), *Coral Reefs: Research Methods*. Paris: UNESCO, Monographs on Oceanographic Methodology, No. 5, pp. 413-19.

Webb, K. L., W. D. DuPaul and C. F. D'Elia, 1977. Biomass and nutrient flux measurements on *Holothurai atra* populations on windward reef flats at Enewetak, Marshall Islands. In D. L. Taylor (ed), *Proceedings: Third International Coral Reef Symposium*. Miami: University of Miami, vol. 1, pp. 409-15.

Webb, K. L., W. D. DuPaul, W. Wiebe, W. Sottile and R. E. Johannes, 1975. Enewetak (Eniwetok) Atoll: Aspects of the nitrogen cycle on a coral reef. *Limnology and Oceanography* 20, 198-210.

Westrum, B. L. and P. A. Meyers, 1978. Organic content of seawater from over three Caribbean reefs. *Bulletin of Marine Science* 28, 153-8.

Wiebe, W. J., 1978. Flux of bacteria. In D. R. Stoddart and R. E. Johannes (eds.), *Coral Reefs: Research Methods*. Paris: UNESCO, Monographs on Oceanographic Methodology, No. 5, pp. 433-8.

Wiebe, W. J., 1979. Anaerobic benthic microbial processes: Changes from the estuary to the Continental Shelf. In R. J. Livingston (ed.), *Ecological Processes in Coastal Marine Systems*. New York: Plenum Publishing Corporation, pp. 469-85.

Wiebe, W. J., R. E. Johannes and K. L. Webb, 1975. Nitrogen fixation in a coral reef community. *Science* 188, 257-9.

Williams, P. J. le B., 1975. Biological and chemical aspects of dissolved organic matter in seawater. In J. P. Riley and G. Skirrow (eds), *Chemical Oceanography*. London: Academic Press, vol. 2, pp. 301-63.

6 Reef Growth

Peter J. Davies

Bureau of Mineral Resources, P.O. Box 378, Canberra City, A.C.T. 2601

INTRODUCTION

Research over the past decade has resulted in significant advances and important new insights into the processes, products and parameters affecting the growth of coral reefs. No longer are Atlantic reefs considered to be poorly developed expressions of their Pacific counterparts. Algal ridges have been described from the Atlantic equal to those described elsewhere, and rates of reef growth are reported to be two or three times faster than in the Pacific. Views on reef structure and growth are therefore changing radically, especially in the identification and measurement of growth, and in the understanding of processes affecting growth. In this review, these three aspects are considered in terms of both the world-wide data base, and that generated for One Tree Reef, the most intensively studied reef in the Great Barrier Reef. Future directions of reef research are examined in the light of present knowledge and its extension to an organized view of coral reef development.

GROWTH PROCESSES

Reef growth involves the increase of both organic material and calcium carbonate. Organic production has recently been the subject of an extensive review (Lewis 1977), so that the present contribution concentrates on carbonate production.

Biologic Accretion

Growth of reef-building organisms during the Holocene has added only a thin veneer of modern reef rock over pre-existing fossil reefs. Recent growth is of the order of only 3–33 m (Davies et al. 1977b; Macintyre et al. 1977; Thom et al. 1978); in the Great Barrier Reef the thickest recorded sequence is 23 m (Harvey et al. 1979). Boreholes through reefs provide comprehensive information on framework development in the Atlantic, Pacific and Indian Oceans. In the Atlantic, cores from the Virgin Islands (Adey et al. 1977; Adey 1978), Campeche Bank (Macintyre et al. 1977) and Galeta Point, Panama (Macintyre & Glynn 1976) show reefs beginning as head coral assemblages and changing vertically into branching assemblages dominated by *Acropora palmata* and/or *Acropora cervicornis*. Some reefs are dominated by either branching (*Acropora palmata* in Florida, Lighty et al. 1978) or massive forms in Curacao (Focke 1978). In the Pacific, branching *Pocillopora* dominates in the Gulfs of Panama and Chiriqui (Glynn & Macintyre 1977) and head corals dominate at Bikini (Wells 1954). Little data is available from the Indian

Ocean, although a borehole at Reunion Island shows branching *Acropora, Porites, Montipora and Pocillopora* dominant in the upper 18 m of the section (Montaggioni 1977).

In the Great Barrier Reef, little is known about the vertical succession of framework builders in the Holocene. Hopley (1977) obtained 1–2 m cores from Carter Reef which show *Porites* spp. dominating within 200 m of the reef margin while both *Acropora* spp. and *Goniopora* spp. occur in the top 2 m further to leeward. Corals were unfortunately not identified in the holes drilled at Bewick and Stapleton Reefs (Thom *et al.* 1978). A head coral assemblage also dominates in the cores at Hayman Island (Hopley *et al.* 1978). On the windward margin of One Tree Reef, in the southern Great Barrier Reef, a head coral assemblage dominates throughout the sequence *(Porites* spp., *Symphyllia* spp. and *Platygyra* spp.), whereas a branching *Acropora* assemblage constitutes much of the 18 m of Holocene section at the leeward margin.

The role of coralline algae in the construction of coral reefs was until recently considered an Indo-Pacific phenomenon (Glynn 1973; Adey 1978). Algal ridges have, however, been described from Bermuda (Ginsburg & Schroeder 1975), the Virgin Islands (Adey 1975) and throughout the Lesser Antilles (Adey & Burke 1976; Adey *et al.* 1977). The taxonomy of the calcareous algae in the Atlantic has been thoroughly described (Adey & Vasser 1975), as has the role of environment in the control of morphology (Steneck & Adey 1976). These same authors report thick crusts of *Lithophyllum congestum* and *Porolithon pachydermum,* which accrete at rates of 1 to 5.2 mm yr^{-1}, at St Croix, Virgin Islands. *Lithophyllum congestum* grows best in a zone 10 cm above and below mean low water.

Algal ridges have been described as major constructional features of Pacific reefs for many decades; their origin having been attributed to the protective veneer of corallines in the surf zone on windward margins. A key to the coralline genera found in the Pacific (Adey & Macintyre 1973) shows that those concerned with reef construction are rigid and inflexible: they include *Porolithon onkodes* and various species of *Hydrolithon, Lithophyllum, Sporolithon* and *Neogoniolithon.* Various studies have shown that coralline species inhabit specific habitats on windward margins (Tracey *et al.* 1948; Doty 1974). In the Great Barrier Reef, studies of coralline algae have been generally neglected, although *Lithophyllum mullucense* is reported as 'important' (Cribb 1973). Cribb states that 'taxonomic work on these algae has not reached a stage where the distribution of the various species in the seaward platform can be considered profitably' (Cribb 1973, p. 51). However, recent collections made by members of the Smithsonian Institute may soon correct this situation.

The 'algal rim' on reefs of the Great Barrier Reef has been described as an elevated 'lithothamnion' ridge (Maxwell 1968, 1969) consisting of a hard, wave-resistant, stepped pavement of encrusting coralline algae; predominantly *Lithothamnium* and *Porolithon.* Fleshy algae form an extensive turf over the coralline pavement (Cribb 1973). It has been suggested by Adey and Burke (1976) that the Great Barrier Reef pavements are akin to the fleshy algal pavements described from Galeta Point, Panama and St Croix, U.S. Virgin Islands. However, studies by Veron & Hudson (1978), on the 'ribbon reefs' in the northern Great Barrier Reef suggest that an algal ridge of the kind described above does not exist. Instead, the reef crest consists of a hard, flat, denuded limestone of solid coral rock. Shallow cores on Carter Reef (Hopley 1977) show *in situ* coral and 'algal cement' approximately 0.5 m thick, confirming Veron & Hudson's observations on the surface morphology of ribbon reefs.

Recent studies (Davies & Marshall 1980) in the southern Great Barrier Reef indicate that the coralline crusts forming the smooth, hard, stepped pavements on windward margins are thin, and do not extend very deep into the subsurface. However, much thicker accumulation of coralline crusts occurs between 4 and 8 m below the reef surface. These deeper crusts are associated with *Millepora,* and probably formed in shallow water environments similar to those described for *Lithophyllum congestum* and *Porolithon pachydermum* in the present-day Atlantic.

In the southern Great Barrier Reef, stepped coralline-encrusted surfaces are best seen a little to leeward of reef fronts, especially on windward margins. Progressing around the reef perimeter, the smooth stepped surfaces are sometimes replaced laterally by a dense low coral cover with little encrusting corallines. At Redbill Reef, in the Pompey Complex, the stepped coralline crusts occur around all margins of the reef. On the western leeward-facing margin of One Tree Reef, coralline activity is prolific, forming both extensive crusts and rhodoliths.

Sedimentologic Accretion

In the Great Barrier Reef, most studies have concentrated on the description of surface sediments, particularly those occurring in lagoons or islands (Maiklem 1970; Davies *et al.* 1977a; Flood 1977; Flood & Scoffin 1978; McLean & Stoddart 1978). Flood *et al.* (1978) claim that faunal and floral contributions to sedimentation are size-related, i.e., very coarse sands and gravels are dominantly coral (especially *Acropora)* and coralline algae; coarse and very coarse sands are mainly composed of foraminifera; medium to fine sands are compositionally mixed; and fine sands are composed of coral fragments, *Halimeda* dust, sponge spicules and planktonic foraminifera. Thus, it seems that little coralline debris appears in sediment fractions other than very coarse sand and gravel, which in turn suggests that coralline material finer than very coarse sand is either not produced, or is lost from the system. *Halimeda* flourishes in most reef environments (Orme *et al.* 1974), but predominantly inhabits the reef flat and outer slopes. In the former environment, it may form greater than 5% of the sediments. Calcification estimates of 50 g $CaCO_3$ m^{-2} yr^{-1} have been made for shallow water non-reefal areas in Bermuda (Wefer 1980).

Inorganically precipitated sediments have apparently contributed little to the sedimentary development of reefs in the Great Barrier Reef. At Lizard Island, ooids are reported as forming up to 10% of intertidal lagoonal sediments (Davies & Martin 1976).

Little information is available, within the Great Barrier Reef, relating to subsurface sediment variation. Lagoonal sediments have been described as coarse, ill-sorted, at least 2 m thick, sometimes containing up to 40% *Halimeda* and showing little evidence of bioturbation (Frankel 1977). On the other hand, punch cores obtained from the bottom of blue holes in the Pompey Reefs indicate extensive bioturbation (Backshall *et al.* 1979).

Destruction

Reef growth is the sum of both constructive and destructive processes operating over a significant period of time. Boulder tracts, eroded reef flats, cay sediments and lagoonal sediments are visible reminders that destructive processes are continually operative and are substantially affecting reef growth. It is a paradox, but nevertheless true that destructive processes have received little attention compared with their constructive counterparts. Destruction of reefs by all forms of erosion has been estimated by relatively few workers. Estimated rates range from 0.05 to 4.0 mm yr^{-1} (table 1). The most recent,

Location	Rate (mm yr^{-1})	Reference
Bikini	0.5	Tracey & Ladd 1974
Guam	0.05	Revelle & Emery 1957
Aldabra	0.5-4.0	Trudgill 1976
Bermuda	1.2	Bromley 1978
Western Australia	0.5-1.0	Hodgekin 1964
Great Barrier Reef	0.25 (solution)	Davies 1974
	0.2-2.9 (bioerosion)	Trudgill, in press

Table 1 Gross estimates of marine erosion of limestones

Agent	Rate	Location	Reference
Sponges) Parrot fish) *Diadema antillarum*)	0.67 cm yr^{-1}	Florida	Hudson 1977
Cliona sp.	0.17 kg m^{-2} yr^{-1}	Florida	Hein & Risk 1975
Cliona sp.	6-7 kg m^{-2} 100 d^{-1}	Bermuda	Neuman 1966
Sponges	0.25-3 kg m^{-2} yr^{-1}	Bermuda	Rützler 1975
Cliona sp.	8-382 g m^{-2} yr^{-1}	Barbados	Stearn & Scoffin 1977
Sponges	1-1.8 kg m^{-2} yr^{-1}	St Croix (U.S.V.I.)	Moore & Shed 1977
Diadema antillarum	9.7 kg m^{-2} yr^{-1}	Barbados	Hunter 1977; Stearn & Scoffin 1977
Echinometra	24 g yr^{-1}	Barbados	McLean 1964
Eucidaris thouarsii	1.7 kg m^{-2} yr^{-1}	Galapagos	Glynn *et al.* 1979
Holothuria difficilis	1 kg m^{-2} yr^{-1}	Pacific	Bakus 1973
Nerita tesselata	154 g m^{-2} yr^{-1}	Barbados	McLean 1967
Acmaea	2.4 g yr^{-1}	Barbados	McLean 1964
Amphopleura sp.	0.2-2.9 mm yr^{-1}	G.B.R.	Trudgill, in press
	18 cm^3 d^{-1} chiton^{-1}	G.B.R.	McLean 1974
Lithophaga sp.	0.911 cm yr^{-1}	Aldabra	Trudgill 1976
Lithotrya sp.	0.844 cm yr^{-1}	Albadra	Trudgill 1976
Tridacna crocea	0.14 kg m^{-2} yr^{-1}	G.B.R.	Hamner & Jones 1976
Mytilus	0.18 kg m^{-2} yr^{-1}	Florida	Hein & Risk 1975
Sipunculids/sponges	0.13-1.67 mm yr^{-1}	G.B.R.	Trudgill, in press
Spionid worms	0.04 kg m^{-2} yr^{-1}	Florida	Hein & Risk 1975
Parrot fish	40-170 g m^{-2} yr^{-1}	Barbados	Frydl & Stearn 1978
Grazing & browsing fish	230 g m^{-2} yr^{-1}	Bermuda	Bardach 1961
Parrot fish	210 g m^{-2} yr^{-1}	Bermuda	Gygi 1975
Browsing fish	1.3 mm yr^{-1}	Bermuda	Bromley 1978
Parrot fish	490 g m^{-2} yr^{-1}	Panama	Ogden 1977
Grazing & browsing fish	40-60 g m^{-2} yr^{-1}	Pacific Atolls	Cloud 1959

Table 2   Rates of bioerosion in intertidal reef environments

and probably the most reliable, study was conducted at Aldabra where rates of up to 4 mm yr^1 were measured on lithified intertidal platforms (Trudgill 1976). Erosion is thirty times greater in intertidal compared with the subtidal environments. The Aldabra study also showed bioerosion was by far the most important instrument of destruction, a conclusion in accord with studies elsewhere. It is therefore pertinent to consider the reported destructive activities of particular reef dwellers (table 2).

The primary eroders in many coral reefs are the sponges which account, on some Barbados reefs, for 90% of the total borings in most coral heads (MacGeachy 1977). In Florida, Hudson (1977) showed boring sponges to be capable of reducing a one metre high *Montastrea annularis* head at the rate of 14-67 mm yr^{-1} (3.0-13.4 kg CaCO$_3$ m^{-2} yr^{-1}). Sponge chips are reported as comprising 30% of the total lagoon sediment at Fanning Atoll (Fütterer 1974) and 40% of silt-sized sediments within patch reefs off Belize (Halley *et al.* 1977). Experimental studies on *Cliona lampus* have shown potential erosion rates by this sponge of 21-25 kg CaCO$_3$ m^{-2} yr^{-1} (Neuman 1966). However, field studies by Rützler (1975) showed average erosion rates by sponges of 0.25-3 kg CaCO$_3$ m^{-2} yr^{-1}, while Rützler & Rieger (1973) earlier showed that *Cliona lampus* produced regularly-shaped chips, approximately 50μm in diameter. Moore & Shed (1977) calculated similar erosion rates (1.0-1.8 kg CaCO$_3$ m^{-2} yr^{-1}) by sponges in Jamaica.

Echinoids are also major eroders on coral reefs and in some instances are growth limiting. In the Galapagos, Glynn *et al.* (1979) have shown that *Eucidaris* commonly range from 10-50 individuals per square metre, and graze heavily on live corals, especially *Pocillopora*. Most fragments in the gut were found to be around 2 mm in size, and coral particles defecated were in the same size range. Coralline algal particles ingested averaged 0.85 mm in size. Major sediment concentrations, especially around the leeward margin, are composed of coral and coralline particles in these same size ranges, indicating the importance of *Eucidaris* in generating reef-flanking sediment. In the areas of high coral cover (60-80%), where the production of CaCO$_3$ is 6.5-8.5 kg m^{-2} yr^{-1}, sea urchins reduce this production by 11-20%. Significantly, along the reef edge where coral cover is 30%, and where there are high densities of sea urchins, the net accretion of CaCO$_3$ is zero. Clearly, therefore, sea urchins not only cause substantial erosion, but also limit coral growth. On Barbados fringing reefs, *Diadema antillarum* average twenty-three individuals per square metre, with a total number of 422 000 on one reef. Budget figures indicate erosion rates of about 9 kg CaCO$_3$ m^{-2} yr^{-1}. which is equivalent to vertical erosion rates of 6 mm yr^{-1} (Hunter 1977; Stearn & Scoffin 1977). There is a wide range in estimates of erosion by other macro-borers, including bivalves (table 2). However, their effects appear to be much less than those of sponges and echinoids. In the Great Barrier Reef, Hamner & Jones (1976) estimated erosion rates by *Tridacna crocea* as 0.14 kg CaCO$_3$ m^{-2} yr^{-1}.

Little quantitative information is available for other borers, such as polychaetes and other worms. Hutchings (1974) considers that in terms of numbers they are the dominant invertebrate group. At Heron Reef, a record number of 1441 specimens of 103 species were extracted from one single *Porites* head (Grassle 1973). The numbers, however, vary with the habitat, as shown at both One Tree Reef by Hutchings (1974) and in the Bahamas by Vittor & Johnson (1977), where different types inhabit growing or eroding reefs. Carnivorous polychaetes like *Hermodice corunculata* are indirect destroyers of corals, opening the way for algal infestation, and have been reported in Florida and the Great Barrier Reef (Shinn 1976). Combined sipunculid/sponge erosion rates at One Tree Reef (Trudgill, in press) average 0.13-1.67 mm yr^{-1} on cemented intertidal rocks.

Damage to corals by fish has been reported by many workers. Randall (1974) concluded that fish that eat coral are not very abundant although Hiatt & Strasburg (1960) found considerable predation of corals by scarids in the Marshall Islands. In the Caribbean, parrot fish are estimated to erode between 40-490 g m^{-2} yr^{-1} (0.02-0.36 mm yr^{-1}) Gygi 1975; Ogden 1977; Frydl & Stearn 1978). Sediment turnover in lagoons by scarids and other browsers and grazers is variously reported as between 40 and 980 g m^{-2} yr^{-1} (Cloud 1959; Bardach 1961; Gygi 1975; Ogden 1977; Frydl & Stearn 1978).

In the Great Barrier Reef, little information is available on the destructive effects of fish and destruction rates have not been measured. Choat (1966) states that parrot fish do not graze on corals and Stephenson & Searle (1960) conclude that scarids in general are rock scrapers. Goldman & Talbot (1976), however, report that scarids occasionally feed on corals. The author has frequently observed scarids feeding on algal turfs on One Tree Reef, a locality for which quantitative information may soon he available.

In most of the above mentioned studies, little attention has been given to the character and size of the sediment produced by boring. The importance of such studies is illustrated by Hunter's calculations for Barbados, which indicate that 65% of the sediment produced by echinoderm borings is very fine sand and silt and is probably lost from the system. Subtidal and intertidal reservoirs of carbonate rubble and sand dominate on many windward margins, in lagoons and off leeward margins. This material has probably been affected by biological erosion, most of it on more than one occasion and in more than one environment.

MEASUREMENT OF GROWTH

Many workers are currently attempting measurements of coral reef growth. The parameters measured reflect the two disciplines most closely concerned with reef studies.

The Biologists' Parameters

Biologists have traditionally approached the problem of coral reef growth through estimates of primary production using the technique of flow respirometry (Sargent & Austin 1954; Odum & Odum 1955; Kohn & Helfrich 1957; Gordon & Kelly 1962; Kinsey 1972) and more recently the alkalinity anomaly method (Smith 1973). The amount of carbon dioxide used in photosynthesis is a measure of growth. Thus, if photosynthesis exceeds respiration, a biomass increase is indicated; if respiration exceeds photosynthesis, respiratory oxidation of more organic material than is being produced is indicated. Table 3 shows the total metabolic turnover of CO_2 by three reef systems. Overall, there is little or no net gain or loss of organic carbon. They are in apparent equilibrium; consuming by oxidation all the organic matter they create by photosynthesis. However, in the zonal data for a single system (table 4), three other conclusions emerge: 1. the reef flat is the major

| Reef | Latitude | Metabolic Rate ($mmoles\ CO_2\ m^{-2}\ d^{-1}$) | | |
		Gross photosynthesis	Gross respiration	Net gain
One Tree Reef	23°S	190	195	−5
Lizard Island	13°S	270	270	0
Canton Island	3°S	500	495	+5

(Data from Smith & Kinsey 1976; Kinsey & Davies 1979a.)

Table 3 Overall metabolic turnover of CO_2 by coral reefs

Reef zone	Gross photosynthesis	Metabolic rate (mmoles CO_2 m^{-2} d^{-1}) Gross respiration	Net gain
Seaward slopes	?	?	?
Algal pavement	170	45	125
Coral flat	600	617	-17
Prograding sand flat	75	117	-42
Reticulated lagoon	260	260	0
Lagoon with small submerged reef	120	145	-25
Leeward flat	?	?	?
Overlying waters	5	17	-12
Weighted mean	190	195	-5

(Data from Smith & Kinsey 1976; Kinsey & Davies 1979a.)

Table 4 One Tree Reef: metabolic turnover of CO_2 by reef zone

Coral	Location	Rate (cm yr^{-1})
Porites porites	Barbados	3.7
	Jamaica	4.1
Porites lobata	G.B.R.	0.4-1.3
Montastrea annularis	Barbados	1.9
	Florida	1.07
	Bermuda	0.81
	Jamaica	2.5
Acropora cervicornis	Bahamas	4.5
	Florida	11.5
	Barbados	14.5
	Jamaica	26.6
Acropora sp.	G.B.R.	10-20
Favia sp.	India	0.5
Solenastrea hyades	North Carolina	0.15
Stephanocoenia sp.	Jamaica	0.5

(Data from Shinn 1966; Ginsburg & Schroeder 1973; Glynn 1973; Moore & Krishnaswani 1974; Isdale 1977; Jell 1977.)

Table 5(a) Estimates of linear growth rates of corals

site of organic activity, 2. the reef flat is, however, a net consuming zone requiring an input of organic carbon and 3. the algal flat is a net producing zone and probably exports its excess production downstream.

The Geologists' Parameters

Geologists have historically recorded linear measurements of growth, usually by *in situ* measurements of growing corals or by radiocarbon dating of cores. Estimates of linear growth rates of corals have been made in many locations (table 5a). Atlantic species vary from 1.5 mm yr^{-1} to 11.5 cm yr^{-1}. In the Great Barrier Reef, *Acropora* species grow up to 20 cm yr^{-1} while *Porites lobata* varies in growth from 4-13 mm yr^{-1}.

Numerous studies describe variations in the rates of reef growth throughout the Holocene (table 5b, fig. 1). At St Croix, Virgin Islands, Adey (1978) reports rates of 1-12

Location	Rate (m 10^{-3}/yr)
Atlantic	
Florida	0.38–4.85
Bermuda	1.2
Jamaica	1.2
Virgin Islands	1–12 (fringing reef)
	1.75–8 (bank barrier reef)
Curacao (N.A.)	1–4
Alacran (Mexico)	1.25–12
Galeta Point (Panama)	1.3–4.2 (but up to 7.5)
Pacific	
Gulf of Chiriqui (Panama)	3.1–3.9
Enewetak	up to 10
Great Barrier Reef	0.2–6
Indian Ocean	
Reunion Island	1–6

Data from sources referenced in text.

Table 5(b)   Estimates of coral reef growth rates from ^{14}C dating

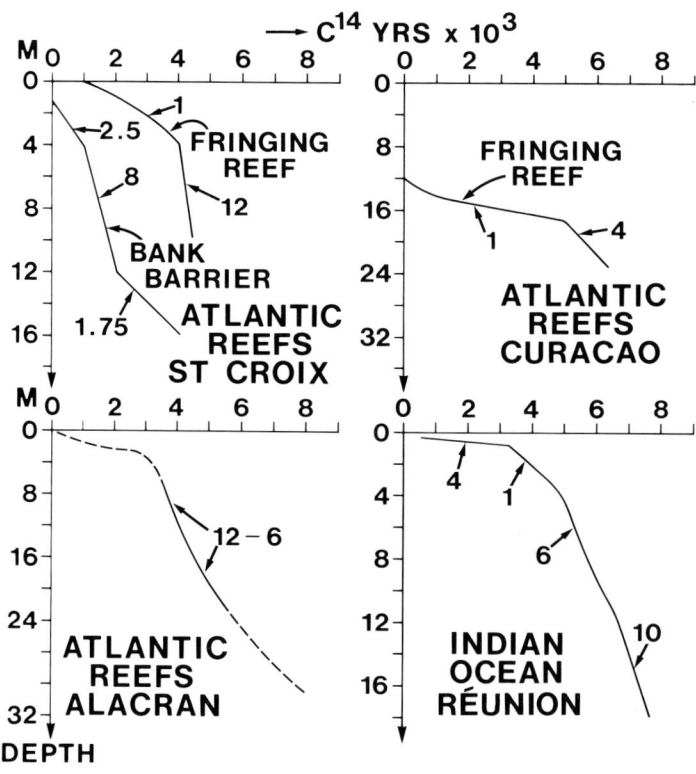

Figure   1   Rates of growth for some Atlantic and Indian Ocean coral reefs. Data from Adey 1978; Focke 1978; Macintyre *et al.* 1977; Montaggioni 1977

m 10^{-3} yr for fringing reefs, which contrast with rates of 1.75-8 m 10^{-3} yr within bank barrier reefs (fig. 1). Fringing reefs at Curacao (Focke 1978) show variations of 1-4 m 10^{-3} yr (fig. 1), and at Alacran, rates of 1.25-12 m 10^{-3} yr reported by Macintyre *et al.* (1977) compare well with the Virgin Islands data (fig. 1). In the Indian ocean rates of 1 -6 m 10^{-3} yr (Montaggioni 1977) suggest slower growth than normally operative in the Atlantic, as do rates published for the Great Barrier Reef. Data from the Great Barrier Reef (fig. 2) suggest that growth rate variations in the Holocene approximate a sine curve with slow rates of growth during reef initiation and again when the reef reaches the surf environment. At St Croix, in the Virgin Islands, Adey (1978) attributes slow growth of reefs, and their eventual demise, to turbid water resulting from erosion during the transgression. Fast rates of 10 m 10^{-3} yr in the early Holocene occurred at Enewetak (J. Tracey, pers. comm.). Growth rates in Pacific Panama and One Tree Reef (table 5b) are instructive because they are the only places where rates have been measured both at the present time and in cores. At Panama, current rates of 1.3-4.2 m 10^{-3} yr (Glynn & Macintyre 1977) contrast with past rates of 7.5 m 10^{-3} yr. At One Tree Reef, present rates of 3 m 10^{-3} yr (Kinsey 1979a) compare with past rates of 3-7 m 10^{-3} yr (Davies & Marshall 1980).

Sedimentation rates have been only infrequently measured in coral reef environments. Lagoonal sedimentation resulting from fish grazing is estimated at 1-3 m 10^{-3} yr (Cloud 1959; Bardach 1961). In the Great Barrier Reef, radiocarbon dating of subsurface samples shows sedimentation rates around 1.5 m 10^{-3} yr (Frankel 1977). Calcium carbonate budget studies at One Tree Reef (Davies 1977) also suggest lagoonal sedimentation rates of 1.5 m 10^{-3} yr. Sedimentation at Reunion Island (0.2 m 10^{-3} yr) reported by Montaggioni (1977) are much less than those described elsewhere, and may represent minimum estimates because the sediments are intertidal and probably subject to erosion.

Calculation of sedimentation rates by ^{14}C dating requires that bioturbation has not destroyed the original depositional fabric. This is by no means certain. An added complication is that the ^{14}C data records the time of formation of the biogenic fragment and not the time that it was deposited. Sedimentation rates calculated by radiocarbon dating should therefore be treated with caution and probably represent minimum estimates.

The Geobiological Parameter

Calcium carbonate is the currency that links geological and biological processes on coral reefs and the alkalinity anomaly method is admirably suited to studies of these links. The method assumes that alkalinity variations in seawater over coral reefs are stoichiometrically linked to changes in carbonate expressed as changes in calcium carbonate (Smith 1973). The method estimates net production of calcium carbonate and the results obtained have been substantiated by experimentation (Kinsey & Domm 1974; Smith & Harrison 1977).

Very few investigations have been made into the quantitative flux of carbonates in reefal systems. However all reefs examined show a daily gain of carbonate (table 6). The increment on different reefs varies by a factor of 3 to 4, but the gain clearly implies accretion of the reefs in contrast to the apparent lack of organic growth reported earlier for similar systems.

Calcification rates for reef flats and seaward zones (table 7) on different reefs are remarkably similar. Variations in community structure seem to have no effect on this

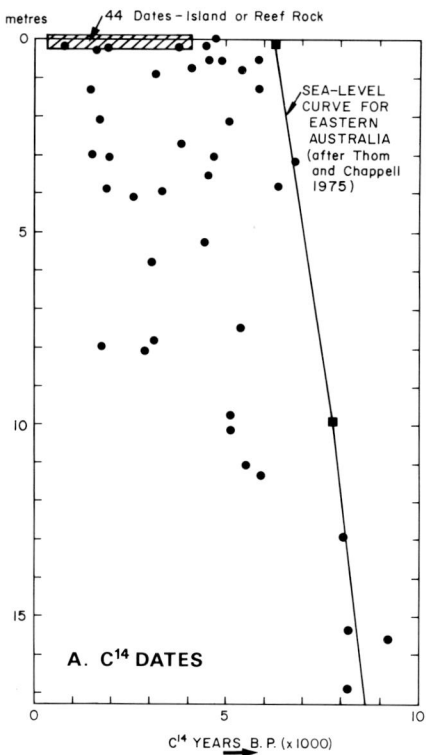

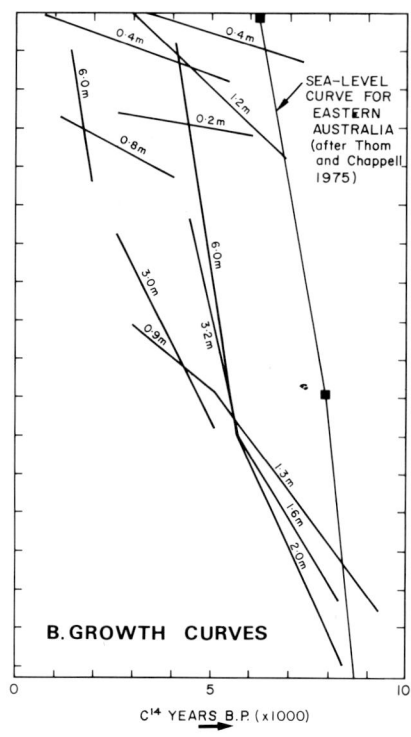

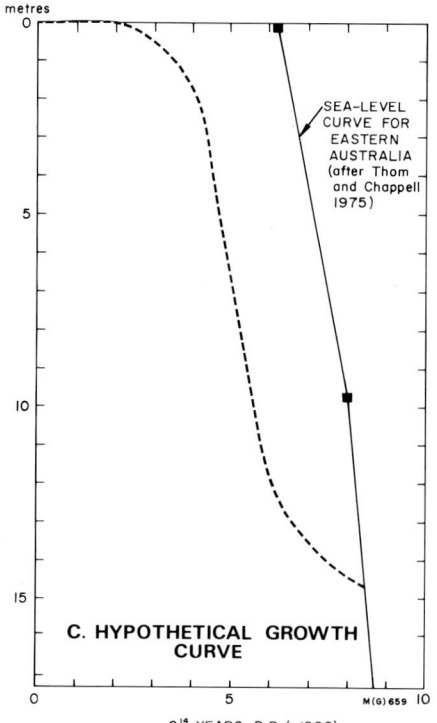

Figure   2   Radiocarbon dates and growth rates
for reefs in the Great Barrier Reef
(from Davies and Marshall 1979)
(A) Published radiocarbon dates
(B) Growth rates
(C) Hypothetical growth curve

Reef	Calcification rate	
	(mmoles CO_2 m^{-2} d^{-1})	(kg $CaCO_3$ m^{-2} yr^{-1})
Canton Atoll	14	0.5
Fanning Island	27	1.0
One Tree Reef	41	1.5
Lizard Island	50	1.8

(Data from Smith & Kinsey 1976; Kinsey & Davies 1979a.)

Table 6   Net growth of coral reefs in terms of calcium carbonate

Reef		Calcification rate
	(mmoles CO_2 m^{-2} d^{-1})	(kg $CaCO_3$ m^{-2} yr^{-1})
One Tree Reef		
Algal flat	110	4.0
Coral flat	123	4.5
Lizard Island		
Seaward pinnacle	101	3.7
Algal flat	104	3.8
Coral flat	99	3.6
Enewetak		
Algal flat	110	4.0
Coral flat	110	4.0

(Data from Smith & Kinsey 1976; Kinsey & Davies 1979a.)

LINES 4

Table 7   Calcification on seaward margins of coral reefs

activity. It has been suggested that some factor, as yet not understood, sets a ceiling on the calcification rates in peripheral seaward zones regardless of the community structure (Smith & Kinsey 1976).

Considering the calcification in different zones of a single reefal system (table 8), it is clear that the general distribution pattern of net carbonate deposition follows the pattern of photosynthetic and respiratory activity. A notable exception is the high level of calcification relative to the low level of organic activity in the algal pavement. The perimeter margins, however, are clearly the sites of major accretion, although positive accrual is characteristic of all the zones.

Growth rates calculated from the alkalinity anomaly method (table 8) appear to be at odds with the rates obtained by radiocarbon dating of cores, reported in Table 5B. This is particularly so when comparison is made between Atlantic rates obtained on reef cores dominated by *Acropora palmata* and *A. cervicornis* which are two to three times faster than modern calcification rates reported for Pacific reefs determined by the alkalinity method. Although these inter-ocean differences have yet to be resolved, they surely suggest variations in coral reef growth rates throughout the Holocene and Recent. This is substantiated by studies at One Tree Reef where Holocene rates of up to 8 m 10^{-3} yr (Davies & Marshall 1981) compare with modern rates of 3 m 10^{-3} yr (Kinsey & Davies

| Reef zone | Calcification rate | | Vertical growth |
	(mmoles CO_2 m^{-2} d^{-1})	(kg $CaCO_3$ m^{-2} yr^{-1})	(mm.yr^{-1})
Seaward slope	?	?	?
Algai pavement	110	4.0	2.8
Coral flat	123	4.5	3.1
Sand flat	8	0.3	0.2
Reticulated lagoon	41	1.5	1.0
Subtidal patch	14	0.5	0.3
Reef lagoon			
Leeward flat	?	?	?
Weighted mean	41	1.5	1.0

(Data from Smith & Kinsey 1976; Kinsey & Davies 1979a.)

Table 8   Calcification at One Tree Reef by reef zone

Location	Subsidence Rate (mm yr^{-1})	Total Potential Subsidence in last 8000 yr (cm)
Caribbean carbonate		
platforms	0.02–0.03	16–24
Enewetak	0.02–0.04	16–32
Midway	0.02	16
Australian shelf		
ridge	0.05	40
basin	0.1–0.2	80–160

(Data from Menard 1964; Adey 1978; B.M.R. unpub.)

Table 9.   Subsidence rates in some coral reef areas

1979a). Further, Kinsey's high calcification estimates for patch reefs at Johnstone Atoll (Kinsey 1979) suggest that different modern environments exhibit substantially different calcification rates, the range equalling that developed in the Holocene.

PARAMETERS AFFECTING GROWTH

Certain factors fundamentally influence the growth of coral reefs. The effects of four such parameters of primary interest to geoscientists (subsidence, sea level, substrate and ocean turbidity/chemistry) are discussed briefly below.

Subsidence

Deep borings show that coral reefs similar to modern reefs have been growing spasmodically over the past 10 to 15 million years. Tectonic movements which began 80 million years ago have caused marked changes in the position of many reefs relative to sea level. Reefs have been 'drowned' or 'exposed' by downward or upward movements of their foundations. In areas of gradual subsidence, reefs have grown fast enough to maintain themselves in the photic zone. Subsidence rates for some reef areas have been calculated (table 9). There is no reason to believe that such subsidence has not continued to operate

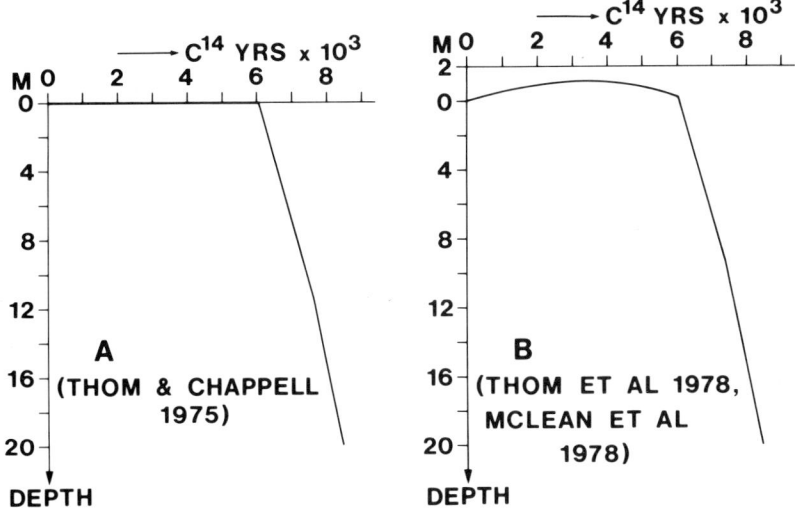

Figure 3 Course of Holocene sea level rise in the Great Barrier Reef
(A) After Thom & Chappell 1975
(B) After McLean *et al.* 1978

in the Holocene. On continental shelves, subsidence is likely to be both complex and variable because of the underlying structure of basins and highs. The Great Barrier Reef represents such a situation. Considering that at least half of the subsidence shown in table 9 would have occurred after most reefs reached sea level, anything from 20-30 cm relief on present reef surfaces must be considered potentially due to subsidence effects. Currently, however, the total absence of accurately determined tidal datums make accurate measurements of reef surfaces to within one metre very difficult. Any subsidence effects are therefore well hidden within this datum error.

Sea Level

Changes in sea level have influenced reef growth more than movements of reef foundations. Holocene sea level history in the Great Barrier Reef is shown in Figures 3a and 3b from which the following conclusions can be drawn: 1. sea level reached its present level by 6200 yr BP, and thereafter; 2. it either remained at present level, or rose to + 1 m between 4-5000 yr BP, stood at this elevated level for one thousand years and fell gradually to its present level.

Evidence for the elevated sea level includes radiocarbon dating of cay surfaces, cemented rubble platforms, micro-atolls, coral shingle ridges, reef flats and mangrove swamps on reefs to the north of Cooktown (McLean *et al.* 1978). A consequence of the elevated sea level, if indeed it did occur, is that reef tops have been areas of marked erosion for 2000 years, as has been suggested for Enewetak (Buddemeier *et al.* 1975; Tracey 1978). A further consequence is that the dating of modern reef flats will not indicate the time at which reefs reached sea level. It also suggests that there may be little hope of resolving reef/sea level relationships to within one metre.

Substrate

A significant body of information has accumulated about the substrate from which the modern Great Barrier Reef has grown. The salient points are summarized below.

1. It is of Pleistocene age. Preliminary ^{230}Th/^{234}U dates (B.M.R., unpub.) indicate that it is last interglacial in age (approximately 125 000 yr BP), similar to previous age determinations at Enewetak and Mururoa atoll (Thurber *et al.* 1965; Lalou *et al.* 1966).

2. The depth of the Pleistocene surface varies from 7 to 23 m below present reef top level (Thom *et al.* 1978; Harvey *et al.* 1979).

3. The gross morphologic features of the Pleistocene surface closely resemble the outlines of the modern reefs (Harvey *et al.* 1979).

4. Studies in the southern Great Barrier Reef show the Pleistocene unconformity in cores to be an altered coral limestone, a soil approximately 20 cm thick, or a coralline encrusted coral surface (B.M.R., unpub.).

5. The Pleistocene surface is a subaerially eroded former reef. The extent to which it has been modified by subaerial erosion probably varies.

The extent to which the Pleistocene surface affected Holocene colonization is unknown. A lag effect, probably due to slow early growth, is suggested by the data in the Hayman Island borehole (Davies & Marshall 1979). There is, however, no doubt that the depth and morphologic variations of the Pleistocene surface have critically affected the growth of the Holocene reef.

Oceanographic Variables

Turbidity, temperature and sea water chemistry have been suggested as having effected the growth of reefs in the Holocene.

In the eastern Caribbean, high turbidity induced by erosion of soil cover formed during low sea level periods is thought to have prevented continuous Holocene reef growth on wide shelves below 20 m, and to have killed an *Acropora palmata* reef growing on the shelf edge at St Croix (Adey 1977; Adey *et al.* 1977). An early Holocene *Acropora palmata* reef at a depth of 10 m off Florida is thought to have been killed by a cominbation of high turbidities induced by erosion of platform muds and low temperatures resulting from atmospheric cooling of expanses of shallow shelf water, together with intense upwelling (Lighty *et al.* 1978).

World-wide concurrence of radiocarbon dates for the initiation of Holocene coral reef growth (approximately 9000 yr BP) may be due, in part, to the inhibition of calcification resulting from slightly higher nutrient levels, particularly phosphate, in ocean waters, following the Wisconsin glaciation (Kinsey & Davies 1979). More intense circulation and increased upwelling, particularly along the equatorial divergence, seems to have occurred around 18 000 yr BP (Climap 1976) and presumably for some considerable time before and after. Experimental evidence strongly suggests that elevated nutrient levels are capable of suppressing coral reef calcification directly (Kinsey & Domm 1974).

In the Great Barrier Reef, the early Holocene growth phase was slow compared with the sea level rise. This little understood period of slow growth may have arisen as a result of any of the oceanographic factors discussed above.

ONE TREE REEF — A CASE STUDY

In 1938, J. A. Steers visited the reefs of the Capricorn and Bunker Groups in the southern Great Barrier Reef, and concluded that One Tree Reef (fig. 4a) was by far the most interesting reef in the province. In 1965, F. Talbot and D. McMichael set a similar course,

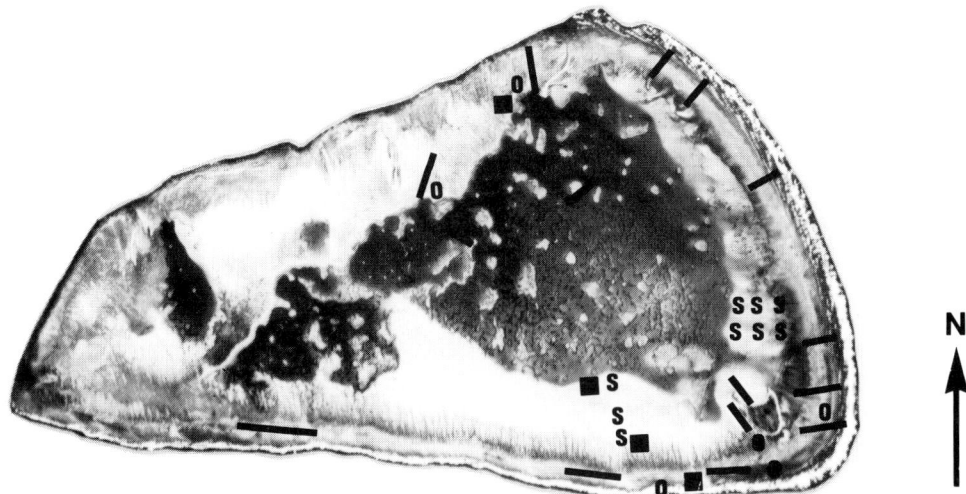

Figure 4a   One Tree Reef. Aerial photograph showing sites of data collection for geological
database. O = drill sites; S = sediment cores; full squares = current and sediment
monitoring sites; unbroken lines = seismic refraction profiles

reached an identical conclusion, and transformed their excitement into a field research
station that has operated continuously since 1968. Results of the studies conducted at One
Tree Reef currently represent the largest modern data base for any reef in the Great
Barrier Reef and the most complete case study of growth. Studies of modern calcification
date from the early days of the research station and studies of carbonate evolution have
been conducted over the past five years. The complete geological data base is shown in
figure 4a, a summary of growth characteristics in figure 4b, and the present-day
sedimentological and biological zonation of the reef is shown in figure 5. It has been
assumed that sea level history in the area approximates the results of Thom & Chappell
(1975) with sea level stabilizing around 6000 yr BP (fig. 3a).

Growth Prior to Reaching Sea Level

Substrate effects on Growth

One Tree Reef has grown on a karst-modified Pleistocene reef, the depth of which varies
from 10.5 to 23 m below the surface of the modern reef. It is at its shallowest around the
perimeter margins and deepest beneath the central lagoon and off the leeward margin.
Holocene growth was first initiated on the perimeter and on highs within the Pleistocene
lagoon. Detailed morphologic variation of the Pleistocene substrate are reported in
Harvey *et al.* (1979).

Holocene accretion

The Holocene development of One Tree Reef has been determined from the core obtained
from six drill holes; the interpretation of four of the holes are shown in figure 6 (Davies
& Marshall 1980). Growth on the windward margin (fig. 6) is dominated by massive head
corals *(Porites* spp., *Symphillia* spp., and *Platygyra* spp.), less common branching forms,
and a coralline algae-*Millepora*-vermetid assemblage especially in the bottom 5 m of the

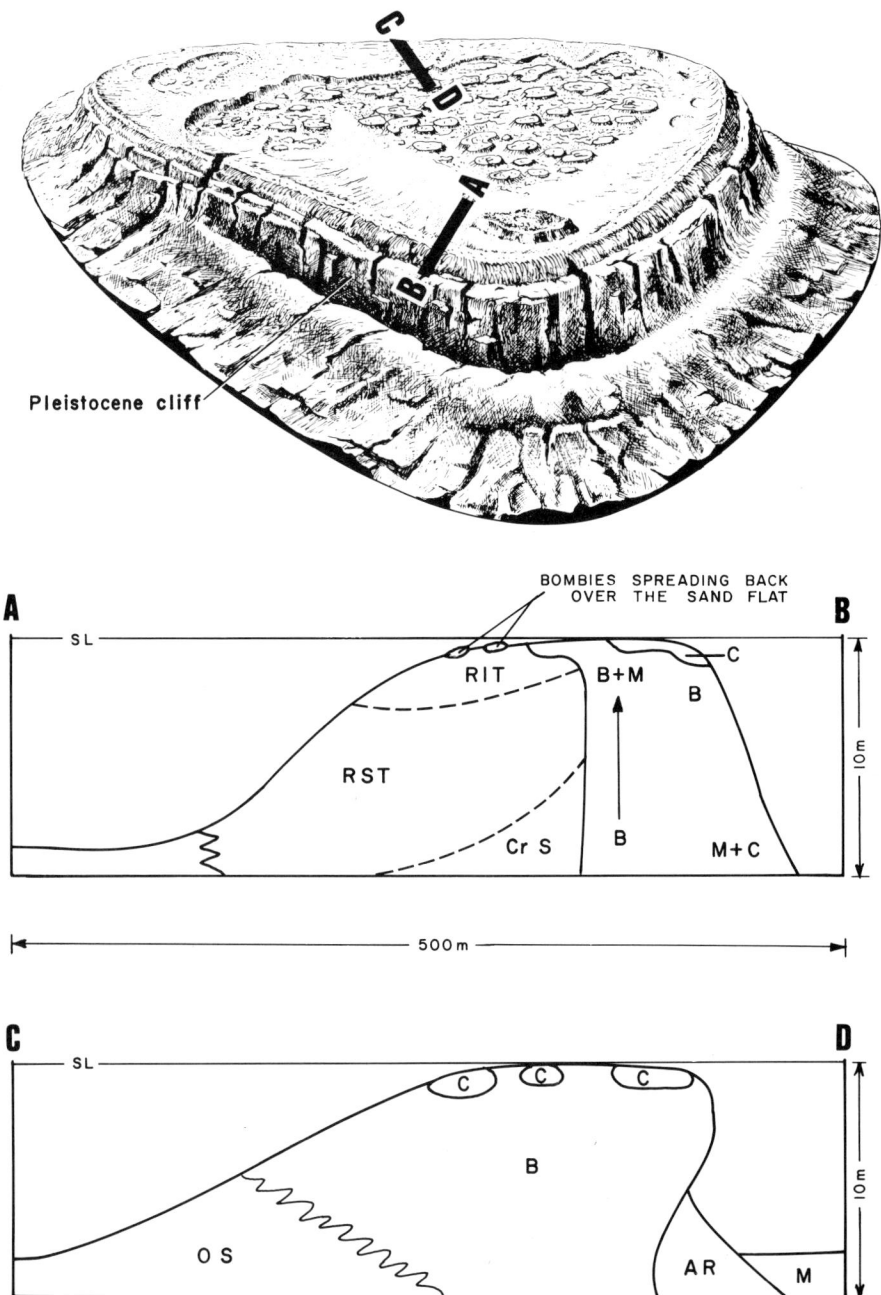

Figure 4b   One Tree Reef. Growth characteristics at profiles of the windward margin (A-B) and leeward margin (C-D). The block diagram is an artistic impression drawn from available physical data showing growth of the modern reef on top of a Pleistocene foundation. Facies variations are: B = branching corals; M = massive corals; C = coralline algae; CrS = coarse subtidal sediments; RIT = rhythmic intertidal sediments; RST = rhythmic subtidal sediments; OS = off-reef sands; AR = autochthonous rubble

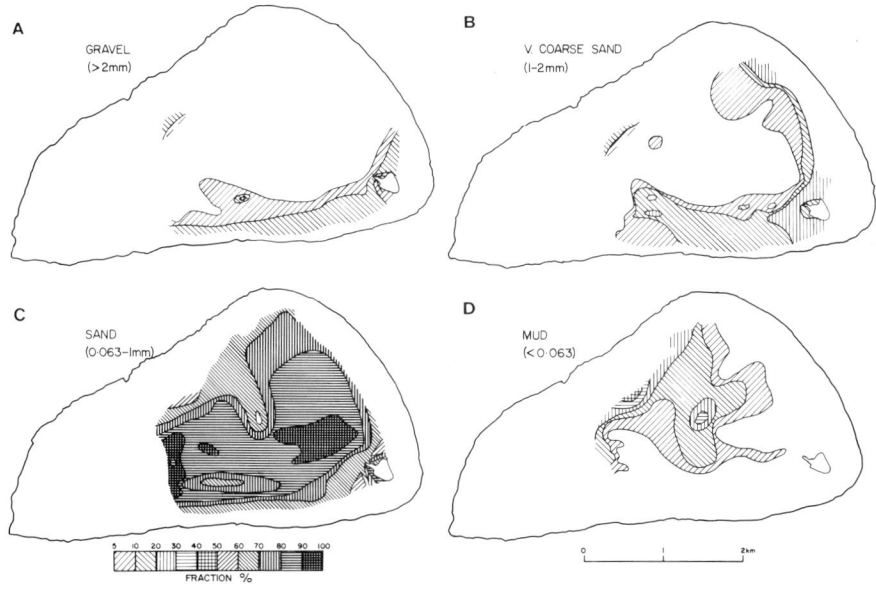

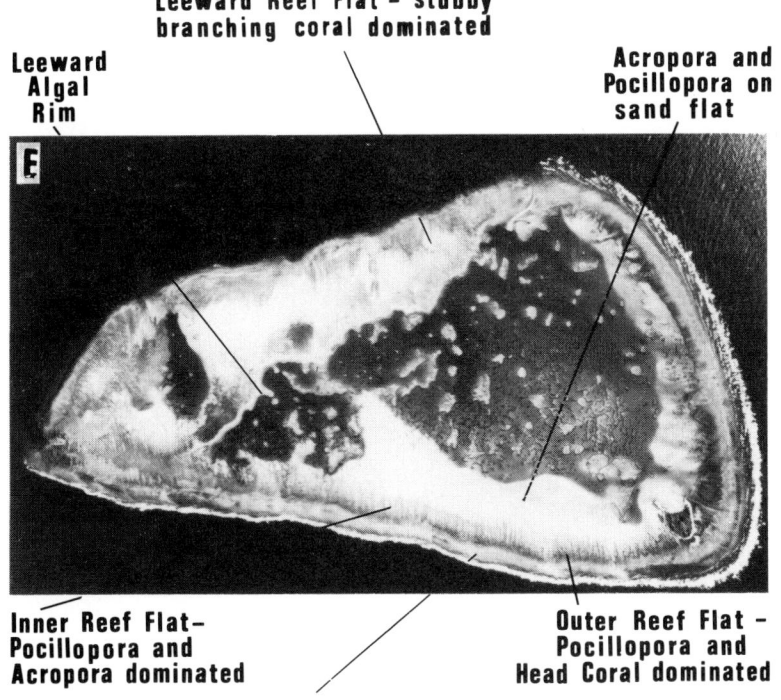

Figure 5 One Tree Reef. Sediment distribution and gross biological zonation
(A-D) Distributions of gravel, very coarse sand, sand and mud
(E) Gross biological zonation

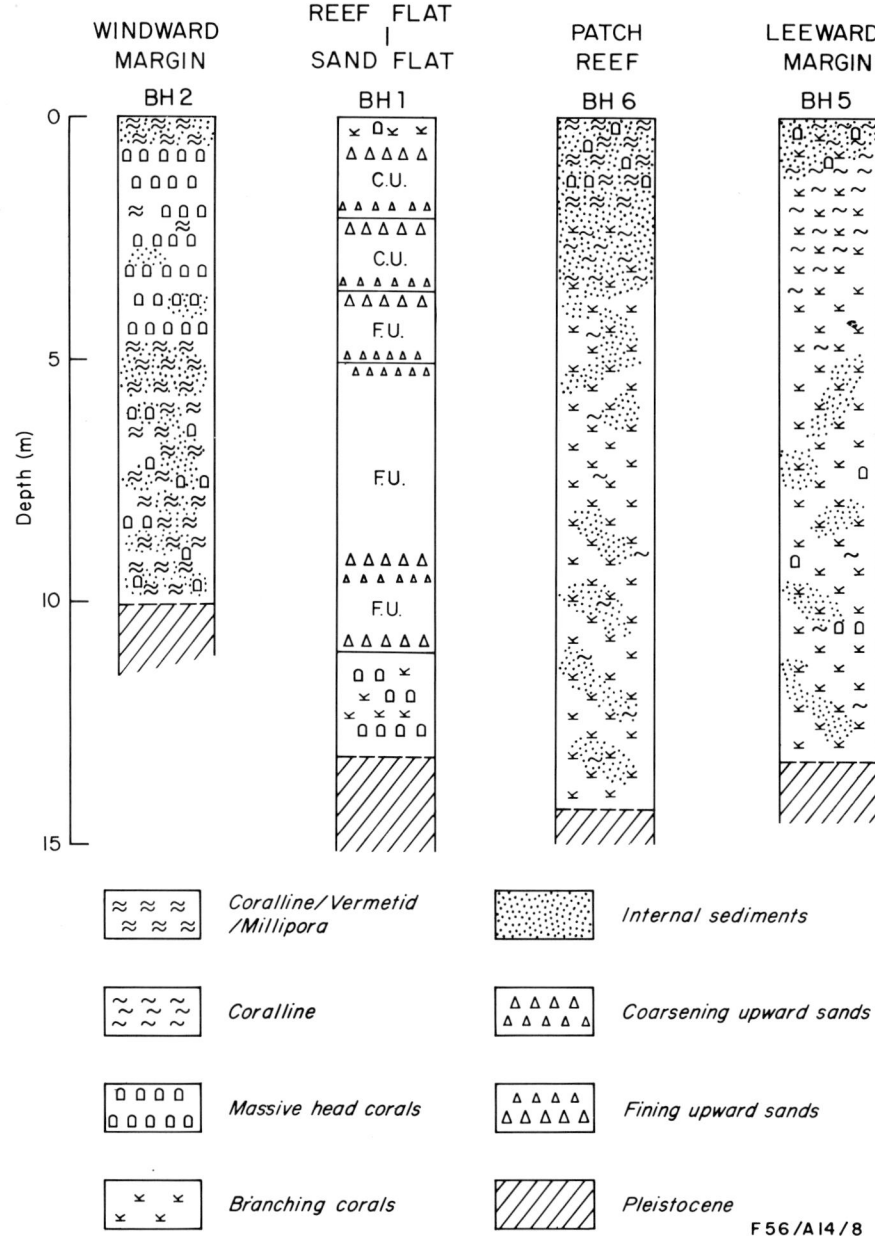

Figure 6   One Tree Reef. Interpretation of four drill holes (from Davies and Marshall 1980)

core. The corallines are largely *Porolithon onkodes*, *Lithophyllum* spp., and *Neogoniolithon* spp. (Davies *et al.*, in press). An identical, but much thinner, assemblage occurs at the top of the section and forms the present day reef surface. The modern algal ridge is therefore thin. Surface faunas immediately leeward of the windward drill hole site are dominated by a coral flat assemblage, i.e., *Acropora* spp., *Pocillopora damicornis*,

Porites spp., *Goniopora* spp., *Goniastrea* spp., and *Favia* spp. Clearly the modern biological reef flat assemblage is very different from that seen in the bore hole.

A drill hole through the reef flat (fig. 6) shows only a thin coral sequence, dominated by branching forms. Much of the section is composed of a thick sand sequence compositionally similar to, and clearly a part of, the prograding sand wedges shown in figure 4. The lower half of the sand sequence exhibits finding upward sediment rhythms indicative of episodic storm supply from the reef margin and representing the earliest deeper subtidal stages of lagoon infill. The upper portions of the sand sequence show a series of coarsening upward sediment rhythms representing shallow subtidal to intertidal progradation. Rhythmic sequences similar to those in the borehole are seen in the southern intertidal and northern subtidal sand sheets currently prograding into the lagoon. The earliest Holocene growth on the Pleistocene surface is at a depth of 12.5 m and is a mixed assemblage of branching and massive corals.

The drill core from the patch reef at the leeward end of the lagoon shows a thick, coralline, algal-*Millepora*-vermetid assemblage above a dominantly branching coral assemblage. The coralline assemblage is similar to that seen on the windward margin but is thicker. Radiocarbon dating of the core indicates that the patch reef reached sea level before the windward margin and must, therefore, have been subject to a high energy surf environment.

On the leeward margin (fig. 6), the borehole sequence is again seen to be composed primarily of a branching coral assemblage which changes into a massive coral-low branching coral-coralline algal assemblage at the surface. The corallines are *Porolithon onkodes, Neogonoilithon fosliei* and *Lithophyllum kotschyanum.*

It is clear from figure 6, and the above brief descriptions, that Holocene growth is thin, that the faunas and floras are environment specific, and that successive changes occurred in growth as the reef struggled upwards to a stablized sea level. This relationship between sea level and growth will be returned to later.

Holocene Growth Rates

The estimated rates of growth for the windward margin, the windward reef flat, and the leeward reef flat are shown in figure 7. Large variations occur both within and between these environments. The most rapid and prolonged growth occurred on the leeward margin. The windward margin was the first to reach sea level (around 5000 yr BP) whereas the leeward margin, which grew faster, was last to reach sea level (1000 yr BP), a clear indication that the depth of the Pleistocene substrate affects reef growth. Biological growth throughout much of the Holocene (as seen in cores 2 and 5) has been equal to or greater than the potential vertical growth ($3-5$ m 10^{-3} yr) currently occurring. The accretion rate of the sand sheet below the windward reef flat compares well with other independent estimates of sand accumulation (Davies 1977; Frankel 1977). Comparison of curves 1 and 2 in figure 7 shows that sand accumulation had not begun until the windward margin had reached sea level, i.e., erosion and shedding of sediment did not occur when the reef was growing upwards but began only when the reef grew into shallow water.

Growth at the Stabilized Sea Level Position

Growth is the positive balance between construction and destruction, and both biologic and abiologic processes play their part.

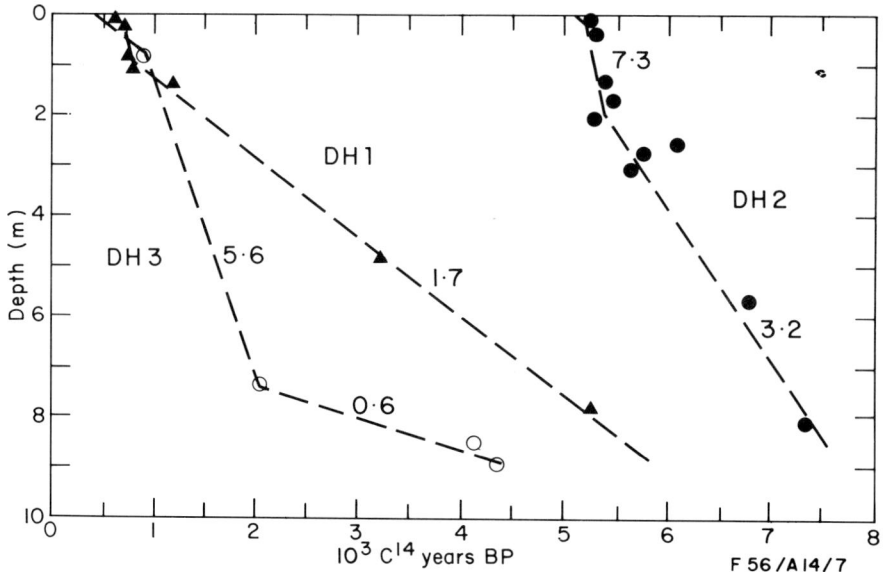

Figure 7 One Tree Reef. Rates of growth as seen in the windward margin (DH2), windward coral flat (DH1) and the leeward margin (DH3)

Biological Accretion

The distribution of coral and algal dominated zones on One Tree Reef is shown in figure 5e. This figure also indicates the net calcification associated with particular environments. SCUBA examination of the slopes supports the suggestion that maximum accretion is occurring on leeward ends (Davies *et al.* 1977) which are dominated by flourishing and flamboyant coral growth, often as spurs growing over a gently sloping sand substrate. Such spurs coalesce with patch reefs growing from a depth of 20 m. Windward slopes on the other hand, although often supporting *Acropora* forests in deeper water, do not exhibit the diversity seen on the leeward margin. In addition, the windward edge is fronted by a smooth, gently sloping, hard-rock surface on which growth is patchy or non-existent. The lateral accretion of the windward margin over the past 8000 years is markedly less than that which has taken place on its leeward counterpart.

Along the southern side of the reef, the spread of the coral flat backwards across the sand flat exemplifies the response of growth to environment (fig. 8). The sand flat is first colonized by *Acropora* spp. which were transported as living debris from the reef flat. Such *Acropora* tips easily colonize sand substrates and grow into small colonies (50 cm high), usually anchored at one point. These first colonizers are quickly followed by other branching forms such as *Pocillopora damicornis*. The second wave of migrants uses niches within the framework created by Acropora, but eventually take over and expand the colony to some metres in diameter. The rate of growth probably depends upon proximity to the reef flat since the smallest patches occur furthest from it. A solid patch reef structure is finally formed after settlement and growth of larvae of massive corals such as *Goniastrea, Favites, Goniopora* and *Porites*. The heart of the coral patch has often by this time died so that growth is confined to the perimeter; upward growth being limited by water depth at low tide. Coralline algae encrust the dead coral skeletons (which

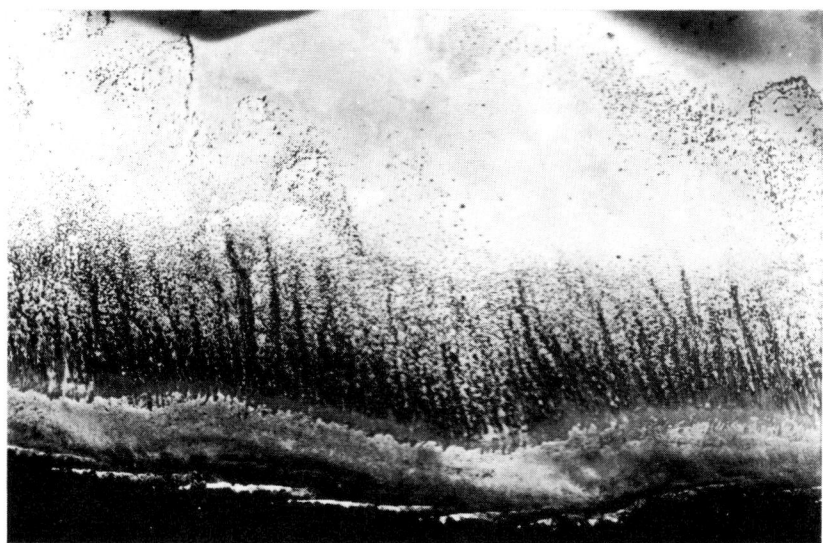

Figure 8 One Tree Reef. Aerial photograph of the southern margin showing spread of the coral flat backwards across the sand flat

are primarily the early colonizers, *Acropora* and *Pocillopora)* and sand accumulates in the centre of the patch. Such patches may elongate by growth at upstream and downstream margins so that water is channelled between adjacent patches, thus increasing and localizing the flow. Consequently, backward growth of the coral flat across the sand flat is a result of coalescence of the patch reefs which colonized the shallow sand flat.

A similar mode of growth to the above also occurs on both lagoonal patch reefs, and the 10 m wall at the leeward edge of the lagoon. Overhangs frequently break off and fall or slide into deeper water. Arborescent species of *Acropora* (e.g. *A. formosa)* spread across the lagoon floor, and form the primary substrate for other branching and massive forms. If the lagoon floor is relatively shallow (<3 m), the staghorn *Acropora* spreads rapidly and forms an impenetrable forest.

Physical Accretion

Physical accretion on a coral reef results from interactions between the reef and the prevailing hydrologic regime. At One Tree Reef, sediment distribution patterns, together with computer simulation experiments of the reef/hydrologic interactions, delineate major accretional sites and define the windward margin as the principal sediment source. A number of environments are recognized where temporary or permanent accretion has, or is, occurring (fig. 4).

On the upper fore-reef slope, extensive coral rubble accumulations occur along the south-eastern and eastern slopes, and locally elsewhere as on the floors of hanging valleys and grooves. Such material is locally derived from branching and plate-corals growing on the reef front. It is, therefore, an ephemeral autochthonous accumulation, and is the source for much of the rubble accumulation on the reef flat. This upper fore-reef slope is clearly a temporary sediment holding zone. Little is known of the volume of material held in the zone, the length of time it remains in the zone, or at what rate it accumulates.

Rubble banks, spits and sheets (fig. 4a) are characteristic features of the eastern windward reef flat. Such bodies are composed of massive coral heads, plate-like coral heads (up to 0.5 m) and sticks of coral debris. Marked imbrication points to their source being the windward margin, and the clasts themselves are identical to those in the fore-reef holding zone. Most are coated by filamentous algae and the number of living colonies tossed up in high energy events is small. The banks and spits are often laterally and vertically graded, and are raising the eastern side of the reef into the supratidal environment. Such deposits along the eastern side of One Tree Reef are currently estimated to contain 2.4×10^6 tonnes of calcium carbonate. Their probable development sequence, dating from the time at which the reef approached stabilized sea level (approximately 5000 yr BP), is firstly, subtidal sheet deposits smothering the earlier living coral flat, followed by intertidal and supratidal bank and spit growth with coalescence of adjacent spits by intertidal spreading of the ridge material. Currently, multiple shingle spits are accreting into the lagoon. Intertidal and supratidal spits also occupy the western margin testifying to the energy at this end of the reef.

Sediment accumulation in the reef flat rubble zone is linked to accumulation in, and transportation from, the reef front. A bank, similar to that in figure 9a, and containing an estimated 4272 tonnes of calcium carbonate would require 100 years to produce. The amount of material currently held in the reef flat holding zone (2.4×10^6 tonnes) necessitates a 4000 tonne transference from the reef front every 6 to 10 years. The frequency of such movements was probably greater in the past when the reef flat was subtidal. The reef-front holding zone, which is receiving material less frequently, is itself undergoing erosion, the products of which are shed downstream. These two environments must therefore be seen as progressing from cyclical changes, perhaps operating on about 2000 year periods.

Intertidal sand sheets occur on both northern and southern margins. The intertidal southern sand sheet currently occupies an area of 822×10^3 m^2 and contains a likely sand volume of 14×10^6 tonnes. A section through the sand sheet (fig. 6) shows that it consists of a lower, subtidal sand sequence overlain by an upper, intertidal sequence. Both sand sequences are composed of coral-coralline-foraminiferal debris; corals in growth position occur spasmodically within the section. The lower subtidal section is characterized by upward fining sediment rhythms while the upper intertidal sequences are dominated by coarsening upward rhythms.

Subtidal, deltaic sand deposits separate the rubble banks from the lagoon along the northern lagoon margin. Surface sediments show a decrease in grain size towards the lagoon while cores show fining upward rhythms. These sand sheets occupy an area of 538×10^3 m^2, probably contain 9×10^6 tonnes of sand, and are accreting into the lagoon.

Lagoonal sediments are made up dominantly of sand, with mud more common in the lagoon centre and at its leeward edge. Grain size variations confirm this distribution (fig. 5), and follow the general variations in bathymetry. Localized rubble concentrations are associated with patch reefs. The total volume of sediment trapped in the lagoon, excluding the patch reefs, approximates 22×10^6 tonnes.

The leeward sand/gravel wedge forms a prominent depositional feature. Little is known of the grain size variations within it, but visual inspection suggests a decrease in grain size away from the leeward reef margin. Rubble is concentrated closest to the reef.

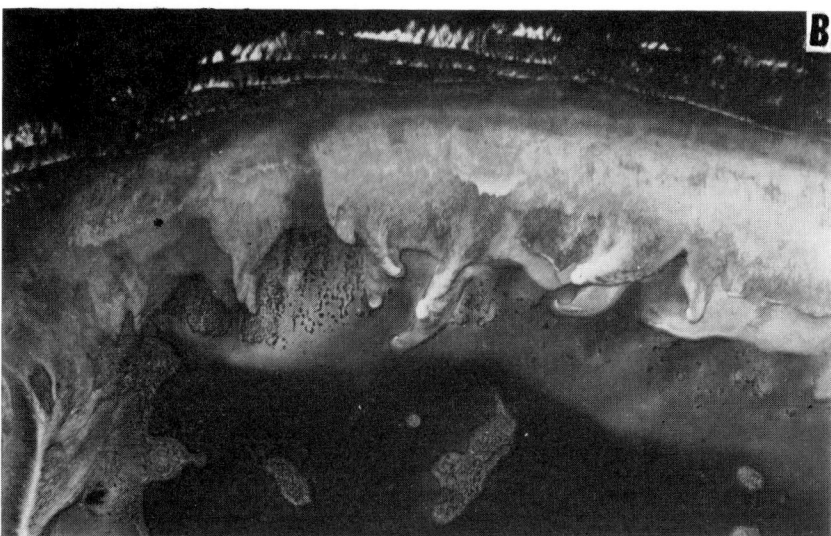

Figure   9   One Tree Reef. (A) Intertidal rubble banks on the windward reef flat
             (B) Subtidal and intertidal rubble banks along the northern margin of the lagoon

Seismic reflection studies confirm that the sand sheet is wedge-shaped. It is approximately 5 to 10 m thick, adjacent to the reef, and 1 km wide, and contains approximately 35 x 10^6 tonnes of sediment (fig. 10).

Monitoring of water movements and sediment suspension and traction flux of different reef environments has thrown some light on sediment sources and mechanisms, and rate of sedimentation of the sand sheets described above. Data obtained during tradewind conditions have led to the following conclusions.

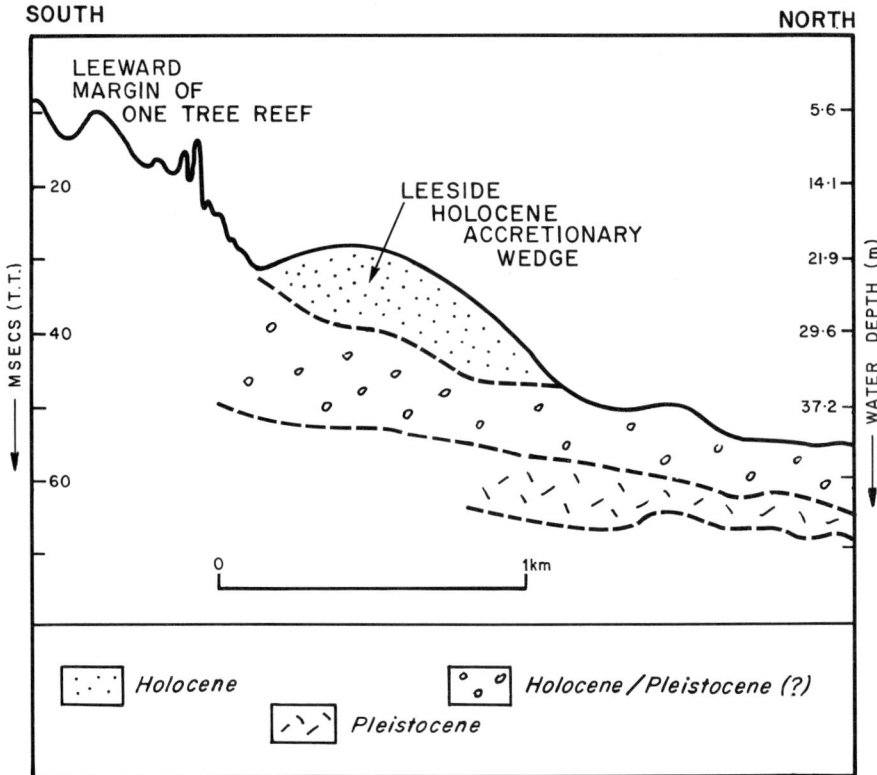

Figure 10   One Tree Reef. Interpretation of seismic reflection profile of the leeward margin. The leeward accretionary wedge is up to 1 km wide and 8 m thick

1. On a rising tide, water enters the lagoon over all margins.

2. During a falling tide, water mostly flows over the leeward margin, however, water also spills through topographic lows in all margins.

3. Surface flow into the lagoon over the leeward margin on the rising tide penetrates only 400 m into the lagoon. Such water may penetrate further southwards at depth by dipping below the warmer lagoonal water. It is, however, unlikely that penetration is great.

4. At low tide, the lagoon is effectively isolated from the ocean for a period of 6 hours. The level of the ponded water is highest on the windward side.

5. Sediment movement reflects dominant water movement. Sediment generated on windward margins is either trapped in the lagoon or by-passes the lagoon and is transported over the leeward margin.

Suspended sediment entering the lagoon over the leeward margin is trapped in the lagoon only during neap tide conditions; sediment entering during spring tides is flushed out.

The hydrological/sedimentological interactions described above explain both the sedimentary patterns shown in figure 5, and the leeward sediment wedge seen in seismic section (fig. 10). Based on the gross movements of sediment first order approximate sedimentation rates of 0.5 m 10^{-3} yr, and 0.75–1.00 m 10^{-3} yr may be calculated for the leeward end of the lagoon and off the leeward margin.

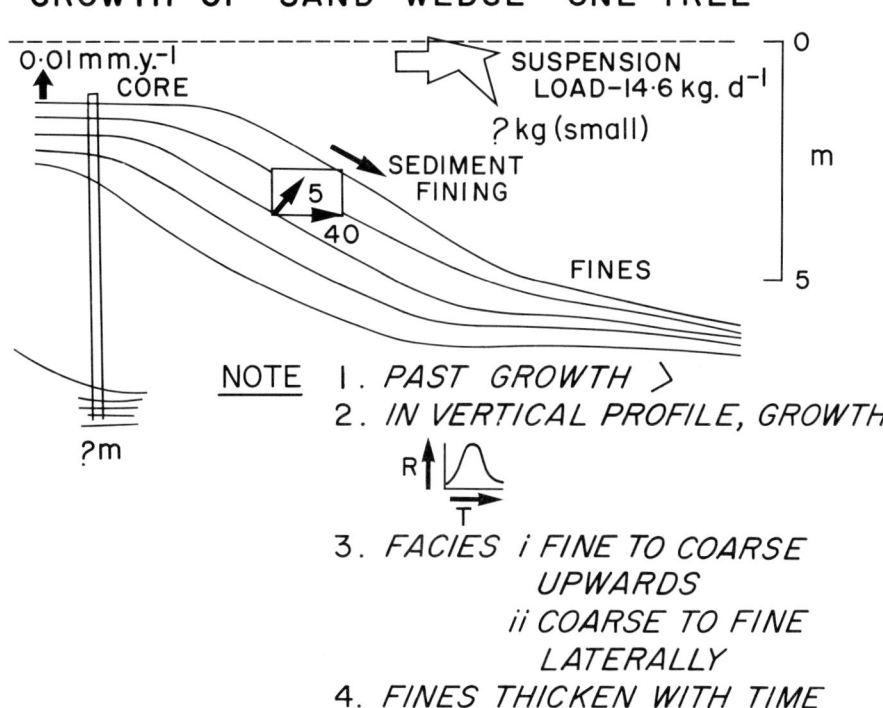

Figure 11   One Tree Reef. Growth of the southern sand sheet summarized as a section through the junction of the sand sheet and the lagoon. Lateral growth in the past was greater than at present, while the rate of vertical accretion approximates a sine curve

Of greater importance than lagoon infill by suspended sediment transport, is the growth of the prograding sand sheets. Sections through the nose of the southern sand sheet suggest its structure to be similar to that shown in figure 11. Cores taken across the transition between the sand sheet and lagoon, together with traction and suspension load monitoring, suggest that the sand sheet has grown as a prograding deltic edge. The foreset slope shows fining of sediment into the deeper water. Therefore, as the edge of the sand sheet migrates into the lagoon, sediments will become coarser upwards. The sand wedge is currently prograding into the lagoon at the rate of about 40 m 10^{-3} yr, and must be slowing down as the prograding edge moves away from the dominant sediment source.

Reef Destruction

Very little is known about the mechanisms or rates of destruction operating at One Tree Reef. That destruction is occurring is evident not only in the sedimentary constructional features that are its by-products, but also in the erosive forms which are the thumb-prints of its progressive occurrence. These are most evident on the intertidal platform rocks and reef flats of the windward margins, where razor-sharp rock edges, under-cut notches, interconnected pool systems, and bizarre shaped upstanding outcrops (fig. 12) testify to the combined efforts of solution, mechanical abrasion and biological erosion. No comprehensive quantitative data, comparable to that for Aldabra (see table 1) are

Figure 12   One Tree Reef. (A) Eroded platform rock along the south-east ridge
            (B) Eroded reef flat with undercut pinnacles of platform rock
            (C) Bizarre-shaped coral remnants indicating erosion of coral flat

available for One Tree Reef, although some data have been obtained for the windward margins. Davies (1974) calculated that destruction of the windward flats by all processes affecting solution amounts to 0.25 mm m^{-2} yr^{-1}. Trudgill (in press) shows that chiton-grazing on the upper surface of the platform rocks removes 0.2–2.9 mm of calcium carbonate per year, whereas sipunculid and sponge borers are removing 0.13–1.67 mm yr^{-1}. No data are available for mechanical erosion. However, if rates comparable to those calculated by Davies (*op. cit.*) and Trudgill (*op. cit.*) are representative of intertidal destruction, then it is sobering to realize that the 4272 tonnes of calcium carbonate contained in the rubble bank shown in figure 9, would be destroyed in less than 1000 years. Such destruction invariably leads to a decrease in grain size. Sediments are then removed to depositional sites in the sand sheets, in the lagoon, on the leeward edge, or alternatively, they are lost from the reef system.

Summary of growth (fig. 4)

Over the last 8000 years, 156 x 10^6 tonnes of calcium carbonate have accumulated at One Tree Reef. The manner in which this material was deposited is largely controlled by the morphology of the pre-existing Pleistocene reef. Vertical growth has varied from 0.5 m 10^{-3} yr. The windward margin reached sea level around 5000 yr BP, while the leeward margin reached sea level only 1000 to 3000 years ago. On reaching sea level, interactions between the reef and the high energy surface environment began the destructive-constructive growth cycles that elevated the perimeter margins as intertidal and supratidal rubble deposits, and led to the development of backward prograding sand sheets, backward growth of the reef flats over the sand sheets, lagoon infill, and leeward extensions of the western margin. Sand sheet progradation is currently occurring at 40 m 10^{-3} yr and infill of the leeward end of the lagoon at 0.5 m 10^{-3} yr. Lagoon infill is unlikely to be complete within the next 5000 years. After the lagoon sediment trap has been obliterated, leeward progradation of the western margin by sediment accretion will be accelerated past the current rate of 0.75–1 m 10^{-3} yr.

VARIATIONS ON A THEME

Coral reefs are net accumulators of calcium carbonate and three factors have governed the rate and manner of accumulation during the Holocene — substrate, a rise in sea level and a still-stand. Macintyre *et al.* (1977), Adey (1978) and Davies & Marshall (1979) show that the curve of reef growth in the Holocene (fig. 7) is a likely expression of their effects, emphasizing firstly, slow early growth for the reasons postulated by Adey (1978) and Lighty *et al.* (1978), secondly, a rapid growth when rates as high as 8 m 10^{-3} yr occurred in the Great Barrier Reef (B.M.R., unpub.) and 12 m 10^{-3} yr in the Atlantic and thirdly, a period of slower growth, usually 3 m 10^{-3} yr when the reef was within 3 m of present mean low water, clearly a result of biological changes and erosion. Davies & Marshall (1979) suggest that such a curve may have environmental application if a 'normal' growth curve could be identified, deviations from which would then represent perturbations about the norm. Kinsey (1979) suggested a similar concept for calcification estimates on a 'standard reef flat', deviations from which are measurable and interpretable in terms of environmental perturbations. The concept of standard growth curves and standard calcification estimates must await validification in the shape of much more data than is currently available. However, several existing factors prevent their useful adoption at the present time.

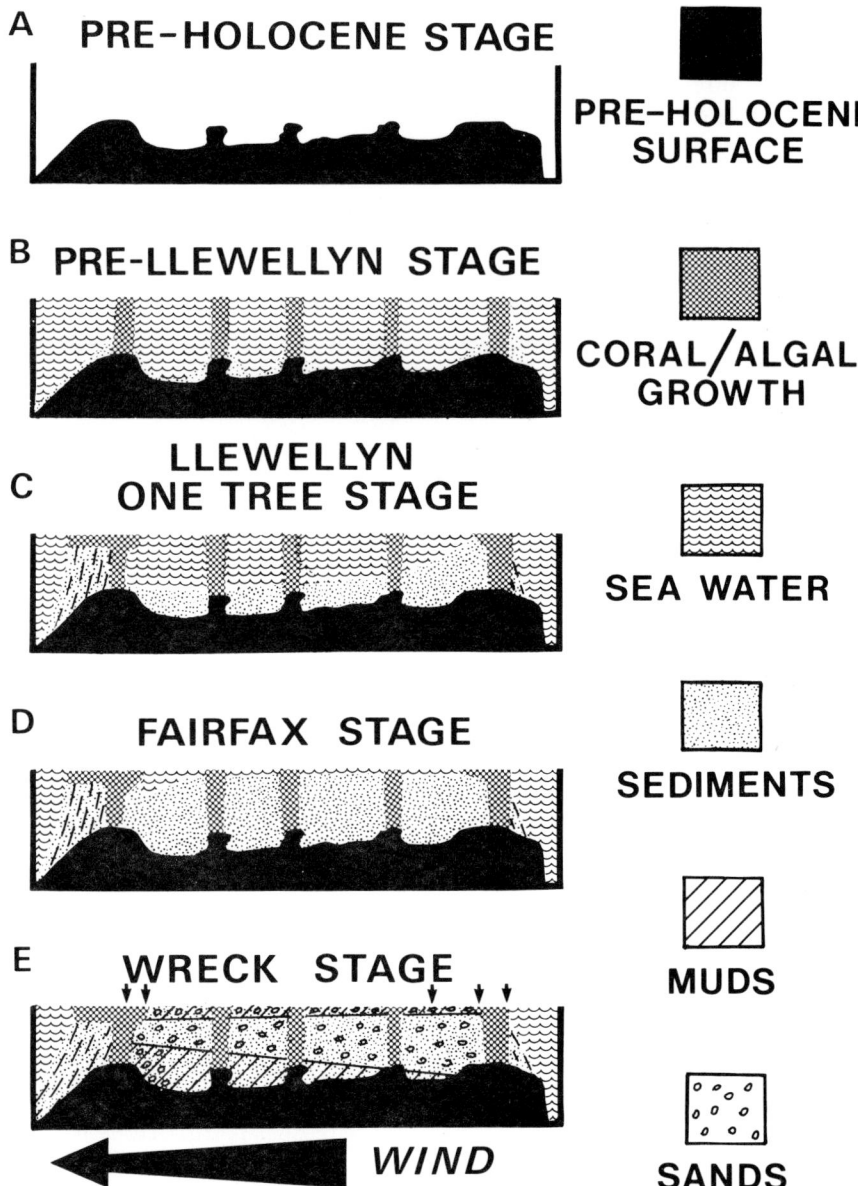

Figure 13   Evolutionary growth stages of reefs in the southern Great Barrier Reef. (A) The
preHolocene limestone surface
(B) Vertical growth phase effected by rising sea level
(C–E) Lateral growth phases during period of relative sea level stability

1. Different environments of the same reef have calcified at different rates during growth to stabilized sea level (fig. 6). Which then constitutes the standard curve for a particular reef?

2. Similar environments on different reefs have calcified at different rates throughout the Holocene, both in the vertically growing and surf-adjustment phases of growth. For example, the rapid growth rate for the One Tree Reef windward reef flat contrasts with the slower rate for the same environment at Fairfax Reef (B.M.R., unpub.).

3. Essentially similar environments on a reef are currently calcifying at different rates. For example, at One Tree Reef, the biological character of the southern coral flat changes appreciably from east to west. Concomitant calcification changes are to be expected solely as a result of changes in the ratio of coral cover to sand. Which part of the reef flat constitutes the norm?

4. The cause of the change in calcification rates between the transgressive and still-stand growth phases are not clearly understood. Such changes may have arisen from a reduced calcification capacity affected by biological changes, or from destruction of some part of the record by erosion, or both. Without understanding the reason for the effect, it is difficult to define the norm.

Perhaps of greater importance than the definition of standard curves or standard calcification estimates, is the recognition, in the southern Great Barrier Reef, that reefs have evolved through a series of overlapping stages (fig. 13) termed juvenile, mature and senile (Davies & Marshall 1980). Such reefs have the characteristics shown in table 10. Their evolutionary gradation, expressed as biological and sedimentological reactions to the three controlling factors, demands that the growth performance of a reef be assessed against its projected position in the evolutionary scheme. Thus, juvenile reefs will exhibit growth characteristics (biological-sedimentologic expression, calcification rates) significantly different from mature and senile reefs. In any assessment of performance therefore, it is necessary to know much more about coral reefs than is known at present, and also to move away from the dominantly descriptive studies of patterns towards analyses of process, controls and rates. One way of achieving this is via a simple and organized view of the pathways of calcium carbonate transfer in the reef system (fig. 14a). The pathways should represent likely flow patterns between the principal reef compartments and between compartments and hydrosphere. Figure 14a is drawn for a five-year period which is sufficiently long to make quantitative data gathering meaningful. More than one agent may be responsible for any one of the mega-processes shown in figure 14a, so that a flow line is in fact a series of lines representing different or similar processes operating at different rates, e.g., biologic reef destruction is affected by many organisms, and different mechanisms, operating at different rates. Calcium carbonate flux is the totality of the processes. In constructing figure 14a, one is therefore forced to examine all aspects of the system so that gaps in knowledge immediately flow from the inspection. Some of the gaps that pertain to controls and rates of mechanisms in this system are shown in table 11. This is not an exhaustive list but if answers to these questions were obtained, then, the quantitative calcium carbonate flux diagram shown in figure 14b would have real meaning: it would be an index of the ecosystem at work, allowing the assessment, measurement and even simulation of change. Further, a series of such diagrams depicting juvenile, mature and senile reefs at work would illustrate the turnover or replacement time within an evolving coral reef system. Much more than standard curves or rates, such diagrams would have immense value in both scientific and managerial applications in coral reef development.

Stage	Example	Relation to sea level	Major diagnostic features
Juvenile	Shoals of northern Capricorns	Subtidal; growing vertically	1. Interspersed patches of dominantly branching corals.
Adolescent	Llewellyn Reef	Perimeters are intertidal having recently reached sea level	1. Incipient algal flat 2. Wide well developed coral flat 3. Deep, open lagoon 4. Absence of appreciable rubble 5. Little sediment on leeward margin 6. Incipient prograding subtidal sand sheets
Mature	One Tree Reef	Perimeters are intertidal–supratidal. At sea level for some appreciable time	1. Reef perimeters exposed at low tide 2. Rubble ridges on perimeter margins 3. Sand and rubble cays unstable 4. Windward flats are bare pavements 5. Coral flats spreading backwards over sand sheets 6. Lagoons moated 7. Lagoons open or partially filled with patch reefs 8. Extensive sand sheet progation and partial lagoon infill 9. Extensive sediment accretion on leeward margin 10. Profuse coral development on leeside reef flat
Senile	Wreck Reef	Most of the reef surface is intertidal–supratidal	1. Windward margins eroding 2. Original high rims eroded; erosional remnant blocks on windward and leeward margins 3. Lagoon infilled 4. Secondary coral/soft algal colonization of upper reef surface. *Halimeda* prolific 5. Island progradation both to windward and leeward 6. Prograding leeward margins 7. Leeward lagoon edge is a marked remnant feature

Table 10   Reef attributes as a function of evolutionary development

First order process	First order problem	Secondary working problems
Biological and abiological construction	1. Is biologic production of $CaCO_3$ environmentally controlled?	1. Is form and rate energy related? 2. Is $CaCO_3$ a limiting nutrient?
	2. What part does abiological construction play in the economics of reef growth?	1. Under what conditions and at what rates does abiological construction occur? 2. To what extent do such products remain a stable part of the reef system?
Biological and abiological lithification	1. To what extent is lithification an inorganic, organic or biologically induced process?	1. In what ways are form, growth rates and abundance related to energy, substrate type and formational process?
	2. What fraction of $CaCO_3$ economics does lithification represent?	1. What percentage of total $CaCO_3$ flux does lithification represent? 2. Is this environment specific?
Biological and abiological destruction	1. What are the major biological elements effecting reef destruction?	1. How do such agents operate and what are the destructive mechanisms? 2. At what rates do they operate, and are they environment specific or dominant? 3. What are the destructive products?
	2. To what extend does such activity secondarily affect the biological development of the reef?	1. By what processes is such destruction growth limiting or growth promoting? 2. To what extent, by what processes and at what rates does biological destruction promote abiological construction and lithification?
	3. How does the reef and hydrosphere interact in terms of energy distribution, dispersal and retention?	1. What is the relation of $CaCO_3$ flux within the system as opposed to between the system and the hydrosphere? 2. What are the energy thresholds of particular environments? 3. How does energy transfer reflect $CaCO_3$ mobility? 4. To what extent does total growth rate and form reflect total energy?

Table 11   Problems requiring urgent attention

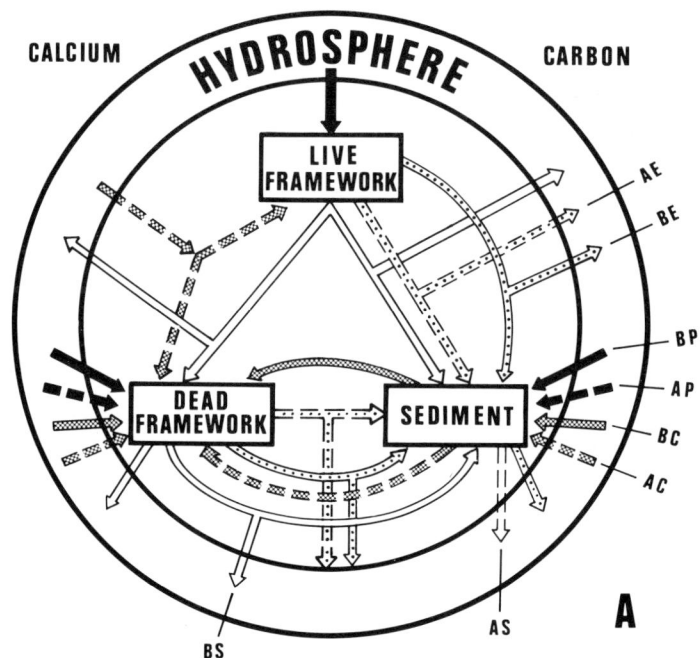

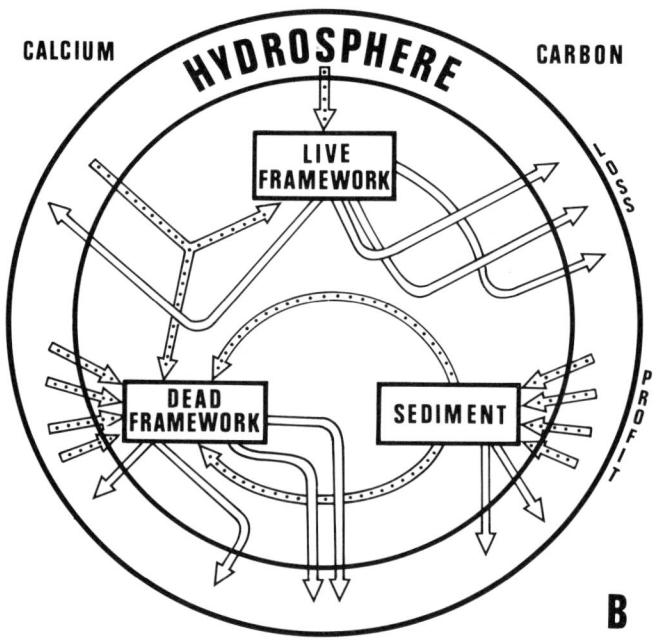

Figure 14   Coral reef calcium carbonate budget

     (A) Pathways of CaCO₃ transfer between major compartments of a reef system, i.e., hydrosphere, living framework, dead framework and sediments.

     (B) Diagramatic representation of the quantitative CaCO₃ flux between reef compartments and the hydrosphere.

     Live framework — calcifiers representing both primary framework builders and squatters. *Dead framework* — mainly dead equivalents of live framework together with products of lithification. *Sediments* — destructional products of both live and dead framework. *Hydrosphere* — seawater around and within reef. The mega-processes which move CaCO₃ within and between these four compartments are as follows. *Constructional* — (a) Biologic calcification by the live framework. (b) Abiologic processes. (i) Fixing of CaCO₃ in abiologic precipitates, for example, ooid shoals. (ii) Abiologic construction from biologically produced materials, for example, beach rock, storm ramparts, cays and all sedimentary units. *Biological and abiological lithification* — all processes effecting encrustation, and primary and secondary cavity infill. In some instances it is unknown whether the dominant process of biologic or abiologic, for example, peloid formation. *Biological and abiological destruction* — all processes responsible for breakdown of the three major reef compartments. More than one agent may be responsible for any of these mega-processes. A flow line in figure 14A represents the totality of the processes operating

LITERATURE CITED

Adey, W. H., 1975. The algal ridges and coral reefs of St. Croix their structure and Holocene development. *Atoll Research Bulletin* 187, 1-67.

Adey, W. H., 1977. Shallow water Holocene bioherms of the Caribbean Sea and West Indies. In D. L. Taylor (ed.), *Proceedings: Third International Coral Reef Symposium.* Miami: University of Miami, Vol. 2, pp. xxi-xxiv.

Adey, W. H., 1978. Coral reef morphogenesis: A multidimensional model. *Science 202, 831-57.*

Adey, W. H., P. J. Adey, R. B. Burke and L. Kaufman, 1977. The Holocene reef systems of eastern Martinique, French West Indies. *Atoll Research Bulletin* 218, 1-40.

Adey, W. H. and R. Burke, 1976. Holocene bioherms (algal ridges and bank-barrier reefs) of the eastern Caribbean. *Geological Society of America Bulletin* 87, 95-109.

Adey, W. H. and I. G. Macintyre, 1973. Crustose coralline algae: A re-evaluation in the geological sciences. *Geological Society of America Bulletin* 84, 883-904.

Adey, W. H., I. G. Macintyre and R. Stuckenrath, 1977. Relict barrier reef system off St. Croix: Its implications with respect to Late Cenozoic coral reef development in the western Atlantic. In D. L. Taylor (ed.), *Proceedings: Third International Coral Reef Symposium.* Miami: University of Miami, Vol. 2, pp. 15-22.

Adey, W. H. and J. M. Vassar, 1975. Colonization, succession and growth rates of tropical crustose coralline algae (Rhodophyta, Cryptonemiales). *Phycologia* 14, 55-69.

Backshall, D. G., J. Barnett, P. J. Davies, D. C. Duncan, N. Harvey, D. Hopley, P. J. Isdale, J. N. Jennings and R. Moss, 1979. Drowned dolines — the blue holes of the Pompey Reefs, Great Barrier Reef. *BMR Journal of Australian Geology and Geophysics* 4, 99-109.

Bakus, G. J., 1973. The biology and ecology of tropical Holothurians. In O. A. Jones and R. Endean (eds.), *Biology and Geology of Coral Reefs.* New York: Academic Press, Vol. 2, Biology 1, pp. 326-67.

Bardach, J. E., 1961. Transport of calcareous fragments by reef fishes. *Science* 133, 98-9.

Bromley, R. G., 1978. Bioerosion of Bermuda reefs. *Palaeogeography, Palaeoclimatology, Palaeoecology* 23, 169-97.

Buddemeier, R. W., S. V. Smith and R. A. Kinzie, III, 1975. Holocene windward reef-flat history, Eniwetak Atoll. *Geological Society of America Bulletin* 86, 1581-4.

Choat, J. H., 1966. Parrot fish. *Australian Natural History* 15, 265-8.

Cloud, P. E., Jr., 1959. Geology of Saipan, Mariana Islands. Part 4. Submarine topography and shoal-water ecology. *United States Geological Survey Professional Paper* 280-K, 361-445.

Climap Project Members, 1976. The surface of the Ice-Age earth. *Science* 191, 1131-7.

Cribb, A. R., 1973. The algae of the Great Barrier Reefs. In O. A. Jones and R. Endean (eds.), *Biology and Geology of Coral Reefs.* New York: Academic Press, Vol. 2, Biology 1, pp. 44-75.

Davies, P. J., 1974. Cation electrode measurements in the Capricorn area, southern Great Barrier Reef Province. In, *Proceedings of the Second International Symposium on Coral Reefs.* Brisbane: Great Barrier Reef Committee, Vol. 2, pp. 449-55.

Davies, P. J., 1977. Modern reef growth — Great Barrier Reef. In D. L. Taylor (ed.), *Proceedings: Third International Coral Reef Symposium.* Miami: University of Miami, Vol. 2, pp. 325-30.

Davies, P. J. and J. F. Marshall, 1979. Aspects of Holocene reef growth — substrate age and accretion rate. *Search* 10, 276-79.

Davies, P. J. and J. F. Marshall, in press. A model of epicontinental reef growth. *Nature.*

Davies, P. J., J. F. Marshall and M. A. Borowitzka, in press. Recent algal nodules (Rhodoliths) from One Tree Reef, Great Barrier Reef. *Geology.*

Davies, P. J., J. F. Marshall, D. Foulstone, B. G. Thom, N. Harvey, A. D. Short and K. Martin, 1977a. Reef growth, southern Great Barrier Reef — preliminary results. *BMR Journal of Australian Geology and Geophysics* 2, 69-72.

Davies, P. J., J. F. Marshall, B. C. Thom, N. Harvey, A. D. Short and K. Martin, 1977b. Reef Development — Great Barrier Reef. In D. L. Taylor (ed.), *Proceedings: Third International Coral Reef Symposium.* Miami: University of Miami, Vol 2, pp. 331-7.

Davies, P. J. and K. Martin, 1976. Radial aragonite ooids, Lizard Island, Great Barrier Reef, Australia. *Geology* 4, 120-22.

Doty, M. S., 1974. Coral reef roles played by free-living algae. In, *Proceedings of the Second International Symposium on Coral Reefs.* Brisbane: Great Barrier Reef Committee, Vol. 1, pp. 27-33.

Flood, P. G., 1977. Coral Cays of the Capricorn and Bunker Groups, Great Barrier Reef Province, Australia. *Atoll Research Bulletin* 195, 1-7.

Flood, P. G., G. R. Orme and T. P. Scoffin, 1978. An analysis of the textural variability displayed by inter-reef sediments of the impure carbonate facies in the vicinity of the Howick Group. *Philosophical Transactions of the Royal Society of London, Series A* 291, 73-83.

Flood, P. G. and T. P. Scoffin, 1978. Reefal sediments of the northern Great Barrier Reef. *Philosophical Transactions of the Royal Society of London, Series A* 291, 55-71.

Focke, J. W., 1978. Holocene development of coral fringing reefs, leeward off Curacao and Bonaire (Netherlands Antilles). *Marine Geology* 28, M31-M41.

Frankel, E., 1977. Previous *Acanthaster* aggregations in the Great Barrier Reef. In D. L. Taylor (ed.), *Proceedings: Third International Coral Reef Symposium.* Miami: University of Miami, vol. 1, pp. 201-208.

Frydl, P. and C. W. Stearn, 1978. Rate of bioerosion by Parrotfish in Barbados reef environments. *Journal of Sedimentary Petrology* 48, 1149-57.

Fütterer, D. K., 1974. Significance of the boring sponge *Cliona* for the origin of fine grained material of carbonate sediments. *Journal of Sedimentary Petrology* 44, 79-84.

Ginsburg, R. N. and J. H. Schroeder, 1973. Growth and submarine fossilization of algal cup reefs, Bermuda. *Sedimentology* 20, 575-614.

Glynn, P. W., 1973. Aspects of the ecology of coral reefs in the western Atlantic region. In O. A. Jones and R. Endean (eds.), *Biology and Geology of Coral Reefs.* New York: Academic Press, vol. 2, Biology 1, pp. 271-324.

Glynn, P. W., G. M. Wellington and C. Birkeland, 1979. Coral reef growth in the Galápagos: Limitation by sea urchins. *Science* 203, 47-9.

Goldman, B. and F. H. Talbot, 1976. Aspects of the ecology of coral reef fishes. In O. A. Jones and R. Endean (eds.), *Biology and Geology of Coral Reefs*. New York: Academic Press, vol. 3, Biology 2, pp 125-54.

Gordon, M. C. and H. M. Kelly, 1962. Primary productivity of a Hawaiian coral reef: A critique of flow respirometry in turbulent waters. *Ecology* 43, 473-80.

Grassle, J. F., 1973. Variety in coral reef communities. In O. A. Jones and R. Endean (eds.), Biology and Geology of Coral Reefs. New York: Academic Press, vol. 2, Biology 1, pp. 247-270.

Gygi, R. A., 1975. *Sparisoma viride* (Bonnaterre), the Stoplight Parrotfish, a major sediment producer on coral reefs of Bermuda? *Ecologae Geologicae Helvetiae* 68, 327-59.

Hamner, W. M. and M. S. Jones, 1976. Distribution, burrowing and growth rates of the clam *Tridacna crocea* on interior reef flats. Formation of structures resembling micro atolls. *Oecologia* 24, 207-27.

Halley, R. B., E. A. Shinn, J. H. Hudson and B. Lidz, 1977. Recent and relict topography of BooBee Patch reef, Belize. In D. L. Taylor (ed.), *Proceedings: Third International Coral Reef Symposium*. Miami: University of Miamia, vol. 2, pp. 29-35.

Harvey, N., P. J. Davies and J. F. Marshall, 1979. Seismic refraction — a tool for studying coral reef growth. *BMR Journal of Australian Geology and Geophysics* 4, 141-7.

Hein, F. J. and M. J. Risk, 1975. Bioerosion of coral heads: inner patch reefs, Florida reef tract. *Bulletin of Marine Science* 25, 133-8.

Hiatt, R. W. and D. W. Strasburg, 1960. Ecological relationships of the fish fauna on coral reefs of the Marshall Islands. *Ecological Monographs* 30, 65-127.

Hodgkin, E. P., 1964. Rate of erosion of intertidal limestone. *Zeitschrift fr Geomorphologie* 8, 385-92.

Hopley, D., 1977. The age of the outer ribbon reef surface, Great Barrier Reef, Australia: implications for hydro-isostatic models. In D. L. Taylor (ed.), *Proceedings: Third International Coral Reef Symposium*. Miami: University of Miami, vol. 2, pp. 23-28.

Hopley, D., R. F. McLean, J. Marshall and A. S. Smith, 1978. Holocene-Pleistocene boundary in a fringing reef: Hayman Island, North Queensland. *Search* 9, 323-5.

Hudson, J. H., 1977. Long-term bioerosion rates on a Florida reef: a new method. In D. L. Taylor (ed.), *Proceedings: Third International Coral Reef Symposium*. Miami: Third International Coral Reef Symposium. Miami: University of Miami, vol. 2, pp. 491-7.

Hunter, I. G., 1977. Sediment production by *Diadema antillarum* on a Barbados fringing reef. In D. L. Taylor (ed.), *Proceedings: Third International Coral Reef Symposium*. Miami: University of Miami, vol. 2, pp. 105-9.

Hutchings, P. A., 1974. A preliminary report on the density and distribution of invertebrates living in coral reefs. In, *Proceedings of the Second International Symposium on Coral Reefs*. Brisbane: Great Barrier Reef Committee, vol. 1, pp. 285-96.

Isdale, P., 1977. Variation in growth rate of hermatypic corals in a uniform environment. In D. L. Taylor (ed.), *Proceedings: Third International Coral Reef Symposium*. Miami: University of Miami, vol. 2, pp. 403-8.

Jell, J., 1977. Workshop on the Great Barrier Reef. In K. R. Marin (ed.), *Sedimentological Newsletter*. Australasian Sedimentologists Group of the Geological Society of Australia, 8, pp. 31-40.

Kinsey, D. W., 1972. Preliminary observations on community metabolism and primary productivity of the pseudo-atoll reef at One Tree Island, Great Barrier Reef. In C. Mukudan and C. S. Gopinadha Pillai (eds.), *Proceedings of the Symposium on Corals and Coral Reefs*. Cochin: The Marine Biological Association of India, pp. 13-32.

Kinsey, D. W., 1979. Carbon turnover and accumulation by coral reefs. Ph.D. dissertation, University of Hawaii, 248 pp.

Kinsey, D. W. and P. J. Davies, 1979a. Carbon turnover, calcification and growth in coral reefs. In P. A. Trudinger and D. J. Swaine (eds.), *Biogeochemical Cycling of Mineral Forming Elements*. Amsterdam: Elsevier Scientific Publishing Company, pp. 131–62.

Kinsey, D. W. and P. J. Davies, 1979b. Effects of elevated nitrogen and phosphorus on coral reef growth. *Limnology and Oceanography* 24, 935–40.

Kinsey, D. W. and A. Domm, 1974. Effects of fertilization on a coral reef environment — primary production studies. In, *Proceedings of the Second International Symposium On Coral Reefs*. Brisbane: Great Barrier Reef Committee, vol. 1, pp. 49–66.

Kohn, A. J. and P. Helfrich, 1957. Primary organic productivity of a Hawaiian coral reef. *Limnology and Oceanography* 2, 241–51.

Lalou, C., J. Labeyrie and G. Delebrias, 1966. Datation des calcaires corelliens de l'atoll de Muroroa (Archipel de Taumotu) de l'epoque actuelle jusqua 500,000 ans. *Académie Des Sciences. Comptes Rendus Hebdomadaires Des Séances. Ser.D* 263, 1946–9.

Lewis, J. B., 1977. Processes of organic production on coral reefs. *Biological Reviews of the Cambridge Philosophical Society* 52, 305–47.

Lighty, R. G., I. G. Macintyre and R. Stuckenrath, 1978. Submerged early Holocene barrier reef, south-east Florida shelf. *Nature* 275, 59–60.

Macintyre, I. G., R. B. Burke and R. Stuckenrath, 1977. Thickest recorded Holocene reef section, Isla Perez core hole, Alacran Reef, Mexico. *Geology* 5, 749–54.

Macintyre, I. G. and P. W. Glynn, 1976. Evolution of modern Caribbean fringing reef, Galeta Point, Panama. *American Association of Petroleum Geologists Bulletin* 60, 1054–72.

MacGeachy, J. K., 1977. Factors controlling sponge boring in Barbados reef corals. In D. L. Taylor (ed.), *Proceedings: Third International Coral Reef Symposium*. Miami: University of Miami, vol. 2, pp. 477–83.

McLean, R. F., 1964. Mechanical and biological erosion of beach rock in Barbados, West Indies. Ph.D. dissertation, McGill University, 265pp.

McLean, R. F., 1967. Measurements of beachrock erosion by some tropical marine gastropods. *Bulletin of Marine Science* 17, 551–61.

McLean, R. F., 1974. Geologic significance of bioerosion of beachrock. In, *Proceedings of the Second International Symposium on Coral Reefs* Brisbane: Great Barrier Reef Committee, vol. 2, pp. 401–8.

McLean, R. F., D. R. Stoddart, D. Hopley and H. A. Polach, 1978. Sea level change in the Holocene on the northern Great Barrier Reef. *Philosophical Transactions of the Royal Society of London, Series A* 291, 167–86.

Maiklem, W. R., 1970. Carbonate sediments in the Capricorn Reef complex, Great Barrier Reef, Australia. *Journal of Sedimentary Petrology* 40, 55–80.

Maxwell, W. G. H., 1968. *Atlas of the Great Barrier Reef*. Amsterdam: Elsevier Scientific Publishing Company, 258pp.

Maxwell, W. G. H., 1969. The structure and development of the Great Barrier Reef. In K. S. W. Campbell (ed.), *Stratigraphy and Palaeontology*. Canberra: Australian National University Press, pp. 353–574.

Menard, H. W., 1964. Marine Geology of the Pacific. New York, McGraw Hill, 271pp.

Montaggioni, L., 1977. Structure interne d'un recif corallien Holocene (Ile De La Renion, Ocean Indien). *France. Bureau de Recherches Geologiques et Minieres Memoires* 89, 456–66.

Moore, C. H., Jr. and W. W. Shedd, 1977. Effective rates of sponge bioerosion as a function of carbonate production. In D. L. Taylor (ed.), *Proceedings: Third International Coral Reef Symposium* Miami: University of Miami, vol. 2, pp. 499–505.

Moore, W. S. and S. Krishnaswami, 1974. Correlation of X-radiography revealed banding in corals with radiometric growth rates. In, *Proceedings of the Second International Symposium on Coral Reefs*. Brisbane: Great Barrier Reef Committee, vol. 2, pp 269–76.

Neuman, A. C., 1966. Observations on coastal erosion in Bermuda and measurements of the boring rate of the sponge *Cliona lampa*. *Limnology and Oceanography* 11, 92–8.

Odum, H. T. and E. P. Odum, 1955. Trophic structure and productivity of a windward coral reef community on Eniwetok Atoll. *Ecological Monographs* 25, 291–320.

Ogden, J. C., 1977. Carbonate-sediment production by parrot fish and sea urchins on Caribbean reefs. *American Association of Petroleum Geologists. Studies in Geology* 4, 281–8. .

Orme, G. R., P. G. Flood and A. Ewart, 1974. An investigation of the sediments and physiography of Lady Musgrave Reef — a preliminary account. In, *Proceedings of the Second International Symposium on Coral Reefs.* Brisbane: Great Barrier Reef Committee, vol. 2, pp. 371–86.

Randall, J. E., 1974. The effect of fishes on coral reefs. In, *Proceedings of the Second International Symposium on Coral Reefs.* Brisbane: Great Barrier Reef Committe, vol. 1, pp. 159–66.

Revelle, R. and K. O. Emery, 1957. Chemical erosion of beach rock and exposed reef rock: Bikini and nearby atolls, Marshall Islands. *United States Geological Survey Professional Paper* 260-T, 699–709.

Rützler, K., 1975. The role of burrowing sponges in bioerosion. *Oecologia* 19, 203–16.

Rützler, K. and G. Rieger, 1973. Sponge burrowing: fine structure of *Cliona lampa* penetrating calcareous substrata. *Marine Biology* 21, 144–62.

Sargent, M. C. and T. S. Austin, 1954. Biologic economy of coral reefs. Bikini and nearby atolls, Marshall Islands. *United States Geological Survey Professional Paper* 260-E, 293–300.

Shinn, E. A., 1966. Coral growth rate an environmental indicator. *Journal of Paleontology* 40, 233–40.

Shinn, E. A., 1976. Coral Reef Recovery in Florida and the Persian Gulf. *Environmental Geology* 1, 241–54.

Smith, S. V., 1973. Carbon dioxide dynamics: a record of organic carbon production, respiration, and calcification in the Eniwetok reef flat community. *Limnology and Oceanography* 18, 106–20.

Smith, S. V. and J. T. Harrison, 1977. Calcium carbonate production of the Mare incognitum, the upper windward reef slope, at Eniwetok Atoll. *Science* 197, 556–9.

Smith, S. V. and D. W. Kinsey, 1976. Calcium carbonate production, coral reef growth, and sea level change. *Science* 194, 937–9.

Stern, C. W. and T. P. Scoffin, 1977. Carbonate budget of a fringing reef, Barbados. In D. L. Taylor (ed.), *Proceedings: Third International Coral Reef Symposium.* Miami: University of Miami, vol. 2, pp. 471–6.

Steers, J. A., 1938. Detailed notes on the islands surveyed and examined by the Geographical Expedition to the Great Barrier Reef in 1936. *Reports of the Great Barrier Reef Committee* 4, 51–96.

Steneck, R. S. and W. H. Adey, 1976. The role of environment in control of morphology in *Lithophyllum congestum*, a Caribbean algal ridge builder. *Botanica Marina* 19, 197–215.

Stephenson, W. and R. B. Searle, 1960. Experimental studies on the ecology of intertidal environments at Heron Island. I. Exclusion of fish from beach rock. *Australian Journal of Marine and Freshwater Research* 11, 241–67.

Thom, B. G. and J. Chappell, 1975. Holocene sea levels relative to Australia. *Search* 6, 90–3.

Thom, B . G. and J. Chappell, 1978. Holocene sea level change: an interpretation. *Philosophical Transactions of the Royal Society of London,* Series A 291, 187–94.

Thom, B. G., G. R. Orme and H. Polach, 1978. Drilling investigations of Bewick and Stapleton Islands. *Philosophical Transactions of the Royal Society of London, Series A* 291, 37–54.

Thurber D. I., W. S. Broecker, R. L. Blanchard and H. A. Potratz, 1965. Uranium-series ages of Pacific Atoll coral. *Science* 149, 55–8.

Tracey, J. I., Jr., 1978. Reef growth, sedimentation, and Holocene sea level. *Tenth International Congress on Sedimentology (Jerusalem),* Abstracts, II, 687.

Tracey, J. I., Jr., and H. S. Ladd, 1974. Quaternary history of Eniwetok and Bikini Atolls, Marshall Islands. In, *Proceedings of the Second International Symposium on Coral Reefs.* Brisbane: Great Barrier Reef Committee, vol. 2, pp. 537–50.

Tracey, J. I., Jr., and H. S. Ladd and J. E. Hoffmeister, 1948. Reefs of Bikini, Marshall Islands. *Geological Society of America Bulletin* 59, 861–78.

Trudgill, S. T., 1976. The marine erosion of limestones on Aldabra Atoll, Indian Ocean. *Zeitschrift fur Geomorphologie,* Supplementband 26, 164–200.

Trudgill, S. T., in press. Intertidal erosion of limestones on One Tree Island, southern Great Barrier Reef.*Zeitschrift fur Geomorphologie.*

Veron, J. E. N. and R. C. L. Hudson, 1978. Ribbon reefs of the Northern Region. *Philosophical Transactions of the Royal Society of London, Series B* 284, 3–21.

Vittor, B. A. and P. G. Johnson, 1977. Polychaete abundance, diversity and trophic role in coral reef communities at Grand Bahama Island and the Florida Middle Ground. In D. L. Taylor (ed.), *Proceedings: Third International Coral Reef Symposium.* Miami: University of Miami, vol. 1, pp. 163–8.

Wefer, G., 1980. Carbonate production by algae. *Halimeda Penicillus* and *Padina. Nature* 285, 323–4.

Wells, J. W., 1954. Recent corals of the Marshall Islands. *United States Geological Survey Professional Paper* 200-I, 385–486.

7 Coral Zonation: Its Nature and Significance

T.J. Done
Australian Institute of Marine Science
P.M.B. No. 3, Townsville M.S.O. Q. 4810
Australia

INTRODUCTION

Of the many definitions of the word 'zone', three are of great interest to reef ecologists. The first definition, listed in the Concise Oxford Dictionary, is 'belt or girdle worn round the body . . . symbol of virginity' — an evocative reflection on the webbing and lead belts worn by diving ecologists. Another definition is 'any well-defined tract of more or less belt-like form' and another, in reference to coral (Wells 1954, p. 396), 'an area where local ecological differences are reflected in the species association and signalized by one or more dominant species'. These definitions provide the basis for the discussion below.

A large part of the 'ecological differences' referred to in Wells' (1954) definition is physical; the range of physical environments in any one reef may be considerable. The physical environment on a coral reef represents both a resource to be partitioned, and a stress to be endured. Light provides energy but ultra-violet light damages living organisms. Water motion brings food and removes waste but it can also destroy corals. Sediment builds substrate but it also abrades and smothers benthic organisms. Evolutionary subdivision of resources and adaptation to stresses bring together different species associations in different places. Finer subdivision of a habitat may result from the inability of some species to live in close proximity to others. One result of this interaction between organisms and environment is zonation of corals and associated communities. A zonation pattern may be unique to a particular reef slope, or may be characteristic of entire reef systems. It is important to distinguish between these because the general body of ecological knowledge about coral reefs at present is built almost exclusively from studies of small areas.

The nature and causes of benthic zonation on reefs have been reviewed by Wells (1954, 1957), Yonge (1963), Stoddart (1969), and Glynn (1973). Recent ecological research has continued in a number of broad areas.

ZONATION OF GROWTH FORMS

The observation that growth forms of corals show a clearer zonation than do species distributions is longstanding (e.g. Wood-Jones 1910; Marshall 1931; Vaughan & Wells 1943). A world wide pattern in growth form zonation has been described by Pichon (1978a).

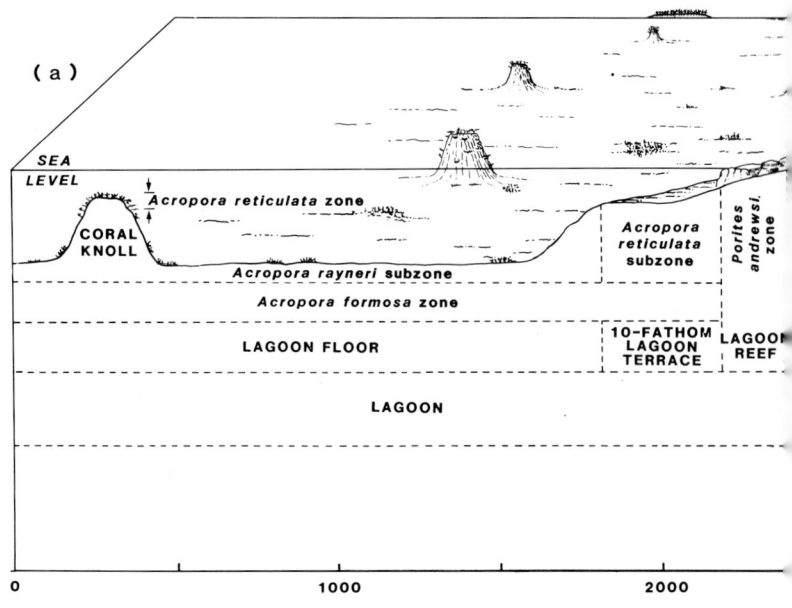

(a)

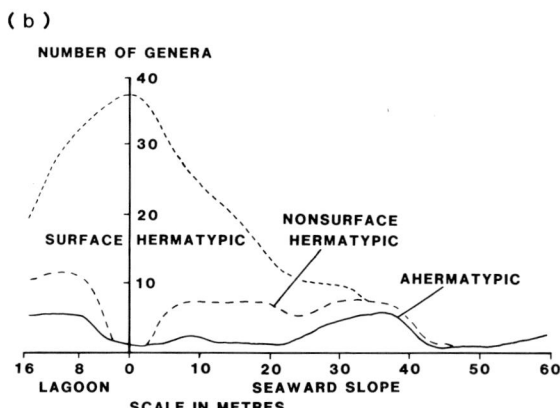

(b)

Figure   1   Zonal analysis of Bikini Atoll (redrawn from Wells 1954).
(a) Zonal subdivisions. Note broadly inclusive zones further subdivided on the basis of localized dominance of species.
(b) Bathymetric distribution of surface and nonsurface genera.
(c) Bathymetric distribution of surface and nonsurface species

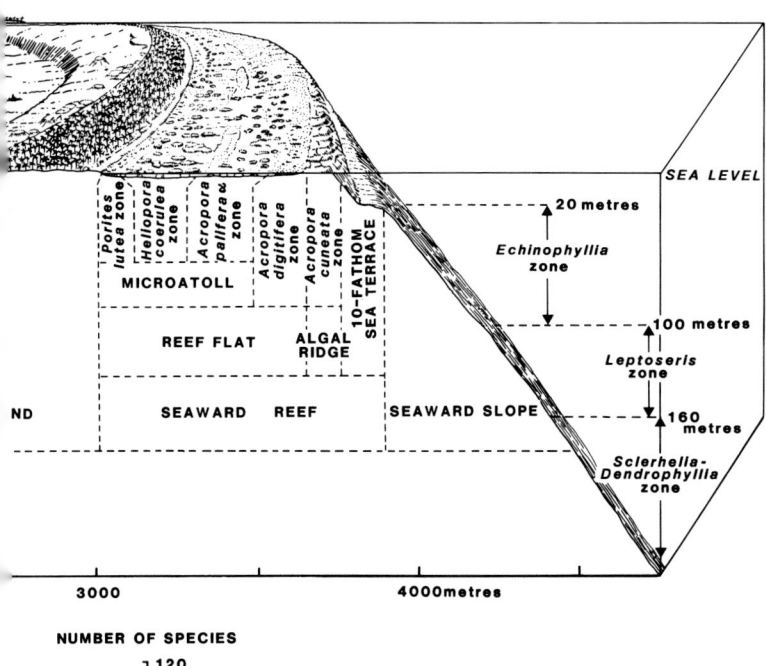

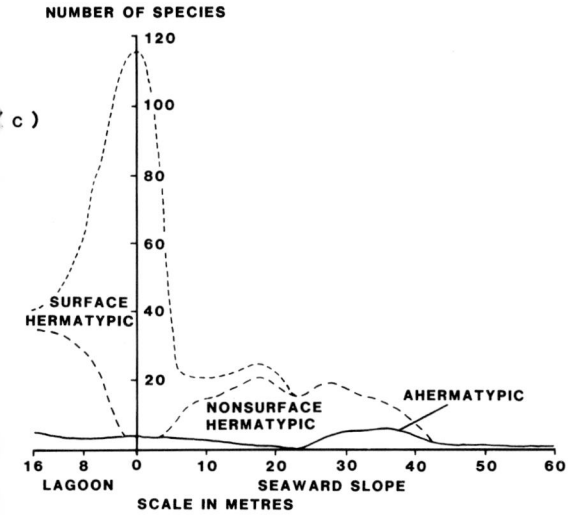

ZONAL PATTERNS AND ENVIRONMENTAL GRADIENTS

Gradients in water movement (waves, currents) and light (with depth and turbidity) have long been seen as determinants of zonation. Ecologists have continued to schematize zonation (of taxa and growth forms) in terms of gradients in water movement and light but have not progressed far in quantifying the gradient-zone relationship, or in identifying causal pathways.

ZONATION, COMMUNITY STRUCTURE AND BIOLOGICAL INTERACTIONS

Community structure (total cover, intra-zonal spatial organization, diversity and dominance, size class distribution) varies between zones in response to a combination of physical controls and biological interactions. At the time of Stoddart's major review (1969), the importance of biological interactions was recognized but had been little studied. Lessons learnt from terrestrial and intertidal ecology suggest that patterns in community structure and zonation may not be a simple summation of the independent responses of species to environmental tolerances and preferences (Connell 1972; Whittaker 1975). The 'realized zone' of a species may, as a result of competition and predation, be a significant contraction of the 'fundamental zone' (Pielou 1977).

TRANSIENCE OF CORAL COMMUNITIES

Recent work raises questions regarding community succession and climax, and the degree to which community composition is a response to prevailing conditions on the one hand and aperiodic, episodic events on the other (Stoddart 1963, 1974; Loya 1976a; Connell 1978; Chappell 1980; Woodley 1980; Woodley & Porter 1980; Pearson, 1981).

Each of these areas is discussed below, with particular emphasis on zonation in scleractinian corals. Also included are sections on zonation schemes, numerical methods of study, environmental zonation and synthetic analyses. It is hoped that works referred to will direct readers to useful sources. A more comprehensive bibliography in the area of coral distribution studies is provided by Sheppard (1982).

NATURE OF CORAL ZONATION

This discussion of coral zonation requires some passing reference to the physical environment: a more detailed consideration of reef environment follows below.

Zonation of Species

Corals and other benthic species distribute themselves in different segments of a reef environment hyperspace which is both broad ranging and stratified. The result is their zonation on the reef, i.e., the occurrence in given reef habitats of predictable but overlapping subsets of the total species complement (see fig. 1a). For logistic reasons, coral zonation has been most often described on intertidal reef flats; for ecological reasons it is most clearly manifested there.

REEF FLATS

Reef flats support belt-like tracts of corals, frequently of low cover and/or diversity and often dominated by one or two characteristic species (Mayer 1918; Stephenson *et al.* 1933; Wells 1954, 1957; Spencer-Davies *et al.* 1971; Morrisey 1980). In this habitat, which is uniformly exposed to high light levels, these tracts reflect zonally-distributed

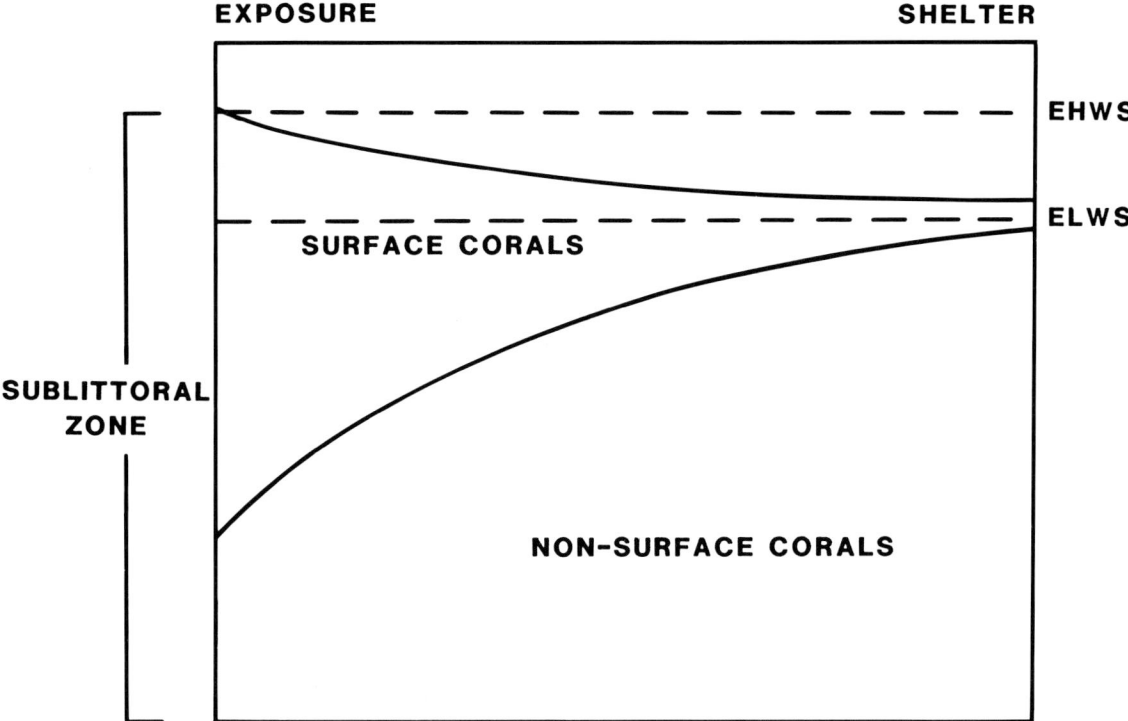

Figure 2 The effect of wave exposure on the bathymetric distribution of surface and nonsurface species. Nonsurface species, which are restricted to depths below the wave base, extend their distributions into shallower depths in sheltered habitats

differences in substrate type, elevation (and thus frequency and duration of subaerial exposure and/or high temperature stress) and exposure to surf, currents or relatively stationary water. It is a habitat wherein an elevational difference of only a few centimetres, or a lateral distance of just a few metres, may be accompanied by major differences in substrate or hydrodynamic regime, and where associated changes in species composition or dominance are marked.

REEF SLOPES

On reef slopes (subjected to gradients of both light and water movement) coral zonation is usually more subtle than on reef flats. The corals are divisible into two suites of species of which the bathymetric distributions, although broadly overlapping, characterize opposite ends of a depth and exposure-related environment gradient. These suites have variously been referred to as 'surface' and 'nonsurface' (Wells 1954), 'reef' and 'subreef' (Pichon 1972, 1978a) and A and B (Sheppard 1981a). The distribution of these suites at Bikini Atoll is indicated in figures 1b and 1c.

The shallower suite of species (of which the reef flat corals are a more or less depauperate subset) has its maximum diversity at various depths depending upon wave exposure, slope and illumination. Wells (1954) noted that the downward extent of the shallower suite was limited by the normal wave base; as a corollary, the upper limit of the deeper suite extends into shallower water on reef slopes with sheltered aspect (see fig. 2).

The deeper suite is less diverse than the shallower, and contains both hermatypic and ahermatypic species. By and large, these species do not have sufficient morphological plasticity to enable them to survive the high mechanical stresses associated with shallower wave-affected areas (except in sheltered refuges).

ZONAL COMMUNITIES

Zonal communities are subdivisions of these suites, particularly of the shallower suite. Subdivision is frequently based on the localized predominance of one or more species (or genera) following Wells' (1954) definition (e.g. Goreau 1959; Loya & Slobodkin 1971; Rosen 1971, 1975; Sheppard 1980a) and the zones may be named after the dominant(s). Alternatively, subdivision may be made on the basis of:

1. the absence of dominant species ('indominate' zones, Sheppard 1980a),
2. quantitatively unimportant but nonetheless 'characteristic' or 'differential' species (Scheer 1978),
3. predominance of characteristic growth forms (Pichon 1972, 1978a; Rosen 1971),
4. the total species complement, using multivariate techniques (e.g. Loya 1972; van den Hoek *et al.* 1975, 1978; Done 1977; Veron & Done 1979; Bradbury & Young, 1981).

These alternative devices for differentiation of zonal communities are particularly applicable on reef slopes, where single species dominance is frequently less marked. The intergrading nature of zonal communities may be seen in the following example.

SPECIES ZONATION: AN EXAMPLE

Zonal subdivision of some Great Barrier Reef ribbon reefs is presented in table 1. The table was derived from a multivariate classification described by Done & Pichon (in prep.) and includes only 'constant' species (*sensu* Sheer 1978); that is, tabulated species are also from time to time found as 'accidentals' in zones other than those indicated.

Single species and groups of species show varying degrees of fidelity to topographic and environmental zones. Species with patterns A, B and E (table 1) by and large have high fidelity to circumscribed sections of the reef (the outer surf, inner surf and back-reef slope respectively). Pattern D demonstrates a low fidelity to any single bathymetric or topographic zone but a high fidelity to a single broad environmental zone (moderate wave energy) found on fore-reef, reef flat and back-reef. Species with pattern C are so ubiquitous that neither individually nor collectively are they indicative of any particular bathymetric, topographic or environmental zone. Pattern C species extend from the calm, poorly illuminated depths (25 to 40 m) to the highly illuminated and surf beaten sections of the reef flat. However, all of group C are extremely polymorphic, and in several cases, there is a strong fidelity of growth form to a particular zone.

In this and many other examples, the high degree of species overlap between topographic zones makes zonal delineation on the basis of species composition difficult. By contrast, *growth forms* of corals are more distinctly zonal in their distribution.

Ecomorph Zonation

Ubiquitous species tend to adopt different growth forms in different zones. Individual species thus exhibit 'ecomorph zonation'; the ecomorphs differing from each other in colony and/or corallite shape and/or size (Veron & Pichon 1976, Chapter 2). The diverse hydrodynamic, photic and sedimentary environments in the different reef zones each favour different morphologies (Morton 1974) and a characteristic sequence of growth forms across reef profiles has been described (Pichon 1972, 1978a; see also below).

	TOPOGRAPHIC ZONE	1	2	3	4	5	6	7	8	9	10
A	*Acropora palmerae*				·	++					
	Acropora sp 1.				+	++				·	
	Acropora rotumana			·	++	++	+				
	Acropora variabilis			++	+	++					
	Pocillopora eydouxi		+	++	++	+				·	
	Acropora digitifera		·	·	++	++	++	·	+	·	·
	Acropora robusta		·	·	++	+	+	+	+		
B	*Gonistrea* c.f. *favulus*					·	++	++			
	Acropora millepora				·	·	++	++	++	+	·
C	*Pocillopora verrucosa*	++	++	++	++	++	++	++	+	+	+
	Acropora humilis	·	++	++	++	++	++	++	++	+	+
	Acropora palifera	+	++	++	++	++	++	++	++	++	++
	Porites (massive spp.)	++	++	++	++	++	++	++	++	++	++
	Stylophora pistillata	++	++	++	++	+	+	++	++	++	++
	Acropora hyacinthus	+	+	++	++	++	+	+	++	+	·
	Millipora platyphylla	++	++	++	+	++	·	+	+·	·	+
D	*Platygyra pini*	+	+	·			+	+	+	+	++
	Acropora cerealis	·	++	++	·	+		++	++	·	++
	Pocillopora damicornis	+	+	·	·	·		+	+	++	++
	Favia pallida	++	+	·			++	++	+	+	+
	Astreopora listeri	++	+				+	++	+	++	+
	Seriatorpora hystrix	++	++	+	·	·			+	++	++
	Cyphastrea serailia	+	·	+	·			++	·	+	+
	Goniastrea pectinata	++	+	·					·	+	++
	Acropora squarrosa	+	+				·	·	+	++	++
	Montipora verrucosa	++	·					·	·	·	+
	Favia favus	+		·					·	++	++
	Fungia spp.	·	+	·					·	+	++
	Millepora tenera	·	+							+	++
	Tubastrea nigrescens	+							·		
E	*Goniastrea edwardsi*	·	·		·	·		·	+	++	+
	Echinopora mammiformis	·							·		++
	Lobophyllia pachysepta		·						·		+
	Fungia echinata	·	·						·		+
	Acropora carduus		·	·					·		+
	Montastrea magnistellata	·							·	++	·
	Rumphella sp.									++	++

Key: blank = absent or present as 'accidental' only
· = present up to 0.3 ave cover grade
+ = present between 0.3 and 1.0 ave cover grade
++ = ave cover grade 1.0 or greater
1, lower slope; 2, 3, mid slope; 4, upper slope; 5, shoulder and outer reef flat; 6, mid reef flat; 7, back reef flat; 8, back reef margin; 9, shallow pinnacle slopes and floor; 10, deep pinnacle slopes and floor.

Table 1   Patterns in the distribution of species on a wave beaten reef of the Great Barrier Reef. For explanation see text

In wave-affected areas, corals exhibit morphological adaptation to mechanical stress (see Shinn 1966; Wainwright *et al.* 1976; Graus *et al.* 1977; Vosburgh 1977). As may be expected, encrusting and streamlined morphs of massive species (e.g. Faviids, *Porites* spp. and, to a lesser extent, Mussiids) are found in surf areas. However, more predominant on both Indo-Pacific and Atlantic reefs are surf resistant ecomorphs of branching species (e.g. Mayer 1918; Wells 1954; Goreau 1959; Rosen 1971, 1975; Giester 1977). In the Caribbean, *Acropora palmata* maintains an open ramose habit in the surf zone and survives mechanical stresses by branch-thickening and by adopting an orientation in which hydrodynamic forces are directed along rather than across the branches (Shinn 1966; Graus *et al.* 1977). Coral branches have low tensile strength but high compressive strength (Wainwright *et al.* 1976; Tunnicliffe 1979).

In the Indo-Pacific, the branching genera *Acropora, Pocillopora* and *Stylophora* are dominant in surf areas (Wells 1954; Rosen 1971, 1975). The strategy of orientation with respect to hydrodynamic forces is dramatically demonstrated by *Acropora palifera*. In calm waters, this species presents several branching growth forms (columns, knobs, micro-atolls). In strong surf it grows as low, ridged colonies with the ridges orientated parallel to water movement. The species dominates this zone (fig. 3). In the Pocilloporidae, compact branching varieties (previously described as separate species) have been shown to be wave-adapted ecomorphs of species which are elsewhere open-branching and fragile (Veron & Pichon 1976). Perhaps the most extreme example of adaptation to the surf zone will prove to be *Acropora palmerae* (Wells 1954), an entirely encrusting coral, which may in fact be a surf ecomorph of one of the ramose *Acropora robusta* group (Wallace 1978; Veron & Wallace, in prep.). It is enigmatic that one of the most delicate *Acropora* species, with fine, close-packed, fragile vertical branches, thrives in most Great Barrier Reef surf zones (fig. 4).

In more sheltered areas, where there is less mechanical stress, colony shapes are more diverse than in surf zones. Patterns of colony shape distributions in such areas may be understood in terms of species-specific nutritional requirements (Porter 1976; Jaubert 1977; Spencer-Davies 1977; Porter *et al.* 1980), competitive abilities (Lang 1973; Porter 1976; Sheppard 1979) and sediment-removing ability (Hubbard & Pocock 1972; Hubbard 1974; Dodge *et al.* 1974; Loya 1976b; Bak 1978).

Many species have been shown to adopt a more flattened habit with increasing depth, thereby maximizing the interceptions of light, e.g. *Montastrea annularis* (Barnes 1973) and many *Acropora* species (Wallace 1978). However, because a flattened surface also accumulates sediment, many other shapes (including branching, massive, bracket and vase) occur in sheltered habitats (Vaughan & Wells 1943; Morton 1974; Porter 1976).

Micro-atolls (Scoffin & Stoddart 1978) form the best known of all zones defined on the basis of coral growth form. In these zones, the tops of colonies die off due to emersion and, except where water is moated, indicate the level of extreme low spring tides (Scoffin & Stoddart 1978). Instances of single colonies covering hundreds of square metres have been described (e.g. the *Heliopora coerulea* micro-atoll zone, Wells 1954). Forty-three Indo-Pacific species have been recorded as micro-atolls (Rosen 1978) although most never attain the size of Wells' *Heliopora*.

Zonation in community structure

Reefs exhibit zonation in coral species diversity, area of cover, colony size, crowding and spatial pattern — those non-taxonomic community attributes on which a great deal of synecological interpretation is based. With few exceptions (e.g. Mayer 1918), the quantitative field work providing these types of data has been published since Stoddart's (1969) observation that 'quantitative work so far has added little to qualitative zonation studies'.

Foremost among recent quantitative studies is that of Loya (1972) at Eilat. Statistical analysis of a large and thorough data base derived from line transects indicated the following trends:

1. a general increase in species richness and diversity with increasing depth to 30 m, even though illumination at 30 m was less than five percent of the surface value,

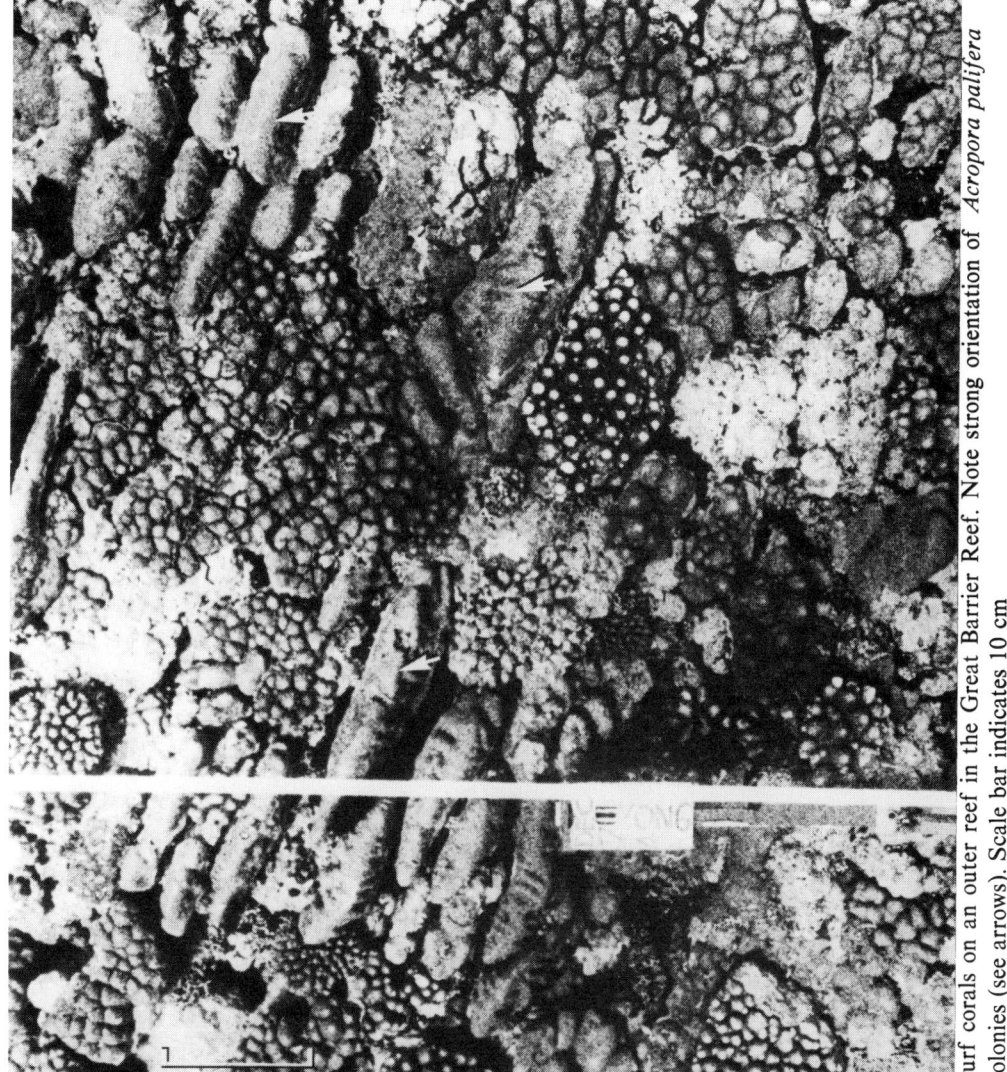

Figure 3   Surf corals on an outer reef in the Great Barrier Reef. Note strong orientation of *Acropora palifera* colonies (see arrows). Scale bar indicates 10 cm

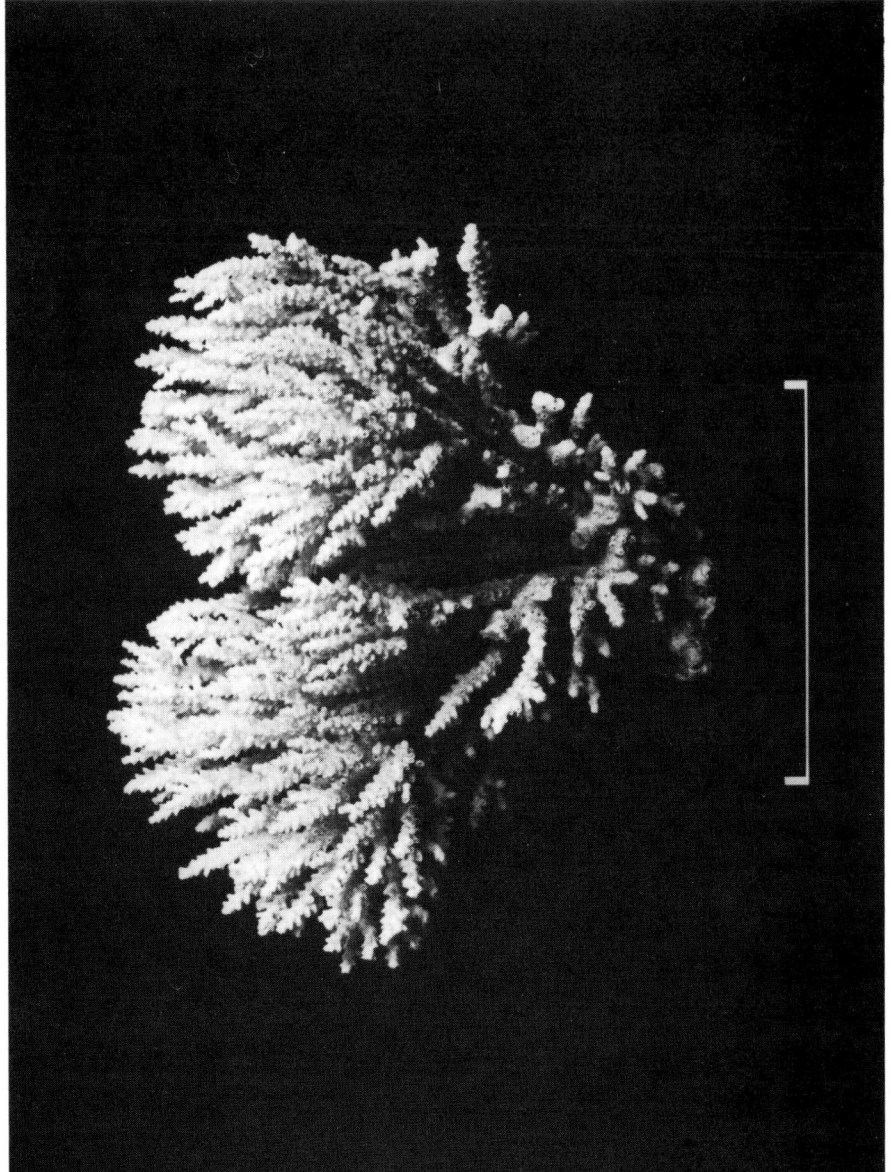

Figure 4   This delicate *Acropora* which is common in surf zones on outer reefs in the Great Barrier Reef resists mechanical damages despite its fragile structure. Scale bar indicates 10 cm. Photograph courtesy of Dr. J. E. N. Veron

2. a smaller average colony size in reef flat and deeper zones than at intermediate depths,

3. a positive correlation between colony size and light intensity down the slope,

4. absence of correlation between species richness and light attenuation,

5. a greater coverage and diversity of living coral on steeper slopes, compared with flat areas.

These trends were attributed to a combination of local factors, especially differential settlement of sediment. The study also led to conclusions about community structure and biotic zonation; to zonal patterns of environmental 'severity' and 'predictability', and of 'physical control' and 'biological accommodation'. These concepts, established in the ecological literature by Sanders (1969), Slobodkin & Sanders (1969) and others, were also developed in relation to reef community structure by Porter (1972a), Connell (1978) and Dana (1979).

Further quantitative studies concerned with zonal patterns in community structure have been provided by Glynn (1976), van den Hoek *et al.* (1975, 1978) Bradbury & Loya (1976), Goodwin *et al.* (1976), Bull (1977), Wallace & Lovell (1977), Ditlev (1978), Morrisey (1980) and Sheppard (1980). Interpretations have been as diverse as the localities in which the studies took place. Gradients and discontinuities in various abiotic factors (such as tides, waves, light, substrate and sediment) and biotic factors (especially grazing pressure and spatial competition) have been suggested as the proximal causes of their observations. Two syntheses concerned with community structure (namely Connell 1978 and Chappell 1980) are considered below.

Zonation in community function

Zonal communities have been shown to differ considerably in their capacity for photosynthesis and calcification (see methods papers by Kinsey 1978; Maragos 1978; Marsh & Smith 1978). The annual mean productivity of zones on One Tree Island reef flat varies twenty-fold (Kinsey 1979) and in four of the five zones studied, mean annual calcification was between 11% to 20% of productivity; in the fifth it reached 60%. These disparities in production and calcification have lead to the concept of upstream 'producer' and downstream 'consumer' zones (LIMER 1976; Kinsey 1979). Overall reef morphology itself is ultimately determined by the rates of carbonate consolidation into the reef framework, rates of dispersion and accumulation as sediment, and in particular, the differences in these rates between zones.

Zonation schemes

Various authors have created schemes to provide synoptic regional views of the ways taxa and growth forms are distributed with respect to reef structure and environment. The schemes are usually presented as subjective ordinations of assemblages against two or three spatio-environmental axes. The assemblages are generally loosely defined and many encompass allied subgroupings with a dominant taxon (or growth form) in common, but with differences at subordinate levels (see fig. 13 and Veron & Done 1979). They thus correspond to the 'alliance' of Beale & Costin (1952) and Scheer (1978) and, since they indicate broader environmental ranges than the more fundamental 'association', they are appropriate for regional synopses.

The authors of the schemes use subjective, ordinal scaling on their environmental axes. This practice is appropriate to the synoptic function of zonal schemes, but it does make inter-region comparisons difficult, since one scheme's 'extreme' may be equivalent to another's 'moderate'.

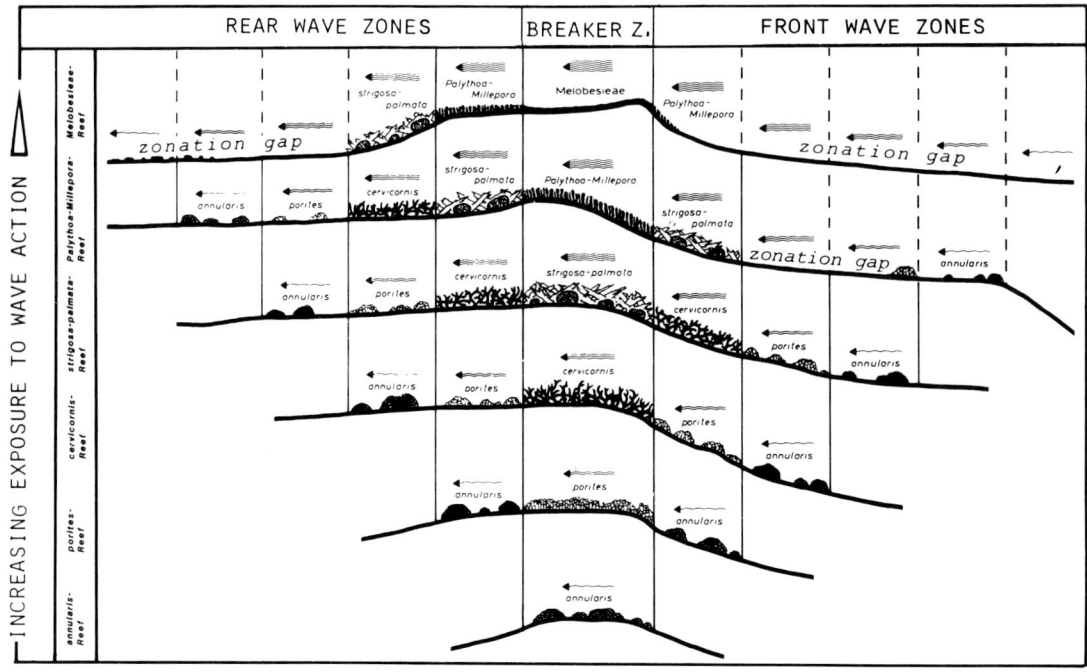

Figure 5 Zonation scheme for Caribbean reefs (reproduced with permission from Giester, 1977). Reefs exposed to maximal wave energy have characteristic coral associations that are not found in sheltered reefs. Shallow water associations on sheltered reefs (e.g. 'annularis' and 'porites') are restricted to deeper and/or sheltered habitats on more exposed reefs

Wave related zonation schemes

Of the two or three dimensions which can comfortably be contemplated or represented diagrammatically, wave action invariably takes pride of place. The comparison of zonation schemes therefore becomes a comparison of the other dimensions in the scheme, and the way the scheme is portrayed diagrammatically.

Giester (1977) provided a scheme of 'climax' breaker and wave zones for Caribbean reefs consisting of a series of reef profiles arranged vertically (fig. 5). The uppermost profile represented a reef exposed to maximum wave energy, the lowest to a minimum. The scheme shows a successive 'dropping out' of high energy alliances with decreasing wave exposure and a shift of deep water alliances to shallower locations on sheltered reefs. A 'zonation gap' caused by seasonal sediment abrasion was not present where surface wave action was reduced. Giester's reef profiles make the zonation trends with exposure and depth readily evident but he stresses that aberrant zonations may result from localized irregularities in bottom topography, substrate change from lagoon to reef facies, seasonal abrasion and unusually shallow reef flat.

Pichon (1978a) provided a 'worldwide' scheme based on very broadly defined growth form alliances (see fig. 6). By including a totally stylized reef shape, Pichon's scheme is more generally representative than Giester's. This scheme emphasizes that a similar sequence of energy conditions and related growth form predominance occurs on reef flat

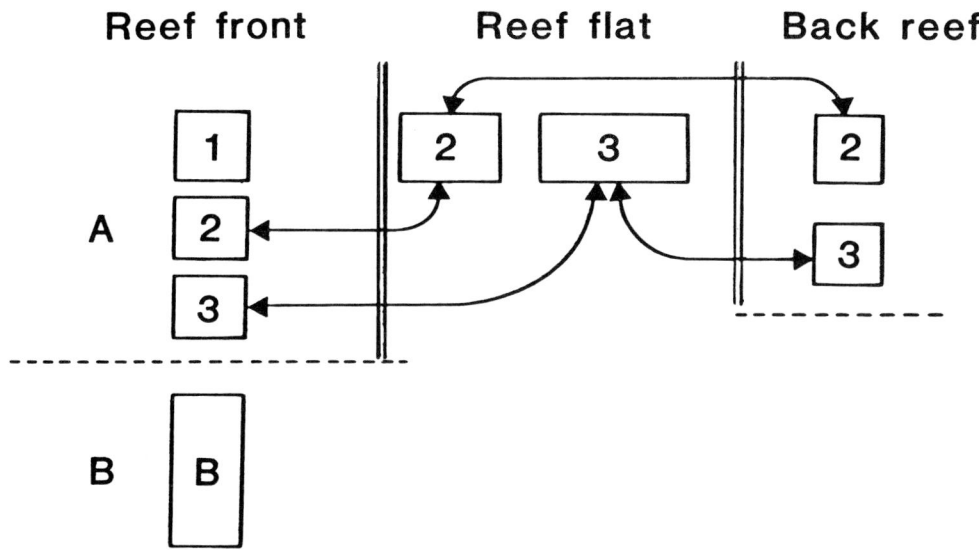

Figure 6 Worldwide coral zonation scheme (redrawn from Pichon 1978). A : reefal biocoenose; B : subreefal biocoenose; 1 : dominant growth forms encrusting, with or without short, thick digitations; 2 : branching forms dominant; 3 : massive forms dominant. The reefal biocoenose is conspicuously zoned, with the sequence across the reef flat a mirror image of the sequence down the reef front. The back reef slope sequence lacks association type 1, which is found only in high wave energy areas. The sub-reefal biocoenose is able to extend into shallower depths on back reef slopes.

and reef slope sites. Zonation was evident only in the reefal biocoenose and it was correlated with wave action, not light. An assemblage occupying sub-reefal depths was not conspicuously zoned.

Rosen (1975) also concerned himself mainly with wave-related zonation. Like Pichon, he noted that water movement attenuated both across the reef flat and down the slope from the surf zone (see fig. 7). He also allowed for longshore attenuation of surf intensity (Roberts 1974) and represented 'water movement' isopleths on a three dimensional block diagram. The five water movement zones thus delineated were occupied by five 'associations' named after the predominant Indo-Pacific taxa. In order of decreasing wave energy they were: the *Porolithon* association, the *Pocillopora* association, the *Acropora* association, the *Faviid* association and the *Porites* association. This scheme is broadly applicable in the Indo-Pacific, but a high energy *Pocillopora* association has not been recognized on the Great Barrier Reef where, in most surf zones in which *Pocillopora* is abundant, *Acropora palifera* (Wells 1954) is significantly more prolific (personal observations; Veron, pers. comm.; Pichon, pers. comm.).

Rosen's scheme emphasizes that 'deep' water movement conditions and associations may occur in shallow depths on a lee shore. Dana (1979) also suggested that deep water environments are so displaced. His scheme (fig. 8) identifies environmental severity, optimately, variability and predictability as primary zoning influences whose actions are reflected in patterns of species coverage, diversity, dominance and successional studies. Although the scheme was derived for a localized study, the ecological concepts it embodies are of universal relevance.

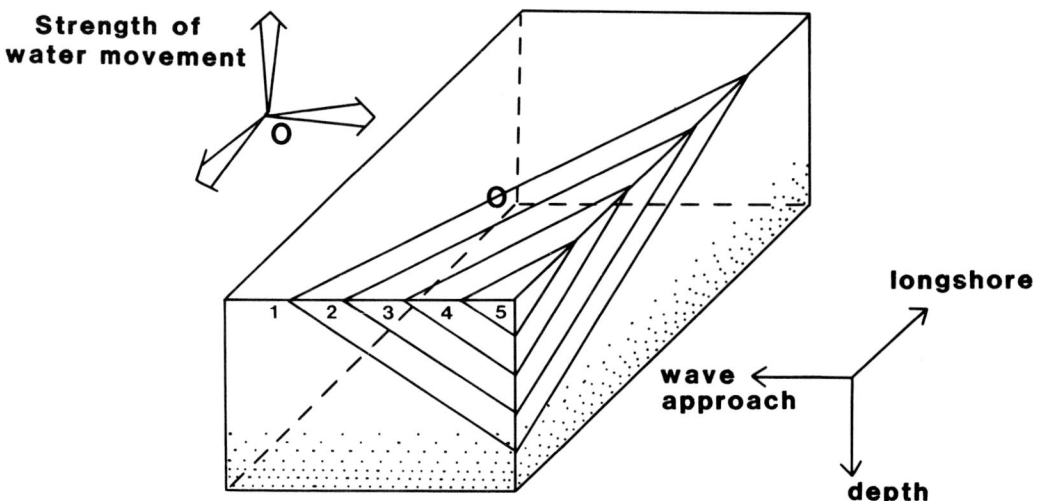

Figure 7 Zonation scheme for Indo-Pacific genera (redrawn from Rosen 1975). The distribution
of five shallow water associations is correlated to the attenuation of wave energy in
three dimensions; with depth, with distance across a reef flat, and with aspect relative
to the direction of wave attack. High energy associations successively 'drop out' as
aspect becomes more sheltered. The five associations are characterized by a
dominance of the following groups; 1 : *Porites;* 2 : Faviids; 3 : *Acropora;* 4 :
Pocillopora; 5 : calcareous algae. The *Porites* association occurs in the most sheltered
habitats and the calcareous algae in the most exposed.

Adey & Burke (1977) also provide a block diagram zonation scheme (see fig. 9). Unlike
Rosen (1975), who uses all three axes to represent wave-attenuating factors, Adey and
Burke use only two: the third axis represents turbidity. This scheme shows a marked
upward zonal shift associated with turbid water (in which light attenuation and its zoning
influence occur in quiet shallow waters). These authors also note similar zonation of the
genera *Porolithon, Millepora* and *Acropora* in both Atlantic and Indo-Pacific reefs.

The relationship between wave action and zonation is thus widely observed and
emphasized. Many vertical, lateral and transverse zonation sequences correlate with
diminution of wave energy from a surf zone which is either moderate or destructively
violent. However, consideration should also be given to those reefs which are rarely, if
ever, exposed to significant surf.

Zonation in low wave areas

Wave-related zonation schemes have little relevance in areas protected from consistent
strong wave action. Rosen's (1975) scheme demonstrates how reef fronts exposed to weak
or no waves are weakly zoned in terms of his broadly defined alliances (i.e. one or two
alliances compared with five on maximally exposed reef fronts). The zonation which
occurs on more sheltered reefs is more at the level of 'association' than alliance
(association defined by Scheer 1978; for examples of this more subtle from of zonation see
Pillai 1969; Loya & Slobodkin 1971; Loya 1972; van den Hoek *et al.* 1975, 1978; Goodwin
et al. 1976; Done 1977; Bouchon 1980).

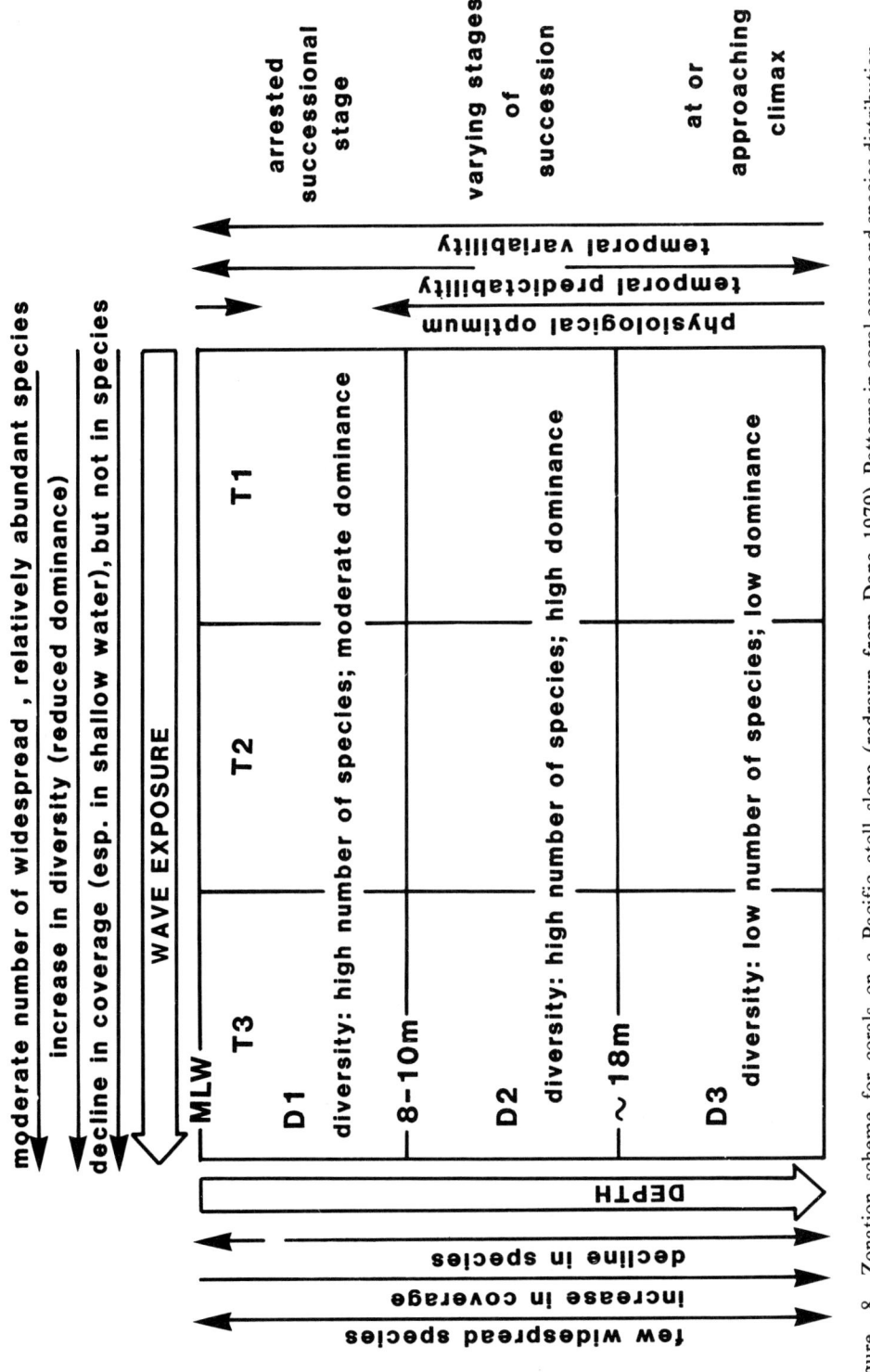

Figure 8 Zonation scheme for corals on a Pacific atoll slope (redrawn from Dana 1979). Patterns in coral cover and species distribution, diversity and dominance are presented as a function of depth and exposure. The scheme includes an interpretation of spatial trends in environmental optimality, predictability and variability to account for the observed patterns

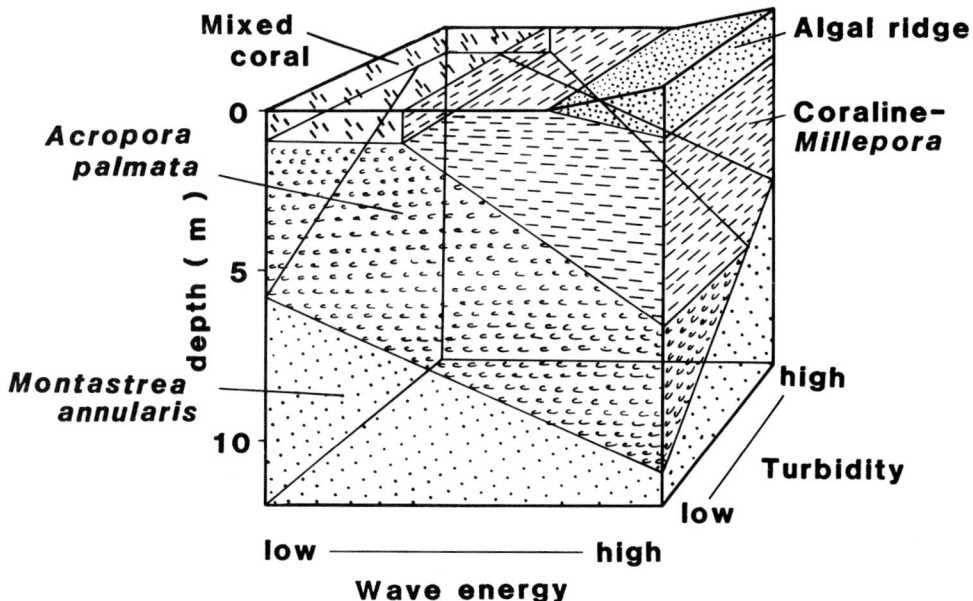

Figure 9   Zonation scheme for dominant Caribbean frame-building corals (redrawn from Adey and Burke 1977). This scheme embodies regional gradients in both wave action and turbidity. High wave energy reefs have the most identifiable zones and high turbidity excludes the *Acropora palmata* zone

Surf on the reefs at Eilat is weak (Mergner 1971) and, although a localized benthic surf zone is described, zonation on most of the reef seems largely a function of longshore current, light levels, substrate and predators. Likewise, Loya & Slobodkin (1971) and Loya (1972) emphasize that a variety of factors is responsible for zonation at Eilat. Bouchon (1980) found difficulty in recognizing zones between the surface and -40 m on a fringing reef in the Gulf of Aqaba which was characterised by low wave action and uniform light attenuation.

In sheltered areas, where biological controls on zonation are obvious, a 'bioenvironment' axis might logically replace the wave action axis. For instance, van den Hoek *et al.* (1978) proposed that a bimodal distribution of algal grazing intensity down a reef slope strongly influenced zonation. The role and effects of biological interactions in determining community structure and zonation are considered below.

NUMERICAL METHODS IN ZONATION STUDIES

The rationale for using numerical methods in ecological studies is discussed in Williams (1976), Clifford & Stephenson (1976) and Whittaker (1978 a,b).

Field methods

Until the time of Stoddart's (1969) review, coral zonation had been studied using two field methods — qualitative survey and quantitative quadrat methods. Stoddart called for optimization of sampling and recording methods for both slopes and reef flats. Several papers in Stoddart & Johannes (1978) refer to recent developments. Several variants of

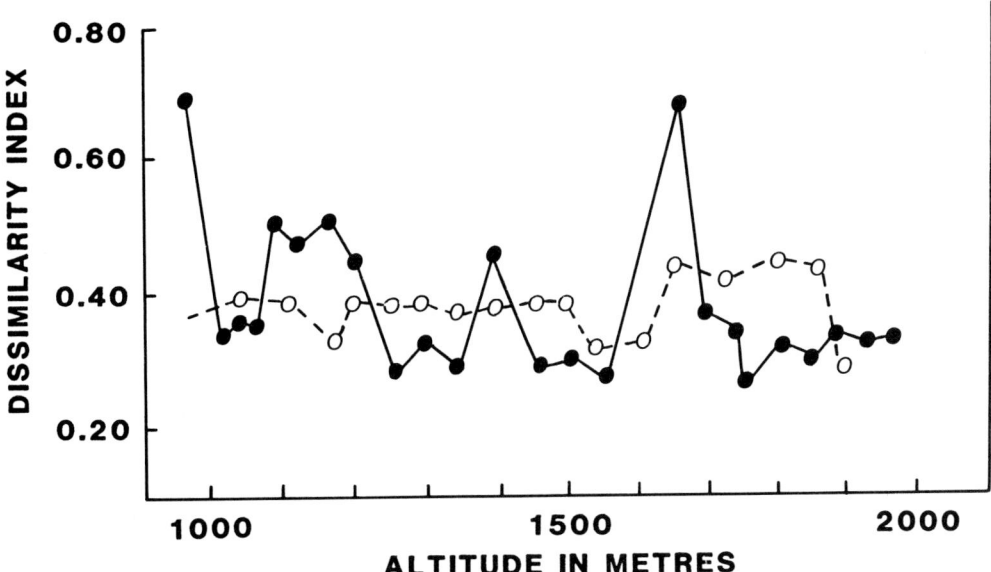

Figure 10   Beals' gradient analysis. Similarity between adjacent samples is markedly discontinuous in strongly zoned areas and more consistent in weakly zoned areas (redrawn from Beals 1969)

the 'line intercept' method introduced by Loya (1972) and Porter (1972) are used in descriptive studies (Pearson 1974, 1981; Wallace & Lovell 1977; Bull 1977); other workers continue to use quadrats (Spencer-Davies *et al.* 1971; van den Hoek *et al.* 1975, 1978; Goodwin *et al.* 1976). Reviews of these and other sampling methodologies are given by Pichon (1978b,c), Scheer (1978), Kinsey & Snider (1978) and Loya (1978). Each method provides a sample of the species present and an estimate of cover and/or abundance of individual species. In general, samples are placed regularly across the study area.

Zone delineation

Ordered samples

A variety of multivariate methods have been used to subdivide sequentially ordered data of this type into groups of samples that may represent zonal communities. All analyses start with a matrix of species (or other attributes such as supraspecific growth form, ecomorph or genus) against sampling units (e.g. individual quadrats or lines). The attributes are entered either qualitatively (presence - absence), semiquantitatively (e.g. importance value or graded cover index) or quantitatively (area cover estimate in quadrats; total line intercept [see Done 1977]).

GRADIENT ANALYSIS

In Beals' (1969) gradient analysis, a similarity index is calculated between spatially consecutive samples and plotted against sample number (see fig. 10). If there are marked discontinuities present, there will be sharp peaks and troughs on the graph, and their position will suggest where zones are separated. If the study area is not markedly zoned, the peaks and troughs will be small or absent. The technique's clarity and simplicity would seem to warrant its use in reef zonation studies. The main disadvantage is its failure to illustrate affinities between non-adjacent samples or groups of samples.

ORDINATION

Wallace & Dale (1977) used a principal co-ordinates analysis in a study of zonation of *Acropora* species and morphological types (fig. 11); lines were drawn between adjacent samples in the ordination. As in Beals' technique, this procedure provides results that indicate whether biotic zones exist and where they may be separated. In addition, it displays between-group affinities. An advantage of ordinations generally is that they may suggest the importance of particular environmental factors causing the zonation (Bradbury & Young 1981). A disadvantage is that sample groups may overlap and that interpretation may be difficult when many points are involved.

Bradbury & Young (1981), improved interpretability by plotting sample position (abscissa) against the scores for each of the first three principal co-ordinates in turn (see fig. 12). The presentation is more easily interpreted than the Wallace and Dale presentation but, as in any ordination, the ecological significance attributed to the axes is hypothetical.

CLASSIFICATION

The most common multivariate technique used is ecological classification (see figs. 13, 14). Ecological classification provides a dendrogram (or dendrograph — Jokiel & Maragos 1978) which suggests a hierarchical relationship between groups of sample units (normal classification) or attributes (inverse classification). Samples have been classified by numerous authors (e.g. Loya 1972; Done 1977; van den Hoek *et al.* 1975, 1979; Jokiel & Maragos 1978). Major groupings in the classifications were invariably associated with major topographic subdivisions and as such do not provide any unexpected insights. However, minor groupings may suggest similarities and dissimilarities that are not self-evident and may be shown to have ecological significance (e.g. Jokiel & Maragos 1978; Veron & Done 1979). Pitfalls for the unwary are indicated by figure 14.

Non-ordered samples

It is the author's opinion that classification and ordinations are of greatest value when there is no single 'correct' order of sample units in the data matrix. Veron & Done (1979) made coral surveys at sixty-six sites, of which half were situated along seven cross reef transects and half were located independently of other stations. Unlike a single cross reef transect, these data did not have an inherent 'correct' order and a classification (fig. 13) produced groups which consisted of sites which were frequently widely dispersed on the reef. Because of the complex morphology of the reef, it was not possible to relate the resulting site groups to a single generalized reef profile, but a plot of mean depth versus mean 'aspect' illustrated an ordered distribution of site groups relative to a subjective assessment of water turbulence. This plot of site groups against spatio-environmental axes (fig. 15) is similar to a 'vegetation chart' used by botanists (e.g. Whittaker 1975, p. 122).

Zonation studies, particularly where the samples are not orderable, may in future use a class of 'data display' programs of the 'minimum spanning tree' type (Prim 1957). This technique is free of the exaggerating tendency of some dendrograms, and is graphically clearer than ordinations. It has been used in a study of mangrove distribution and zonation by Bunt & Williams (1980).

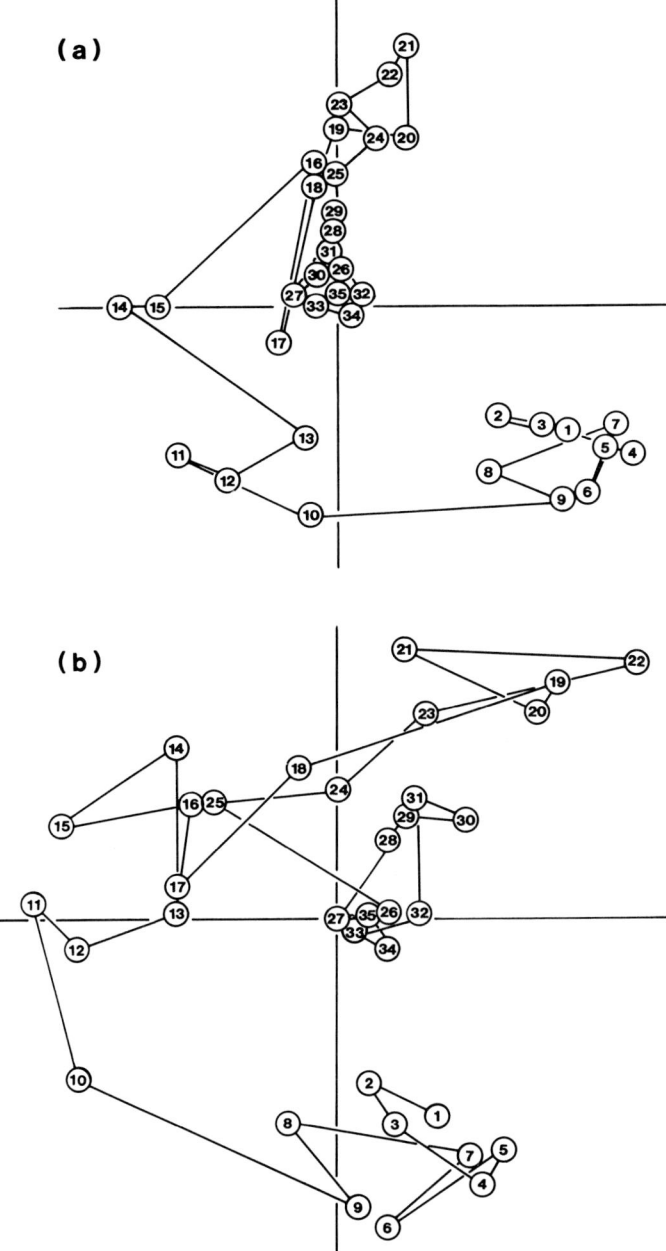

Figure 11 Principal co-ordinates analysis of line transects down a reef slope (redrawn from Wallace and Dale 1977). Lines in the figure join transects (numbered circles) which are adjacent to each other on the reef. Ordination (a) (based on species composition) illustrates a distinct separation between reef crest (transects 1–9), and reef slope (16 –35) with a loosely coherent 'transition zone' (10–15) between. Ordination (b) (based on colony shape) shows the same reef crest zone, but a much broader transition zone and a smaller deep zone. These differences highlight differences in the way species and growth forms divide up the slope

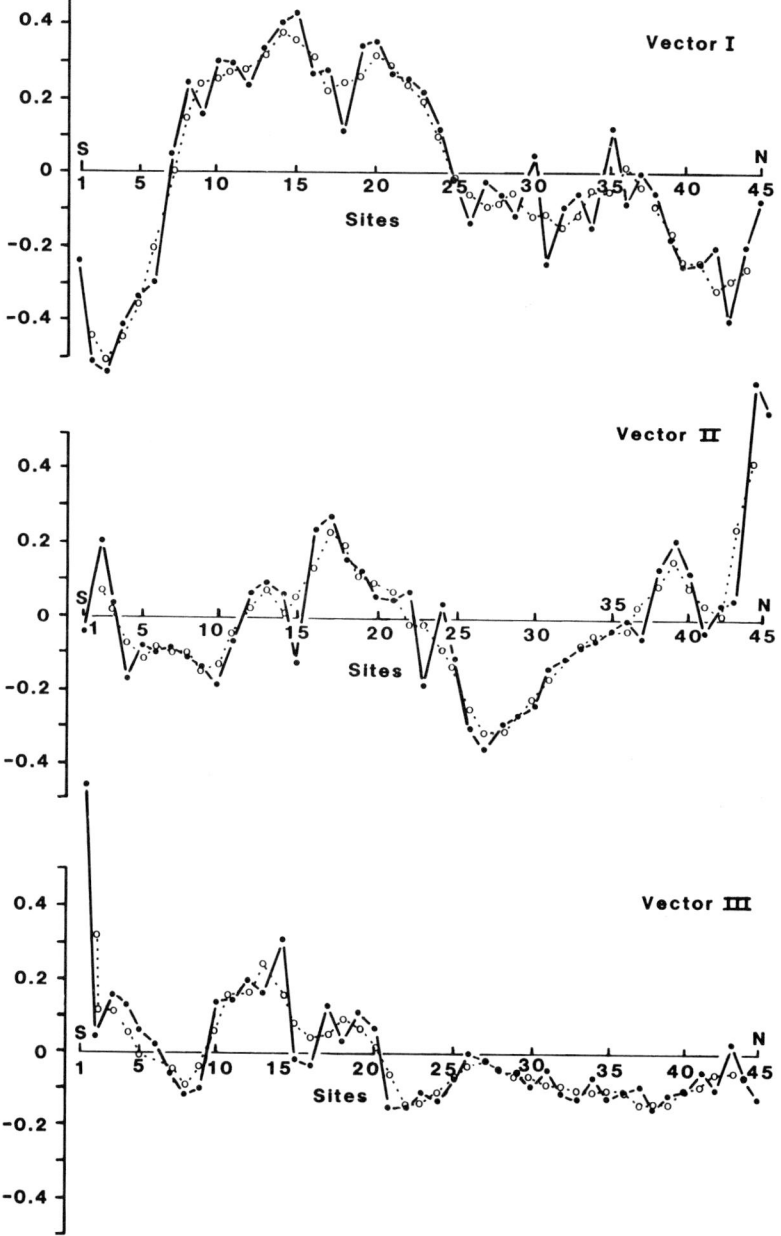

Figure 12   Illustration of structural zonation of a reef flat using principal co-ordinate vectors I, II
and III plotted against transect position (redrawn from Bradbury and Young, 1981a).
Vector I differentiated sand from other substrates, vector II rubble from thin ramose
corals, and vector III one reef crest from the remainder of the transect. The limits of
structural zones are indicated where runs of high positive or negative scores cross the
abscissa; the steepness of the crossing reflects the abruptness of the change. The plot
of running means (dotted lines) reduces irregularities caused by transects which are
dissimilar to neighbours on both sides

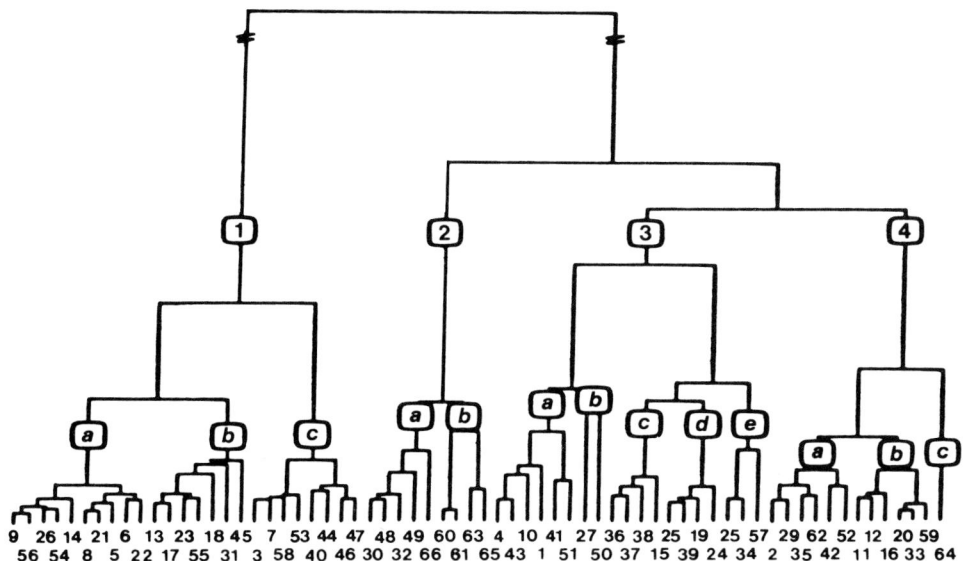

Figure 13   Dendrogram of coral survey data from sixty-six sites widely distributed over Lord Howe Island reef (redrawn from Veron and Done 1979). Groups 1 to 4 reflect the major morphological subdivisions of the reef (flat, lagoon, outer slope, passages); subgroups 'a' to 'e' were composed of faunistically homogeneous sites whose characteristic composition and dominance relationships were tabulated using diagnostic and sorting programs

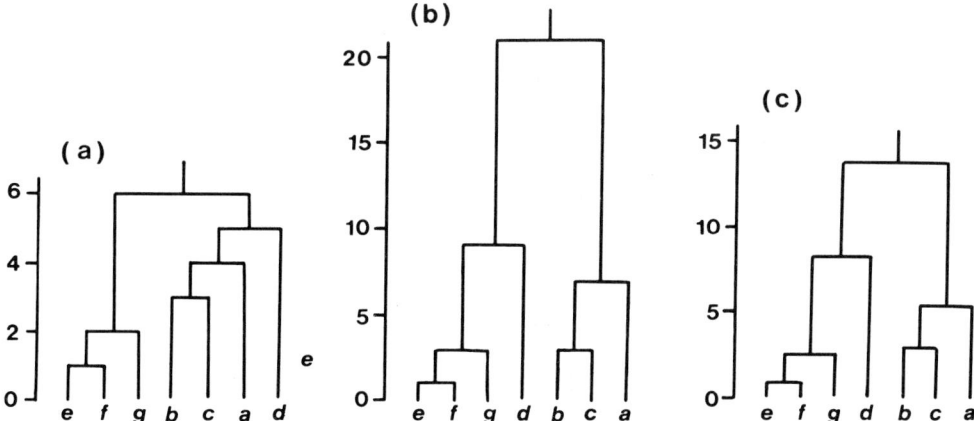

Figure 14   Three classifications of the same hypothetical data matrix using strategies (a) nearest neighbour; (b) furthest neighbour, and (c) group average. Those major patterns and sample groupings which are independent of strategy used may be assumed to have ecological significance. However, the differences (namely, the position of sample 'd' and the details of the branching pattern) have no ecological significance, being due solely to differences in computational characteristics of each sorting strategy

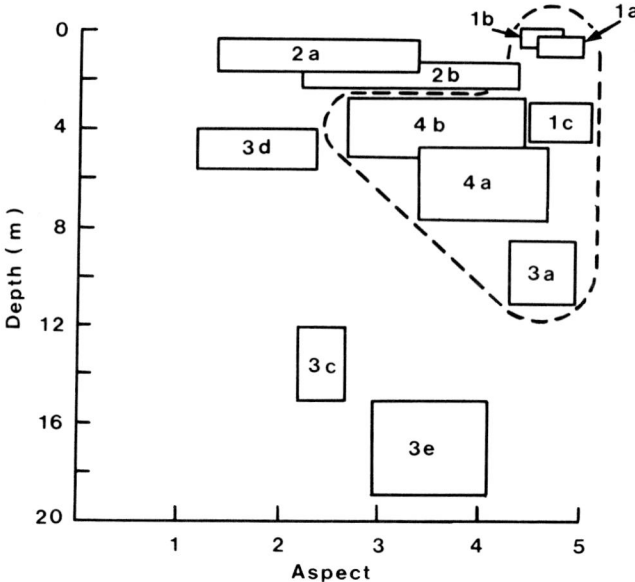

Figure 15   Subjective ordination of site groups (from Fig. 13) against depth and aspect (on a scale
of 0 to 5). The lengths of the boxes are equal to the standard deviations of depth and
aspect; the centres of the boxes are at mean depth and aspect. The broken line
indicates the reef-front series (redrawn from Veron & Done 1979)

Zone composition

A full list of species found in any zone may be reduced to a smaller number which
adequately describe the zone and differentiate it from others. These are the 'constant',
'dominant', 'characteristic' and 'differential' species (Scheer 1978). They may readily be
defined by various package and special purpose computer programs. Topographic data
collected with each sample, such as depth, slope, aspect, substrate composition and total
cover may be similarly diagnosed.

   Constant and dominant species are easily defined using simple programs which sort the
species list for any set of samples in order of descending frequency (number of samples in
which attributes was present) and mean cover, respectively. The dominants or constants
so defined for a set of zones may be displayed using vector roses, as in figure 16 (after
Lamacraft 1979).

   The degree to which a species is characteristic of a zone may be quantified as its T-
value and its F-ratio (ratios of within zone mean cover and variance to overall mean cover
and variance, respectively; Wishart 1975). Species with high T and low F are very
characteristic of a zone (though not necessarily dominant). Species with low T and high
F have distributions which are indifferent with respect to the zone.

Zone differentiation

Zonal communities of greatly overlapping composition may be discriminated from each
other on the basis of the presence or absence of one or more 'differential' species (Scheer
1978). In comparisons of pairs of the sample sets, differential species contribute most to

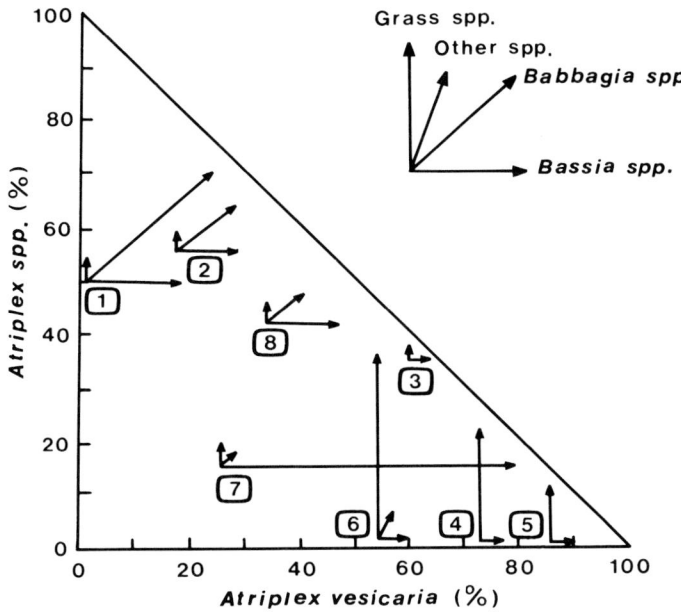

Figure 16   An example of the use of vector roses to display biotic affinities of assemblages (redrawn from Lamacraft 1979). The data are from a study of Australian grasslands

the dissimilarity index. Diagnostic programs 'Grouper' and 'Gowecor' (Lance *et al.* 1968) give each species' contribution to the index, and rank them in descending order. Multiple discriminant analysis (Cooley & Lohnes 1971) has also been used to differentiate zones (Dinesen 1980) using both generic and growth form data.

Zone overlap

The extent and nature of overlap between zones can suggest specific hypotheses regarding the role of abiotic factors and spatial competition in determining the zonation pattern (Sheppard, 1980). Methods for sampling and analysis of overlapping distributions, and the biological interpretation of the results, are presented by Pielou (1977). The approach is suitable for habitats exposed to a single, monotonic environmental gradient, and some reef flats and slopes may fit these criteria. However, studies may be restricted to selected areas because species' distributions in many areas respond to 'non-gradient' factors (such as dicontinuous and patchy substrate) or stress gradients acting in opposition or non-monotonically (see Chappell, 1980).

PHYSICAL DETERMINANTS OF CORAL ZONATION

Light and water movement (waves, currents and tides) are the environmental factors recognized as major determinants of the zonal organization of reef benthos. The latter also determines sediment distribution, which may be the major proximal cause of benthic distribution patterns. The degree to which a zonal sequence at a particular place may be attributed to these factors depends on both their range and their spatial pattern. Ecologically-orientated introductions to light and water movement in the shallow littoral are provided by Jerlov (1971), Riedl (1971a,b), Schwenke (1971) and Drew (in press). A lucid account of a study of waves and currents on a reef is given by Roberts *et al.* (1975) and a highly technical treatment of waves on a reef is given by Lee & Black (1978).

Coral zonation on reefs is not due to environmental variability *per se* but to its stratification on the reef's surface (i.e. environmental zonation). Large and continuous tracts of uniform topography are characterized by photic, hydrodynamic and substrate conditions which encompass only a fraction of total reef variability. Small scale topographic features provide fine grain, perhaps irregular environmetal texture (shadows, eddies, etc.) but the larger scale environmetal pattern, being predetermined by geological history and geographic locality, is, by and large, regular and constant through ecological time.

Zoning influences of abiotic factors may be modified to a greater or lesser degree by competitive and/or predatory interactions, depending on the biota (see below). Notwithstanding such biological interactions (or unpredictable exogenous disturbance), the prevailing physical environment does provide the ultimate deterministic framework in which organismic distribution takes place and so some discussion of the physical environment in relation to its zoning effects is presented here. Catastrophic affects of physical events are considered below.

Water motion

Wave action

Ecological zonation schemes sometimes imply that wave action decreases rather uniformly with depth (on the reef slope) and with distance from the reef margin (on the reef flat) (Rosen 1971, 1975; Pichon 1972, 1978a). This implication aligns reef zonation concepts with terrestrial botany, where vegetation zonation is frequently a response to gradients in one or more major environmental factors (Whittaker 1975; Pielou 1977). However, there is evidence to suggest that the gradient concept of water motion may be misleading over-simplification in many real-life situations.

REEF SLOPE

The water mass on a wave-exposed reef or rocky shore may be subdivided into four vertically intergrading water bodies. The subdivisions made on the basis of water particle dynamics (fig. 17) are referred to as lacerating, oscillating, unidirectionally flowing and two-dimensionally flowing water bodies (Riedl 1971b). Boundaries between the water bodies ascend with decreased wave exposure and, in sheltered areas, the lacerating and unidirectionally flowing water bodies disappear. This scheme indicates that there are significant qualitative changes with depth as well as simple vertical diminution of particle velocity. The scheme also shows the relationship between waves and the currents they generate.

Another instance where a simplistic view of wave attenuation with depth is inappropriate has been described by Roberts *et al.* (1975). Their data indicate that the expected decrease of wave force with increasing depth is counterbalanced by an increase in current force. The resulting 'wave plus current' force curve (fig. 18) is bimodal indicating that the total force on the reef benthos at a depth of 21 m is the same as that experienced at about 3 m, close to the reef margin. These areas of high total force corresponded with zones of prolific coral growth; intermediate depths have lower force and less prolific coral growth.

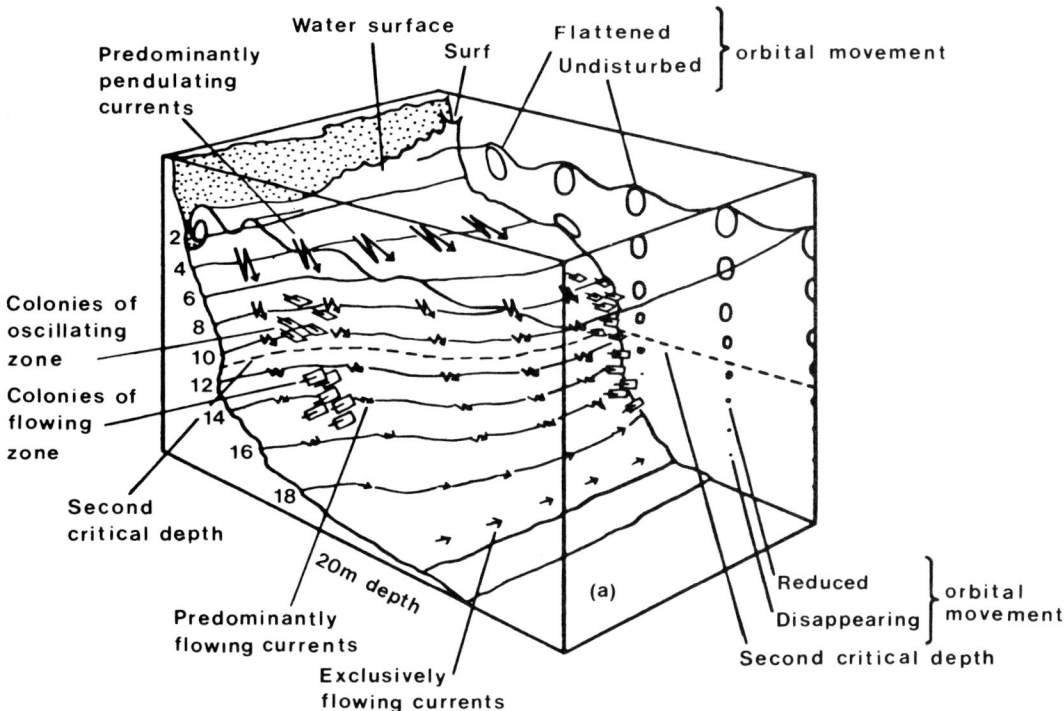

Figure 17 Four intergrading water bodies on a wave-beaten shore. Arrows indicate bottom currents; ellipses indicate orbits of water particles; tags indicate orientation of planar gorgonians (redrawn from Riedl 1971)

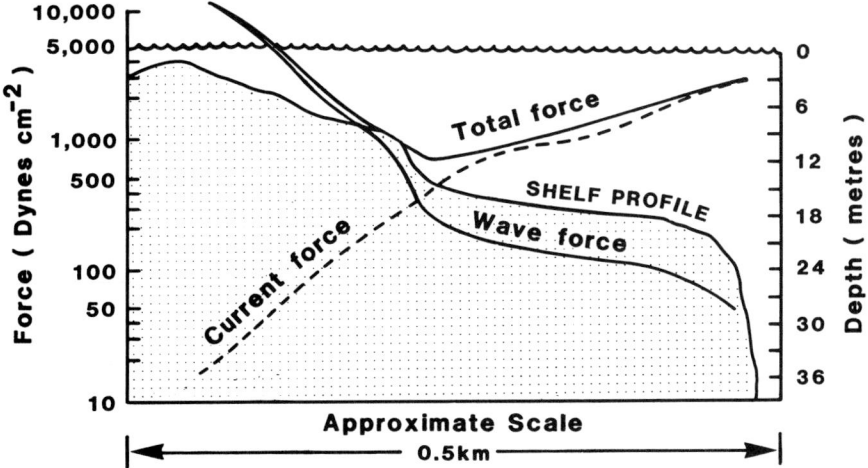

Figure 18 Bimodal distribution of total hydrodynamic force on a reef slope. Although wave and current forces individually attenuate monotonically, their combined affect causes total force at −21 m to equal total force at −3 m (redrawn from Roberts *et al.* 1975)

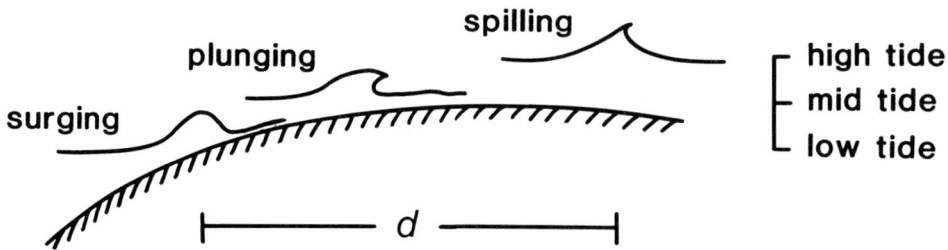

Figure 19   Form and position of wave break for waves of a given height, period and steepness at different tides (based on Wiegel 1964). The distance 'd', occupied by communities adapted to surf conditions, may reach tens to hundreds of metres in macro- to supertidal areas. In micro-tidal areas, surf has only a localized zoning influence

REEF FLAT

Waves moving onto a reef lose energy in three distinct phases — pre-surf, surf and post-surf. A set of waves crossing the fore-reef shelf at Grand Cayman showed a height reduction of about 20%, attributed to the combined effects of shoaling, refraction, reflection, scattering and extremely high frictional attenuation (Roberts *et al.* 1975). The spectral characteristics of the waves only changed significantly after wave break, at which time 75% of the wave energy was dissipated over a narrow crest region. At Ala Moana (Hawaii), small secondary waves were generated after surf break (Lee & Black 1978). Large bottom friction and wave break together caused wave height to reduce by about 50% for every 500 ft (150 m) travelled over the reef.

The shape and position of break for a given wave may vary greatly in macro-tidal areas whereas, in atidal areas, it will remain more constant (fig. 19). Breaker shape (either spilling, plunging or surging) depends on a relationship between wave steepness and substrate slope (Wiegel 1964; Huntley & Bowen 1975). The velocity fields of the water in the three breaker types differs significantly and it may be that zonation of coral species and ecomorphs within surf zones is as much a function of these qualitative differences as of a simple attenuation of total wave forces.

The position and intensity of the surf line may indirectly affect benthic zonation through nutrition. It is likely that the surf line effectively isolates reef slope from reef flat waters, just as it isolates inshore and offshore waters adjacent to beaches (U.S. C.E.R.C. 1977, p. 443). Restriction or channelization, by the surf, of plankton and other nutrients may affect organismic zonation in ways not yet understood (the dependence of downstream 'consumer' zones or upstream 'producer' zones has already been noted; LIMER 1976; Kinsey 1979).

WAVE CLIMATE

The zoning influences of wave action depend ultimately upon long term averages of wave climate (height, period and direction). Distinct weather and leeward slopes are recognizable on reefs with a directional tradewind dominance (Wells 1954, 1957; Stoddart 1971; Roberts 1974). At Grand Cayman Island, an eighty-fold difference in mean annual wave power between weather and leeward shores is reflected in significant differences in reef morphology and zonation (Roberts 1974).

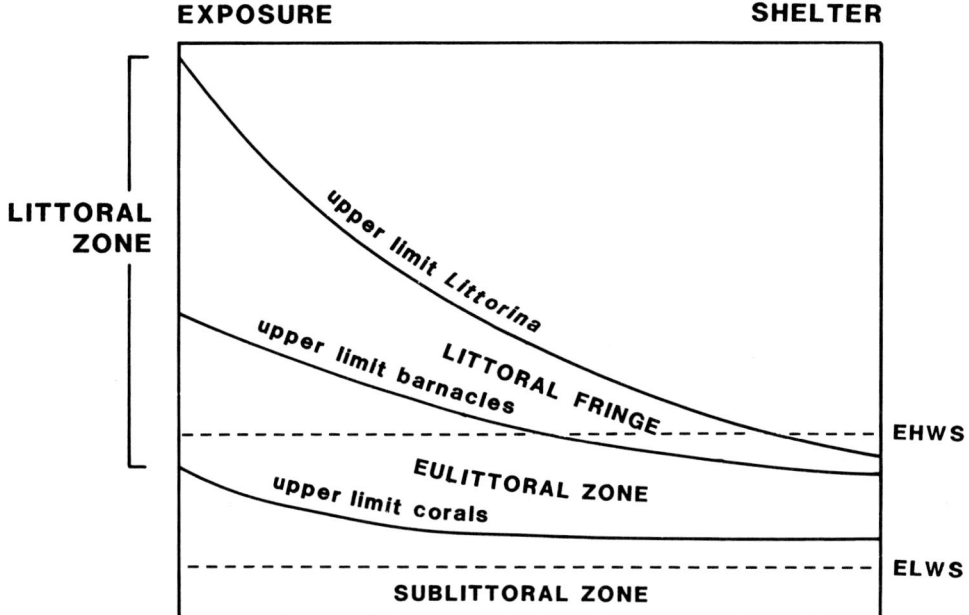

EXPOSURE   **SHELTER**

LITTORAL ZONE

upper limit Littorina

LITTORAL FRINGE

upper limit barnacles

- EHWS

EULITTORAL ZONE

upper limit corals

- ELWS

SUBLITTORAL ZONE

Figure 20   The influence of wave exposure on effective tidal limits and hence, intertidal zonation (redrawn from Taylor, 1978). The vertical limits of intertidal zones on exposed shores are extended beyond the nominal tide limits by wave surge and spray

Data on wave climate are rarely available but may be approximated using hindcasting techniques and numerical wave refraction/friction models (Roberts 1974; Holmes 1975). The approximation may be further stretched to estimate the depth of the wave base (namely, depth of negligible orbital movement) which, for open water waves, is equivalent to half the wave length (Sverdrup *et al.* 1942) or ten times the wave height (Dietrich 1963, Table 3). The complexities of the wave/reef interaction limit this latter approximation to the status of a rough rule of the thumb.

Tides and sea level fluctuations

Reef tops usually have a slight slope or a stepped profile (e.g., Morissey 1980, fig. 3). As a result, the associated reef top communities are subjected to (and limited by) varying frequencies and duration of subaerial exposure and temperature stress associated with both normal tidal fluctuations and longer-term anomalies in mean sea level.

TIDES

Normal tidal fluctuations are responsible for the broadest levels of reef top zonation. A 'tripartite' zonation scheme (Taylor 1978) relates littoral, eulittoral and sublittoral zones with nominal tidal limits, and indicates vertical shifts and extension of zones found in heavily wave-exposed areas (fig. 20).

Coral reefs occur in a wide range of tidal conditions. A regional classification including 'micro-tidal' (0.3-1.0 m), 'meso-tidal' and 'macro-tidal' (>3 m) was used by Stoddart (1971). Reefs in the central Great Barrier Reef (range >6 m; Maxwell 1968) may be termed 'super-tidal'.

Reef flats lie at various depths relative to mean sea level. It is generally held that Caribbean reef tops are deeper than Indo-Pacific reef tops (Stoddart 1969). Periodic tidal emergence and scouring of Indo-Pacific reef flats restricts the coral communities that may persist, or may locally inhibit colonization completely (Stephenson *et al.* 1931; Wells 1954, 1957; Ditlev 1978; Morrisey 1980). In many Caribbean reefs, the combination of deeper reef flat and small tidal range restricts the extent of these conditions. A luxuriant community may therefore develop on many reef flats (Stoddart 1969). However, other Caribbean reefs have been shown to lie shallow with respect to mean sea level, and their biotic communities are subjected to periodic emergence and high temperature stress (Glynn 1968, 1973; Scatterday 1977). In this respect they are similar to Indo-Pacific reefs.

SEA LEVEL FLUCTUATIONS

Aperiodic or long-period fluctuations in sea level may drastically change a zonation pattern which had developed during a preceding period of normal tidal activity. Periods of exceptionally low water levels caused extensive mortality of reef flat benthos at Eilat (Fishelson 1973; Loya 1976a), Guam (Yamaguchi 1975) and Puerto Rico (Glynn 1968). Biotic zonation on reef flats which have been so disturbed then becomes a function of time elapsed since disturbance, as well as of the ongoing ambient conditions.

Currents

Habitat selection on the basis of a preferred range of current velocity contributes to observed zonation patterns. Currents influence sessile organisms directly through hydraulic stresses and by supply and removal of dissolved and suspended material. An indirect influence is exerted through the distribution of reefal sediment which in turn limits biotic distributions.

Excessive hydraulic drag appears to exclude or limit corals in certain habitats, such as surf (discussed above) and narrow passes. For example, current velocities of up to 3.8 m s^{-1} favoured heavy encrustations of coralline algae almost to the exclusion of corals on a highly illuminated channel wall in the northern Great Barrier Reef (Veron 1978).

At mid-range velocities, currents act more as transport media and less as an environmental stress. It has been suggested that, among suspension feeders, there is an optimum range of ambient velocity in which each species feeds most effectively (Wainwright *et al.* 1975). Predominantly autotrophic species and/or morphologically plastic species may be expected to have somewhat broader preferred velocity ranges and hence, distribution ranges. Some of the principles by which species and ecomorphs may adapt to current environments are discussed in Chamberlain & Graus (1975), Wainwright *et al.* (1975) and Jokiel (1978).

Low current velocities may exclude certain species or colony shapes through stagnation. It is questionable whether open water becomes stagnant under normal ecological conditions (Schwenke 1974) but laboratory experiments suggest that a branching coral's own growth could cause a stagnant pocket in its interior (Chamberlain & Graus 1975). Death of the colony could result due to starvation of the interior polyps or poisoning by accumulated metabolic wastes.

Sediments

Currents affect zonation indirectly through the sorting and distribution of sediments across the reef structure. Fine sediments accumulate only in habitats with low current velocities. Sediment depth, grain size, suspended load and deposition rate are known determinants of coral distribution and performance (Hubbard & Pockock 1972; Loya 1972, 1976b; Aller & Dodge 1974; Faure 1974).

The transport of sediment itself can affect zonation. The abrasive effect of suspended sand may restrict or inhibit coral growth in high energy areas (Barnes *et al.* 1971; Bak 1975; Giester 1977; van den Hoek *et al.* 1978). Deposited sediment that moves *en masse* or is 'worked' by waves does not, in general, provide a long term substrate for macro-benthos.

Light and Turbidity

Most reef corals occur in the euphotic zone, the upper stratum of the water column in which photosynthetically active radiation (PAR) attenuates to 1% of its surface value (Yentsch 1966). This attenuation is a major zoning influence on corals and other benthos. Lower bathymetric limits for autotrophic species are determined by their photosynthetic efficiency at reduced irradiance levels, whereas upper limits for subtidal benthos may depend on species specific adaptation against potentially damaging irradiation by visible or ultra-violet radiation. Reviews of utilization of light by marine plants (including symbiotic zooxanthellae) are provided by McCloskey *et al.* (1978) and Drew (in press).

Lower limit

A complete autotroph is limited to depths in which P/R (the ratio of mean daily photosynthesis to mean daily respiration) exceeds 1.0 (McCloskey *et al.* 1978). The irradiance at which compensation occurs (P/R = 1.0) varies between species (Spencer-Davies 1977) and ecomorphs (Jaubert 1977). However, photoadaptation within some species results in a lowering of compensation irradiance in individuals found in habitats with reduced illumination (Wethey & Porter 1976; Porter *et al.* 1980). The total population is therefore characterized by a range of compensation irradiance. Survival below the lower limit of this range requires some degree of heterotrophism (McCloskey *et al.* 1978). Therefore, in principle, the depth at which a species' lowest compensation irradiance exists is a critical determinant of its bathymetric distribution (but see cautionary comments in McCluskey *et al.* 1978).

(Chalker, pers. comm.). This means that photosynthesis in these corals cannot begin to become limiting at a shallower depth than that at which irradiance is about 250 μ Einsteins m^{-2} s^{-1} (commonly 10% of its surface value). This depth varies greatly between and within reefs depending on water transparency (see below).

Upper limit

The shallow limit of many species may be determined by photoinhibition in high light intensities, or vulnerability to ultra-violet radiation, which penetrates clear ocean waters almost as well as visible light (Smith & Baker 1979; Jokiel 1980). Many shallow water corals produce compounds which protect their tissues from ultra-violet light (Maragos 1974). Green pigments emit photosynthetically useful wavelengths when exposed to ultra-violet (Kawaguti 1969). Species-specific differences in the degree of ultra-violet light protection or utilization may set different upper limits to their distribution. Ultra-violet light has been shown to cause death in shallow water transplants of shade-loving bryozoans, sponges and tunicates (Jokiel 1980). This finding gives reason to believe that non-surface corals (*sensu* Wells (1954), including hermatypes and ahermatypes) may similarly be upwards-limited by ultra-violet radiation.

Regional and 'within reef' variability

Vertical light attenuation (and hence, depths at which ultra-violet light is damaging, and depth to compensation points) varies from reef to reef and within reefs, depending upon transparency of the water column to radiation. Reef waters off Townsville, Queensland, range in transparency (measured by Secchi disc) from 1–3 m (nearshore) to 20 m (inner line of reefs) to 34 m (outer line of reefs) (Ikeda *et al.* 1980; T.A. Walker, pers. comm.). These data indicate extreme vertical compression of the euphotic zone in nearshore reefs compared to the inner and outer lines of reefs. Coral zonation on nearshore reefs is both condensed into shallow waters, and truncated at the shallow end (Done 1982).

Depth to 90% extinction is a broad ecological index of transparency. The index has its physiological basis in the finding (referred to above) that many corals are not light-limited until approximately 90% attenuation of surface levels. This index (which may readily be extracted from published accounts) varies greatly in some well described reefs: Eilat, outer slope -8 m (Loya 1972); Tulear, lagoon -10 m (Pichon 1978a); Tulear, outer slope -17 m (Pichon 1978a); Bikini Atoll outer slope -18 m (Wells 1954); Curacao, outer slope -24 m (van den Hoek *et al.* 1978). Compensation depths, photoinhibition depths, depths at which ultra-violet light is damaging and light-dependent expressions of coral zonation may be correlated with these transparency data. However, differences in light measurement techniques and deficiencies in the physiological data presently preclude synthesis of the published results.

In some reef habitats, reflection from sandy floors and/or walls encrusted with crustose algae may elevate irradiance levels at a given depth (Kinzie 1973). The reduced transparency in some lagoons is due to reef-derived suspended material in the water column, and is thus related to flushing rates.

BIOTIC ENVIRONMENT OF CORALS

A coral may be killed by extrinsic biological events over which it has no control or defence. This provides space for colonization by a successor which may not be of the same species. In corals, death may follow coral-coral or coral-algae competition for space, predation upon the colony or boring of the colony. The intensity of biotic interactions many vary between zones; their importance in the determinations of large scale (e.g. zonal) distribution is considered below.

CORAL-CORAL COMPETITION.

Different coral species are not equally effective in occupying space on the reef. When two Atlantic species contest the same space, one will usually kill the other by mesenterial aggression or an overtopping growth. The identify of the victor is predictable (Lang 1970, 1973; Porter 1976). In Indo-Pacific corals, the beginnings of a similar aggressive hierarchy have been described by Sheppard (1979) who attributes a deterministic role to interspecific aggression in the composition of biotic zones. Experimental studies of hard/soft coral interactions are currently in progress (Sammarco *et al.* pers. comm.).

CORAL-ALGAL COMPETITION.

Competition for space between coral and algae may result in the death of the coral. For example, displacement of coral spat by algae has been reported by Pearson (1974), Connell (1976) and Sammarco (1980); overgrowth of living adult scleractinia by encrusting coralline algae has been reported by Grigg & Maragos (1974), Littler & Doty

(1975) and Dana (1979). It is widely held that coral dominance in many areas is possible only because grazing and cropping organisms (especially sea urchins and fish) keep the standing crop of highly productive fleshy algae at low levels (Ogden 1976; Benayahu & Loya 1977; van den Hoek *et al.* 1978; Sammarco 1980; Hatcher, ch.10 this volume).

CORAL PREDATION AND BORING

The impact of predation on a coral community depends upon coral and predator population levels, and the feeding strategies of the predators, which belong to many animal orders (Endean 1976). At normal population levels, *Acanthaster planci* will frequently kill whole colonies but the feeding pattern is patchy (Porter 1972b; Endean 1973; Glynn 1974a). However, plagues of *Acanthaster* kill corals over large areas, and it appears that this form of perturbation may be equivalent in its ecological - geological time scale effects on coral community structure to other extrinsic disturbances such as floods, cyclones and sediment smothering (Endean 1976; Connell 1978; Pearson, 1981).

Numerous organisms bore into scleractinian corals (Scoffin & Garrett 1974; Warme 1977; Bromley 1978; Risk & MacGeachy 1978). Those which bore into the basal regions of colonies can cause them to topple or accelerate dislodgement by wave action (Tunnicliffe 1979). The effect of borers on community structure may thus be great by comparison with the small amount of carbonate they excavate. Moreover, total borer impact may be zonally distributed (Bromley 1978).

CORAL RESPONSES TO THE ENVIRONMENT

Zoning Influences of Biological Interactions

This section considers the relative importance of biological interactions to coral zonation. To what extent are species which are physiologically and mechanically capable of occupying a habitat, excluded by their inability to hold their place in the face of interspecific competition or localized predation? Are these interactions simply a rather inconsequential product of distribution patterns determined by overriding physical factors, or are they themselves determinants of the pattern? These concepts have been discussed by Porter (1972a), Connell (1973, 1978), Glynn (1973) and Dana (1976).

Some of the biological mechanisms for the possible exclusion of some species by others have been discussed above. However, the degree to which these mechanisms propagate upscale from an interaction between two individuals to a determinant of overall community structure and zonation remains to be demonstrated conclusively.

An analogous situation exists in the field of island biogeography. Diamond (1975) has proposed a set of 'assembly rules' for fauna on islands. The rules are based upon some distributional data and some mechanistic biological premises. The empirical bases of Diamond's rules have been attacked by Simberloff and colleagues (e.g. Connor & Simberloff 1979). They indicate that demonstration of a biological mechanism for species exclusion does not provide grounds for believing it to be effective at the level of community organization, a statement which holds for coral communities as well as for islands.

The effect of species interactions in limiting species ranges has been examined in quantitative distributional studies and field experiments. The bathymetric distributions of several markedly zoned coral species at Chagos were found to be strongly truncated

(Sheppard 1980b). With supporting information on the aggressive status of the corals present, these data suggested that competition contributes to determining the nature and extent of coral zones. However, an analysis of coral distribution along a Great Barrier Reef transect suggested that interspecific competition was inconsequential over a similar spatial scale (tens to hundreds of metres; Bradbury & Young, in press).

Grazing by fish and echinoids may significantly affect zonation. For example, the Atlantic echinoid *Diadema antillarum* is seen as functionally analogous to keystone predators in rocky shores (Paine 1969). Its own distribution is strongly zoned on many Caribbean reef slopes (Ogden 1976) with low densities on shallow and deep slopes, and high densities at intermediate depths. Data for algal productivity and consumption rates support the view that the dense mid-slope population of the echinoid indirectly causes the observed high coral cover and diversity by keeping algal standing-crop low (van den Hoek *et al.* 1978). Conversely, absence of the sea urchin from shallow zones (due to a paucity of refuges from wave action) allows high algal standing-crop, which effectively excludes corals. These conclusions are supported by the results of field experiments in which sea urchin density was manipulated (Sammarco *et al.* 1974; Sammarco 1980). The belief that the algal/coral/urchin interaction is a major determinant of community structure is thus well supported in the Caribbean. Echinoid control of algal biomass is suggested on some Indo-Pacific reefs (e.g. Benayahu & Loya 1977) but it is not significant on slopes of several reefs surveyed on the central Great Barrier Reef (Sammarco, pers. comm.).

Age-specific responses

The presence or absence of a given species from a topographic zone may not necessarily be comprehensible in terms of the resource requirements and physiological/mechanical/competitive tolerances of the adult stage alone. Success in a place indicates that the requirements of all life stages (from spat to adult) have been met at the point of settlement. Exceptions are imported mobile coralliths (Glynn 1974b) and adults derived from imported living fragments (Tunnicliffe 1981). Localized barriers to dispersal (such as strongly directional currents or surf barriers) may also cause the absence of a species otherwise capable of surviving and which is, perhaps, abundant nearby.

Little age-specific data is available, although the need is well recognized. Those relevant physiological, behavioural and ecomechanical studies which do exist refer mainly to adult specimens (see sections on water motions, sediment and light in Connell 1973). The deficiency is partly redressed by studies of life history strategies.

Life History Strategies

Life history studies provide species-specific population data which are central to understanding distribution patterns in heterogeneous and zoned environments. Key parameters in the studies include recruitment rate, sediment rejection ability, aggression status, regenerative ability, juvenile and adult survival rates and longevity (Bak & Engel 1979). These authors defined three broadly different life history strategies in terms of these parameters. Similar information was used by Loya (1976c,d), who described an opportunistic life history strategy in one ubiquitous species and drew attention to alternative strategies likely to be found in others. Although life history studies do not necessarily identify the requirements of all life stages, they do identify (and to a degree quantify) key processes and events in the organism/environment interaction.

SYNOPTIC SPATIO-TEMPORAL MODELS

Two recent models together provide a basis for understanding coral community structure and zonation at a synoptic level. These models were devised by Connell (1978) and Chappell (1980).

A decade or so ago, coral cover and diversity in many western Pacific reefs was largely determined by whether or not the community had been infested by *Acanthaster planci* in the past (Pearson 1981). Its appearance depended upon the severity of the infestation, and the stage reached in the secondary succession which that event initiated (Endean 1973, 1976; Pearson 1981).

The successional potential of reef coral communities has been emphasized in a recent synopsis by Connell (1978). Connell reasoned that, because some species are superior to others in occupying space, a succession towards a monospecific stand would result. However, the monospecific end result was not usually attained (i.e. diversity was kept high) because succession was usually interrupted by extrinsic disturbance of varying frequency and intensity. Examples of such disturbances were 'storm waves, freshwater floods, sediments or herds of predators'.

Connell's emphasis on indiscriminate disturbance of extrinsic origin is significant. Work on temperate rocky shores had previously demonstrated a significant role in community organizations for patchy chronic disturbance from within the community (Connell 1961, 1975). It is the opinion of many authors that intrinsic disturbance can similarly maintain a high diversity in coral communities (e.g. sea urchin grazing, Ogden 1976; Sammarco 1980: *Acanthaster* predation, Porter 1972). The question is thus raised as to whether high diversity may be maintained indefinitely without extrinsic disturbance.

Chappell's model has as its basis the premise that a reef is traversed by stress gradients in wave action, subaerial exposure, illumination and sediment (Chappell 1980). Each stress and the sum of the stresses is zonally distributed and provides a predictive basis for the distribution of coral shapes and diversity. However, the successional concepts embodied in Connell (1978) are explicit excluded from the Chappell model.

One may speculate that the next synoptic model will embrace both Connell and Chappell concepts, and thereby rectify their individual deficiencies, namely, zonal differentiation (Connell) and successional concepts (Chappell). Chappell's stress-indicies may be translatable into zone-specific rates and patterns of space production due to intrinsic perturbation. Connell's extrinsic disturbance may be modelled as variables with zone-specific frequencies and intensities. The composite model would thereby embrace the spatio-temperal variables of environmental predictability and severity, which have long been appreciated as zone dependent (Loya 1972; Porter 1972a; Dana 1979).

CONCLUSIONS

Emphasis in reef benthic ecology has changed. Descriptive localized studies of zonation have provided a necessary biological base in the past. Present day time-based process studies, autecological studies and zonal surveys with a broader geographic base are providing an improved basis for understanding reef organization.

Supporting long-term environmental information is essential both for zone/environment correlation, and for process studies. Knowledge of micro-environment at scales from square centimetres to one hectare of reef surface is particularly relevant in analysis of community structure and zonation.

Zonation is the most visible of all manifestations of pattern in the distribution and abundance of organisms. It thus provides a focus for many problems in reef ecology, for it results from a grand natural experiment in which the dose rate of virtually all ecological factors (light, wave energy, currents, sediments, space limitations, diversity and predation) are varied in space and time.

ACKNOWLEDGEMENTS

I thank the following AIMS and James Cook University colleagues for advice and assistance: R. H. Bennett, R. H. Bradbury, J. S. Bunt, B. Chalker, M. Pichon, P. W. Sammarco, C. R. S. Sheppard and W. T. Williams.

LITERATURE CITED

Adey, W. H. and R. B. Burke, 1977. Holocene bioherms of Lesser Antilles-Geologic control of development. *American Association of Petroleum Geologists. Studies in Geology* 4, 67-81.

Aller, R. C. and R. E. Dodge, 1974. Animal sediment relations in a tropical lagoon Discovery Bay, Jamaica. *Journal of Marine Research* 32, 209-232.

Bak, R. P. M., 1975. Ecological aspects of the distribution of reef corals in the Netherlands Antilles. *Bijdragen tot de Dierkunde* 45, 181-90.

Bak, R. P. M., 1978. Lethal and sublethal effects of dredging on reef corals. *Marine Pollution Bulletin* 9, 14-16.

Bak, R. P. M. and M. S. Engel, 1979. Distribution, abundance and survival of juvenile hermatypic corals (Scleractinia) and the importance of life history strategies in the parent coral community. *Marine Biology* 54, 341-52.

Barnes, D. J., 1973. Growth in colonial scleractinians. *Bulletin of Marine Science* 23, 280-98.

Barnes, J., D. J. Bellamy, D. J. Jones, B. A. Whitton, E. A. Drew, L. Kenyon, J. N. Lythgoe and B. R. Rosen, 1971. Morphology and ecology of the reef front of Aldabra. *Symposia of the Zoological Society of London* 28, 87-114.

Beadle, N. C. W. and A. B. Costin, 1952. Ecological classification and nomenclature. *Proceedings of the Linnean Society of New South Wales* 78, 61-82.

Beals, E. W., 1969. Vegetational change along altitudinal gradients. *Science* 165, 981-5.

Bouchon, C., 1980. Quantitative study of the scleractinian coral communities of the Jordanian coast (Gulf of Aqaba, Red Sea): preliminary results. *Tethys* 9, 243-46.

Bradbury, R. H. and Y. Loya, 1978. A heuristic analysis of spatial patterns of hermatypic corals at Eilat, Red Sea. *American Naturalist* 112, 493-507.

Bradbury, R. H. and P. C. Young, 1981. The effects of a major forcing function, wave energy, on a coral reef ecosystem. *Marine Ecology — Progress Series* 5, 229-41.

Bradbury, R. H. and P. C. Young, in press. The race and the swift revisited, or is aggression between corals important? In, *Fourth International Coral Reef Symposium, Manila, 1981.*

Bromley, R. G., 1978. Bioerosion of Bermuda reefs. *Paleogeography, Paleoclimatology, Paleoecology* 23, 169-97.

Bull, G. D., 1977. Comparative analysis of community structure of two fringing reefs of Magnetic Island North Queensland. M.Sc. disseration, James Cook University of North Queensland, 84 pp.

Bunt, J. S. and W. T. Williams, 1980. Studies in the analysis of data from Australian tidal forests ('Mangroves'). I. Vegetational sequences and their graphic representation. *Australian Journal of Ecology* 5, 385-90.

Chamberlain, J. A. and R. R. Graus, 1975. Water flow and hydromechanical adaptations of branched reef corals. *Bulletin of Marine Science* 25, 112-25.

Chappell, J., 1980. Coral morphology, diversity and reef growth. *Nature* 286, 249-52.

Clifford, H. T. and W. Stephenson, 1976. An Introduction to Numerical Classifications. New York: Academic Press.

Connell, J. H., 1961. Effects of Competition, predation by *Thias lapillus,* and other factors on natural population of the barnacle *Balanus balanoides. Ecological Monographs* 31, 61-104.

Connell, J. H., 1972. Community interactions on marine rocky intertidal shores. *Annual Review of Ecology and Systematics* 3, 169-72.

Connell, J. H., 1973. Population ecology of reef-building corals. In O. A. Jones and R. Endean (eds.), *Biology and Geology of Coral Reefs.* New York: Academic Press, vol. II, Biology 1, pp. 205-245.

Connell, J. H., 1975. Some mechanisms producing structure in natural communities: a model and evidence from field experiments. In M. L. Cody and J. M. Diamond (ed.), *Ecology and Evolution of Communities.* Cambridge: Belknap Press, pp. 460-90.

Connell, J. H., 1976. Competitive interactions and species diversity of corals. In G. O. Mackie (ed.), *Coelentrate Ecology and Behaviour.* New York: Plenum, pp. 51-58.

Connell, J. H., 1978. Diversity in tropical rain forests and coral reefs. *Science* 199, 1302-10.

Connor, E. F. and D. Simberloff, 1979. The assembly of species communities: Chance or competition? *Ecology* 60, 1132-40.

Cooley, W. N. and P. P. Lohnes, 1962. *Multivariate Procedures for the Behavioural Sciences.* New York: John Wiley & Sons.

Dana, T. F., 1976. Reef Coral dispersion pattern and environmental variables on a Caribbean coral reef. *Bulletin of Marine Science* 26, 1-13.

Dana, T. F., 1979. Species number relationships in an assemblage of reef-building corals; McKean Island, Phoenix Islands. *Atoll Research Bulletin* 288, 1-27.

Diamond, J. M., 1975. Assembly of species communities. In M. L. Cody and J. M. Diamond (ed.), *Ecology and Evolution of Communities.* Harvard: Belknap, pp. 342-444.

Dietrich, G., 1963. *General Oceanography.* New York: John Wiley & Sons.

Dinesen, Z. D., 1980. Some ecological aspects of coral assemblages in the Great Barrier Reef Province. Ph.D. dissertation, James Cook University of North Queensland, 360 pp.

Ditlev, H., 1978. Zonation of corals (Scleractinia: Coelenterata) on intertidal reef flats at Ko Phuket, eastern Indian Ocean. *Marine Biology* 47, 29-39.

Dodge, R. E., R. C. Aller and J. Thompson, 1974. Coral growth related to resuspension of sediments. *Nature* 247, 574-7.

Done, T. J., 1977. A comparison of units of cover in ecological classifications of coral communities. In D. L. Taylor (ed.), *Proceedings: Third International Coral Reef Symposium.* Miami: University of Miami, vol. 1, pp. 9-14.

Done, T. J., (1982). Patterns in the distribution of coral communities across the central Great Barrier Reef. *Coral Reefs 1,* 95-107.

Drew, E. A., (in press). Light. In D. Eewin and R. Earl (eds.), *The Sublittoral Environment of the British Isles — In Perspective.* London: Oxford University Press.

Endean, R., 1973. Population explosions of *Acanthaster planci* and associated destruction of hermatypic corals in the Indo-West Pacific region. In O. A. Jones and R. Endean (eds.), *Biology and Geology of Coral Reefs.* New York: Academic press, vol. II, Biology 1, pp. 390 -438.

Endean, R., 1976. Destruction and recovery of coral reef communities. In O. A. Jones and R. Endean (eds.), *Biology and Geology of Coral Reefs.* New York: Academic Press, vol. III, Biology 2, pp. 215-54.

Faure, G., 1974. Morphology and bionomy of the coral reef discontinuities in Rodriguez Island (Mascarene Archipelago, Indian Ocean). In *Proceedings of the Second International Symposium on Coral Reefs.* Brisbane: Great Barrier Reef Committee, vol 2, pp. 161-72.

Fishelson, L., 1973. Ecological and biological phenomena influencing coral-species composition on the reef tables at Eilat, (Gulf of Aqaba, Red Sea). *Marine Biology* 19, 183-96.

Geister, J., 1977. The influence of wave exposure on the ecological zonation of Caribbean coral reefs. In D. L. Taylor (ed.), *Proceedings: Third International Coral Reef Symposium.* Miami: University of Miami, vol. 1, pp. 23-29.

Glynn, P. W., 1968. Mass mortalities of echinoids and other reef flat organisms coincident with midday, low water exposure in Puerto Rico. *Marine Biology* 1, 226-43.

Glynn, P. W., 1973. Aspects of the ecology of coral reefs in the Western Atlantic region. In O. A. Jones and R. Endean (eds), *Biology and Geology of Coral Reefs*. New York: Academic Press, vol. II, Biology 1, pp. 271-324.

Glynn, P. W., 1974a. Rolling stones among the Scleractinia: mobile corallith communities in the Gulf of Panama. In, *Proceedings of the Second International Symposium on Coral Reefs*. Brisbane: Great Barrier Reef Committee, vol. 2, pp. 183-98.

Glynn, P. W., 1974b. The impact of *Acanthaster* on coral and coral reefs in the eastern Pacific. *Environmental Conservation* 14, 295-304.

Goodwin, M. H., M. J. C. Cole, W. E. Stewart and B. L. Zimmerman, 1976. Species density and association in Caribbean reef corals. *Journal of Experimental Marine Biology and Ecology* 24, 19-31.

Goreau, T. F., 1959. The ecology of Jamaican coral reefs. I. Species composition and zonation. *Ecology* 40, 67-90.

Graus, R. R., J. A. Chamberlain and A. M. Boker, 1977. Structural modification of corals in relation to waves and currents. *American Association of Petroleum Geologists. Studies in Geology* 4, 135-53.

Grigg, R. W. and J. E. Maragos, 1974. Recolonisation of hermatypic corals on submerged lava flow in Hawii. *Ecology* 55, 387-95.

Hatcher, B. G., ch.10 this volume. Grazing on coral reef ecosystems.

Holmes, P., 1975. Wave conditions in coastal areas. In J. Hails and A. Carr (eds.), *Nearshore Sediment Dynamics and Sedimentation*. London: John Wiley & Sons, pp. 1-14.

Hubbard, J. A. E. B., 1974. Scleractinian coral behaviour in calibrated current experiment: An index to their distribution patterns. In, *Proceedings of the Second International Symposium on Coral Reefs*. Brisbane: Great Barrier Reef Committee, vol. 2, pp. 107-126.

Hubbard, J. A. E. B. and Y. P. Pocock, 1972. Sediment rejection by scleractinian corals: A key to paleao-environmental reconstruction. *Geologische Rundschau* 61, 598-626.

Huntley, D. A. and A. J. Bowen, 1975. Comparison of the hydrodynamics of steep and shallow beaches. In J. Hails and A. Carr (eds.), *Nearshore Sediment Dynamics and Sedimentation*. London: John Wiley & Sons, pp. 69-109.

Ikeda, T., M. Gilmartin, N. Revelanta, A. W. Mitchell, J. H. Carlton, P. Dixon, S. M. Hutchinson, E. Hing Fay, G. M. Boto and K. Isehi, 1980. AIMS data report. I. Biological, chemical and physical observations in inshore waters of the Great Barrier Reef, North Queensland *Australian Institute of Marine Science Technical Bulletin, Oceanography Series* 2.

Jaubert, J. M., 1977. Light, metabolism and growth forms of the hermatypic scleractinian coral *Synaraea convexa* Verrill in the lagoon of Moorea (French Polynesia). In D. L. Taylor (ed.), *Proceedings: Third International Coral Reef Symposium*. Miami: University of Miami, vol. 1, pp. 483-8.

Jerlov, N. G., 1970. Light-general introduction. In O. Kinne (ed.), *Marine Ecology*. London: John Wiley & Sons, vol 1, part 1, pp. 95-102.

Jokiel, P. L., 1978. Effects of water motion on reef corals. *Journal of Experimental Marine Biology and Ecology* 35, 87-97.

Jokiel, P. L., 1980. Solar ultraviolet radiation and coral reef epifauna. *Science* 207, 1069-71.

Jokiel, P. L. and J. E. Maragos, 1978. Reef corals of Canton Atoll. II. Local distribution. *Atoll Research Bulletin* 221, 71-97.

Kawaguti, S., 1969. Effect of the green fluorescent pigment on the productivity of the reef corals. *Micronesica* 5, 313.

Kinsey, D. W., 1978. Productivity and calcification estimates using slack-water periods and field enclosures. In D. R. Stoddart and R. E. Johannes (eds.), *Coral Reefs: Research Methods*. Paris: UNESCO, Monographs on Oceanographic Methodology, no. 5, pp. 439-68.

Kinsey, D. W., 1979. Carbon turnover and accumulation by coral reefs. Ph.D. dissertation, University of Hawaii, 248 pp.

Kinzie, R. A., III, 1973. The zonation of West Indian gorgonians. *Bulletin of Marine Science* 23, 93-155.

Kinzie, R. A., III and R. H. Snider, 1978. A simulation study of coral reef survey methods. In D. R. Stoddart and R. E. Johannes (eds.), *Coral Reefs: Research Methods*. Paris: UNESCO, Monographs on Oceanographic Methodology, no. 5, pp. 231-50.

Lamacraft, R. R., 1979. A method of displaying differences in botanical composition. *Australian Journal of Ecology* 4, 407-409.

Lance, G. N., P. W. Milne and W. T. Williams, 1968. Mixed data classificatory programs. III. Diagnostic systems. *Australian Computer Journal* 1, 178-81.

Lang, J. C., 1971. Interspecific aggression by scleractinian corals. 1. The rediscovery of *Scolymia cubensis* (Milne Edwards & Haime). *Bulletin of Marine Science* 21, 952-9.

Lang, J., 1973. Interspecific aggression by scleractinian corals. 2. Why the race is not only to the swift. *Bulletin of Marine Science* 23, 260-79.

Lee, T. T. and K. P. Black, 1978. The energy spectra of surf waves on a coral reef. *Proceedings of the Sixteenth Coastal Engineering Conference* 1, 588-608.

LIMER 1975 Expedition Team, 1976. Metabolic processes of coral reef communities at Lizard Island, Queensland. *Search* 7, 463-8.

Littler, M. M. and M. S. Doty, 1975. Ecological components structuring the seaward edges of tropical Pacific reefs: The distribution, communities and productivity of *Porolithon*. *Journal of Ecology* 63, 117-29.

Loya, Y., 1972. Community structure and species diversity of hermatypic corals at Eilat, Red Sea. *Marine Biology* 13, 100-23.

Loya, Y., 1976a. Recolonization of Red Sea corals affected by natural catastrophes and man-made perturbations. *Ecology* 57, 278-89.

Loya, Y., 1976b. Effects of water turbidity and sedimentation on the community structure of Puerto Rican corals. *Bulletin of Marine Science* 26, 450-66.

Loya, Y., 1976c. The Red Sea coral *Stylophora pistillata* is an r strategist. *Nature* 259, 478-80.

Loya, Y., 1976d. Settlement, mortality and recruitment of a Red Sea scleractinian coral population. In G. O. Mackie (ed.), *Coelenterate Ecology and Behaviour*. New York: Plenum, pp. 89-100.

Loya, Y., 1978. Plotless and transect methods. In D. R. Stoddart and R. E. Johannes (Editors), *Coral Reefs: Research Methods*. Paris: UNESCO, Monographs on Oceanographic Methodology, no. 5, pp. 197-217.

Loya, Y. and L. B. Slobodkin, 1971. The coral reefs of Eilat Gulf of Eilat, Red Sea. *Symposia of the Zoological Society of London* 28, 117-39.

McCloskey, L. R., D. S. Wethey and J. W. Porter, 1978. Measurement and interpretation of photosynthesis and respiration in reef corals. In D. R. Stoddart and R. E. Johannes (eds.), *Coral Reefs: Research Method*. Paris: UNESCO, Monographs on Oceanographic Methodology, no. 5, pp. 379-96.

Maragos, J. E., 1974. A study of the ecology of Hawaiian reef corals. Ph.D. dissertation, University of Hawaii, 290 pp.

Maragos, J. E., 1974. Coral communities on a seaward reef slope, Fanning Island. *Pacific Science* 28, 257-78.

Maragos, J. E., 1978. Measurement of water volume transport for flow studies. In D. R. Stoddart and R. E. Johannes (eds.), *Coral Reefs: Research Methods*. Paris: UNESCO, Monographs on Oceanographic Methodology, no. 5, pp. 353-60.

Marsh, J. A., Jr. and S. V. Smith, 1978. Productivity measurements of coral reefs in flowing water. In D. R. Stoddart and R. E. Johannes (eds.), *Coral Reefs: Research Methods*. Paris: UNESCO, Monographs on Oceanographic Methodology, no. 5, pp. 361-77.

Marshall, P., 1931. Coral reefs - rough water and calm water types. *Report of the Great Barrier Reef Committee* 8, 64-72.

Maxwell, W. G. H., 1968. Atlas of the Great Barrier Reef. Amsterdam: Elsevier Scientific Publishing Company, 258pp.

Mayer, A. G., 1918. Ecology of Murray Island Reef. *Papers of the Department of Marine Biology Carnegie Institute of Washington* 9, 1-48.

Mergner, H., 1971. Structure, ecology and zonation of Red Sea reefs in comparison with south Indian and Jamaican reefs. *Symposia of the Zoological Society of London* 28, 141-61.

Morrisey, J., 1980. Community structure and zonation of macroalgae and hermatypic corals on a fringing reef flat of Magnetic Island Queensland, Australia. *Aquatic Botany* 8, 91-139.

Morton, J., 1974. The coral reefs of the British Solomon Islands: a comparative study of their composition and ecology. In, *Proceedings of the Second International Symposium on Coral Reefs*. Brisbane: Great Barrier Reef Committee, vol. 2, pp. 31-53.

Ogden, J. C., 1976. Some aspects of herbivore-plant relationships on Caribbean reefs and seagrass beds. *Aquatic Botany* 2, 103-116.

Paine, R. T., 1969. The *Pisaster-Tegula* interaction: Prey patches, predator food preference, and intertidal community structure. *Ecology* 50, 950-61.

Pearson, R. G., 1974. Recolonization by hermatypic corals of reefs damaged by *Acanthaster*. In, *Proceedings of the Second International Symposium on Coral Reefs*. Brisbane: Great Barrier Reef Committee, vol 2, pp. 207-215.

Pearson, R. G., 1981. Recovery and recolonization of coral reefs. *Marine Ecology — Progress Series* 4, 105-122.

Pichon, M., 1972. Les peuplements à base de scléractiniaires dans les récifs de la Baie de Tuléar (Sud-ouest de Madagascar). In C. Munkundan and C. S. Gopinadha Pillai (eds.), *Proceedings of the Symposium on Corals and Coral Reefs*. Cochin: The Marine Biological Association of India, pp. 135-54.

Pichon, M., 1978a. Recherches sur les peuplements à dominance d'anthozoaires dans les récifs coralliens de Tuléar (Madagascar). *Atoll Research Bulletin* 222, 1-447.

Pichon, M., 1978b. Problems of measuring and mapping coral colonies. In D. R. Stoddart and R. E. Johannes (eds.), *Coral Reefs: Research Methods*. Paris: UNESCO, Monographs on Oceanographic Methodology, No. 5, pp. 219-30.

Pichon, M., 1978c. Quantitative benthic ecology of Tulear reefs. In D. R. Stoddart and R. E. Johannes (eds.), *Coral Reefs: Research Methods*. Paris: UNESCO, Monographs on Oceanographic Methodology, No. 5, pp. 163-74.

Pielou, E. C., 1977. Mathematical Ecology. New York: John Wiley & Sons.

Pillai, C. S. G., 1969. The distribution of corals on a reef at Mandapam (Palk Bay), S. India. *Journal of the Marine Biological Association of India* 11, 62-72.

Porter, J. W., 1972a. Patterns of species diversity in Caribbean reef corals. *Ecology* 53, 745-8.

Porter, J. W., 1972b. Predation by *Acanthaster* and its effect on coral species diversity. *American Naturalist* 106, 487-92.

Porter, J. W., 1976. Autotrophy, heterotrophy, and resource partitioning in Caribbean reef-building corals. *American Naturalist* 110, 731-42.

Porter, J. W., G. J. Smith, J. F. Battey, D. G. Dallmeyer, S. Chang and W. Fitt. 1980. Photobiology of reef corals: Photoadaptive mechanisms and their ecological consequences. Abstract, *Third Winter Meeting, American Society of Limnology and Oceanography*, Seattle.

Prim, R. C., 1957. Shortest connection networks and some generalization. *Bell System Technical Journal* 36, 1389-1401.

Riedl, R., 1971a. Water movement — general introduction. In O. Kinne (ed.), *Marine Ecology*. London: John Wiley & Sons, vol. 1, part 2, pp. 1085-9.

Riedl, R., 1971b. Water movement — animals. In O. Kinne (ed.), *Marine Ecology*. London: John Wiley & Sons, vol. 1, part 2, pp. 1123-56.

Risk, M. J., and J. K. MacGeachy, 1978. Aspects of bioerosion of modern Caribbean reefs. *Revision de Biologia Tropical* 26, Supplement 1, 85-125.

Roberts, H. H., 1974. Variability of reefs with regard to changes in wave power around an island. In, *Proceedings of the Second International Symposium on Coral Reefs*. Brisbane: Great Barrier Reef Committee, vol 2, pp. 497-512.

Roberts, H. H., S. N. Murray and J. N. Suhayda, 1975. Physical processes in a fringing reef system. *Journal of Marine Research* 33, 233-60.

Rosen, B. R., 1974. Principal features of reef coral ecology in shallow water environments of Mahe, Seychelles. *Symposia of the Zoological Society of London* 28, 103-83.

Rosen, B. R., 1975. The distribution of reef corals. *Report of the Underwater Association* 1, 2-16.

Rosen, B. R., 1978. Determination of a collection of coral microatoll specimens from the northern Great Barrier Reef. *Philosophical Transactions of the Royal Society of London, Series B* 284, 115-22.

Sammarco, P. W., 1980. *Diadema* and its relationship to coral spat mortality: Grazing, competition, and biological disturbance. *Journal of Experimental Marine Biology and Ecology* 45, 245-72.

Sammarco, P. W., J. S. Levinton and J. C. Ogden, 1974. Grazing and control of coral reef community structure by *Diadema antillarum* Phillipi (Echinodermata: Echinoidea): A preliminary study. *Journal of Marine Research* 32, 47-53.

Sanders, H., 1969. Benthic marine diversity and the stability-time hypothesis. *Brookhaven Symposia in Biology* 22, 71-95.

Scatterday, J. W., 1977. Low water emergence of Caribbean reefs and effects of exposure on coral diversity — Observations off Bonaire, Netherlands Antilles. *American Association of Petroleum Geologists. Studies in Geology* 4, 155-69.

Scheer, G., 1978. Application of phytosociologic methods. In D. R. Stoddart and R. E. Johannes (eds.), *Coral Reefs: Research Methods. Paris:* UNESCO, Monographs on Oceanographic Methodology, No. 5, pp. 175-96.

Schwenke, H., 1971. Water movement - plants. In O. Kinne (ed.), *Marine Ecology* London: Wiley Interscience, vol. 1, part 2, pp. 1091-121.

Scoffin, T. P. and P. Garrett, 1974. Processes in the formation and preservation of internal structure in Bermuda patch reefs. In, *Proceedings of the Second Symposium on International Coral Reefs*. Brisbane: Great Barrier Reef Committee, vol. 2, pp. 429-48.

Scoffin, T. P. and D. R. Stoddart, 1978. The nature and significance of micro-atolls. *Philosophical Transactions of the Royal Society of London, Series B* 284, 99-122.

Sheppard, C. R. C., 1979. Interspecific aggression between reef corals with reference to their distribution. *Marine Ecology — Progress Series* 1, 237-47.

Sheppard, C. R. C., 1980a. Coral cover, zonation and diversity on reef slopes at Chagos Atolls, and population structures of the major species. *Marine Ecology — Progress Series* 2, 193-205.

Sheppard, C. R. C., 1980b. Roles of interspecific and intraspecific competition in coral zonation. *Progress in Underwater Science* 6, 57-60.

Sheppard, C. R. C., 1981a. Reef and soft substrate coral fauna of Chagos, Indian Ocean. *Journal of Natural History* 15, 607-21.

Sheppard, C. R. C., 1981b. The groove and spur structures of Chagos Atolls and their coral zonation. *Estuarine, Coastal and Shelf Science* 12, 549-63.

Shinn, E. A., 1966. Coral growth rate, an environmental indicator. *Journal of Paleontology* 40, 233-40.

Slobodkin, L. B. and H. L. Sanders, 1969. On the contribution of environmental predictability to species diversity. *Brookhaven Symposia in Biology* 22, 82-95.

Smith, R. C. and K. S. Baker, 1979. Penetration of UV-B and biologically effective dose rates in natural waters. *Photochemistry and Photobiology* 29, 311-23.

Spencer-Davies, P., 1977. Carbon budgets and vertical zonation of Atlantic Reef Corals. In D. L. Taylor (ed.), *Proceedings: Third International Coral Reef Symposium*. Miami: University of Miami, vol. 1, pp. 391-6.

Spencer-Davies, P., D. R. Stoddart and D. C. Sigee, 1971. Reef forms of Addu Atoll, Maldive Islands. *Symposia of the Zoological Society of London* 28, 217-59.

Stephenson, J. A., A. Stephenson, G. Tandy and A. A. Spencer, 1931. The structure and ecology of Low Isles and other reefs. *Scientific Reports of the Great Barrier Reef Expedition, 1928-29, British Museum (Natural History)* 3, 17-112.

Stoddart, D. R., 1963. Effects of Hurricane Hattie on the British Honduras reefs and cays, October 30-31, 1961. *Atoll Research Bulletin* 95, 1-142.

Stoddart, D. R., 1969. Ecology and morphology of recent coral reefs. *Biological Reviews of the Cambridge Philosophical Society* 44, 433-98.

Stoddart, D. R., 1971. Environment and history in Indian Ocean reef morphology. *Symposia of the Zoological Society of London* 28, 3-38.

Stoddart, D. R., 1974. Post-hurricane changes on the British Honduras reefs: Re-survey of 1972. In *Proceedings of the Second International Symposium on Coral Reefs*. Brisbane: Great Barrier Reef Committee, vol. 2, pp. 473-83.

Stoddart, D. R. and R. E. Johannes (eds.), 1978. *Coral Reefs: Research Methods*. Paris: UNESCO, Monographs on Oceanographic Methodology, Number 5, 581pp.

Sverdrup, H. U., M. W. Johnson and R. H. Fleming, 1942. *The Oceans, their Physics, Chemistry and General Biology,* New York: Prentice Hall.

Taylor, J. D., 1978. Zonation of rocky intertidal surfaces. In D. R. Stoddart and R. E. Johannes (eds.) *Coral Reefs: Research Methods*. Paris: UNESCO, Monographs on Oceanographic Methodology, no. 5, pp. 139-98.

Tunnicliffe, V., 1979. The role of boring sponges in coral fracture. In C. Levi and N. Boury-Esnault (eds.), *Biologie des Spongiares. Colloques Internationaux du Centre National de la Reserche Scientifique* no. 291 pp. 309-15.

Tunnicliffe, V., 1981. Breakage and propogation of the stony coral *Acropora cervicornis*. *Proceedings of the National Academy of Science* 18, 2427-31.

U.S. Coastal Engineering Research Center, 1975. *Shore Protection Manual*. 2nd ed. Washington: U.S. Department of Defence.

van den Hoek, C., A. M. Breeman, R. P. M. Bak and G. van Buurt, 1978. The distribution of algae, corals and gorgonians in relation to depth, light attenuation, water movement and grazing pressure in the fringing coral reef of Curacao, Netherlands Antilles. *Aquatic Botany* 5, 1-46.

van den Hoek, C., A. M. Cortel-Breeman and J. B. W. Wanders, 1975. Algal zonation in the fringing coral reef of Curacao, Netherlands Antilles, in relation to zonation of corals and gorgonians. *Aquatic Botany* 1, 269-308.

Vaughan, T. W. and J. W. Wells, 1943. Revision of the suborders, families and general of the Scleractinia. *Special Papers of the Geological Society of America* 44, 1-363.

Veron, J. E. N., 1978. Deltaic and dissected reefs of the far Northern Region. *Philosophical Transactions of the Royal Society of London, Series B* 284, 23-37.

Veron, J. E. N. and T. J. Done, 1979. Corals and coral communities at Lord Howe Island. *Australian Journal of Marine and Freshwater Research* 30, 203-36.

Veron, J. E. N. and M. Pichon, 1976. Scleractinia of Eastern Australia. Part I. Families Thamnasteriidae, Astrocoeniidae, Pocilloporidae. *Australian Institute of Marine Science Monograph Series* 1, 1-86.

Veron, J. E. N. and C. C. Wallace, in press. Scleractinia of Eastern Australia. Part V. Family Acroporidae. *Australian Institute of Marine Science Monograph Series* 6.

Vosburgh, F., 1977. The response to drag of the reef coral *Acropora reticulata*. In D. L. Taylor (ed.), *Proceedings: Third International Coral Reef Symposium*. Miami: University of Miami. vol. 1, pp. 477-82.

Wainright, S. A., W. D. Biggs, J. D. Currey and J. M. Gosline, 1976. Mechanical Design in Organisms. Edward A. Arnold, London 423pp.

Wallace, C. C., 1978. The coral genus *Acropora* (Scleractinia: Astrocoeniina: Acroporidae) in the central and southern Great Barrier Reef Province. *Memoirs of the Queensland Museum* 18, 273-319.

Wallace, C. C. and M. B. Dale, 1977. An information analysis approach to zonation patterns of the coral genus *Acropora* on outer reef buttresses. *Atoll Research Bulletin* 220, 95-110.

Wallace, C. C. and E. R. Lovell, 1977. Topography and coral distibution of Bushy and Redbill Islands and surrounding reef, Great Barrier Reef, Queensland. *Atoll Research Bulletin* 194, 1-22.

Warme, J. E., 1977. Carbonate borers — their role in reef ecology and preservation. *American Association of Petroleum Geologists. Studies in Geology 4*, 261-79.

Wells, J. W., 1954. Recent corals of the Marshall Islands. *United States Geological Survey Professional Paper 260-I*, 285-486.

Wells, J. W., 1956. Scleractinia. In R. C. Moore (ed.) *Treatise on Invertebrate Paleontology*. Lawrence, Kansas: University of Kansas Press, part F, pp. 328-444.

Wells, J. W., 1957. Coral reefs. *Memoirs of the Geological Society of America* 67, 609-31.

Wethey, D. S. and J. W. Porter, 1976. Sun and shade differences in productivity of reef corals. *Nature* 262, 281-2.

Whittaker, R. H., 1975. Communities and Ecosystems. Macmillan, New York.

Whittaker, R. H., (ed.), 1978a. *Ordination of Plant Communities*. The Hague: W. Junk.

Whittaker, R. H. (ed.), 1978b. *Classification of Plant Communities*. The Hague: W. Junk.

Wiegel, R. L., 1964. *Oceanographical Engineering*. New Jersey: Prentice-Hall.

Wishart, D., 1975. *Clustan 1C Manual*. London: Computer Centre, University College, University of London.

Woodley, J. D., 1980. Hurricane Allen destroys coral reefs. *Nature* 287, 387.

Woodley, J. D. and J. W. Porter, 1980. Hurricane Allen: Test of the intermediate disturbance hypothesis on the Jamaican reef. Abstract, *Third Winter Meeting, American Society of Limnology and Oceanography, Inc.,* Seattle.

Wood-Jones, F., 1910. Corals and Atolls. Lovell Reeve, London.

Yonge, C. M., 1963. The biology of coral reefs. Advances in Marine Biology 1, 209-60.

Yamaguchi, M., 1975. Sea level fluctuations and mass mortalities of reef animals in Guam, Mariana Islands. *Micronesica* 11, 227-43.

Yentsch, C. S., 1966. Primary production. In R. W. Fairbridge (ed.), *The Encyclopedia of Oceanography*. New York: Van Nostrand Reinhold, pp. 722-5.

8 Geological and Biological Roles of Cavities in Coral Reefs

Robert N. Ginsburg

T. Wayland Vaughn Laboratory for Comparative Sedimentology, Rosenstiel School of Marine and Atmospheric Science, University of Miami, Fisher Island Station, Miami Beach, Florida 33139, U.S.A.

INTRODUCTION

The variety of life on the surface of a coral reef holds one's attention so completely that it takes considerable experience to realize that most of the bulk volume of reefs is empty space. However, the cavities of coral reefs are by no means just empty space. They are an integral, perhaps even essential, element of the reef ecosystem. Furthermore, the organisms and sediments that line or fill the cavities are major contributors to the ultimate geological record. This brief overview will consider the modes of origin of cavities, the organisms that inhabit them (coelobites)and the sediment that fills them.

CAVITIES

Cavity is used here as a general term to include all open spaces regardless of size or origin. Four genetic classes of cavities can be recognized in modern reefs.

1. Intraskeletal cavities (pores).
    a. Cellular — corals, hydrocorallines *(Millepora),* coralline algae, foraminifers. In these skeletons individual cells range from about 20 to 500 μm.
    b. Tubular — worm tubes, gastropods. Tube diameters range from a few millimetres to several centimetres.
    c. Chambers — pelecypods, gastropods. Size ranges from a few centimetres to nearly a metre.

2. Growth, framework and shelter cavities.
    a. Planar, lens, and wedge-shaped cavities — sheet-like spaces, millimetres thick between successive layers of crustose coralline algae; wedge-shaped cavities a few centimetres to tens of centimetres at their thick ends between overlapping or overhanging coral plates.
    b. Shelter cavities — the spaces beneath bivalve shells or plate-like fragments of coral, segments of green algae *(Halimeda)* or large shell fragments. Sizes range from a centimetre to a metre.

c. Caves and networks of irregular cavities — the irregular packing of corals with widely varying shapes — branched, hemispherical, lobate, foliose and plate-like — produces cavities and networks of cavities of widely varying size and shape.

3. Interparticle cavities (interstitial pores of sediments).
   Amoeboid networks, the apertures (pore throats) of which range in size from tens of microns (silt, fine sand-sized sediment) to millimetres (coarse sand and granule-sized grains) to centimetres (coral rubble).

4. Borings.
   a. Cellular — amoeboid networks with apertures of a few millimetres or less (clionid sponges).
   b. Tubules — few to 10 μm; blue-green algae, fungi.
   c. Tubular and vase-shaped — few millimetres to few centimetres: bivalves, polychaete and sipunculid worms, barnacles.
   d. Equant and irregular chambers — few to 10 cm; *(Siphonodictyon sponges)*.

Descriptions and illustrations of some of the types of cavities listed above can be found in Otter (1937), Garrett *et al.* (1971), Scoffin (1972), Zankl & Schroeder (1972), Ginsburg & Schroeder (1973) and Zankl & Multer (1977).

When one looks into a reef through the larger openings that reach the surface, it is clear why there are so few systematic studies of this highly irregular and variable network of cavities. It is possible to recognize some general differences in the volumes and configurations of cavity systems, differences that can be related most readily to the growth forms of the corals. Reefs built principally of massive corals are termed pillar reefs and those in which branched corals predominate are named thicket reefs (Zankl & Schroeder 1972; Zankl & Multer 1977). The variable growth forms of massive corals give complex cavity systems (Garrett *et al.* 1971; Scoffin 1972; Zankl & Schroeder 1972); cavities can arise by the roofing over of spaces between adjacent colonies, and through the overhanging growth (curtain-like) of individual colonies Jackson *et al.* (1971). On the scale of metres, the combination of upward-growing pillars with horizontal growth as mushroom shapes, ledges, and roofs can produce caves and tunnels up to five metres high and tens of metres long (Zankl & Schroeder 1972). Reefs built principally of branched or palmate corals *(Acropora cervicornis, Porites porites, Acropora palmata)* have an intricate, irregular cavity system analogous to that of a pile of loose wooden matches or straw.

It is extremely difficult to develop accurate estimates of the total volume of cavities in a living reef because direct measurements require samples larger than the cavities; moreover, the volume of cavities is constantly changing as new cavities are formed by boring and earlier cavities are infilled with sediment. However, it is possible to make some crude estimates of the cumulative volume of cavities that existed during the history of a reef by considering the porosities of the major elements. The skeletons of corals and green algae have porosities around 50%, and coarse-grained, angular skeletal sand and coral rubble are equally porous. If corals and coarse-grained sediments are assumed to be the principal solid elements of reefs, then the minimum volume of cavities should be near 50%, and any growth or boring cavities would increase the total. Judging from what I have seen in cross-sections of reefs, I expect that the total, cumulative volume of cavities in large masses of reef rock will be more than 75%, and in some examples it may well approach 90%.

COELOBITES

Anyone who has turned over a loose piece of dead coral on a reef has surely been struck by the profusion of organisms attached to the underside and the crabs and ophiuroids scurrying away to find a new refuge. When one digs deeper into coral rubble, or examines freshly-broken pieces of reef rock, it becomes quite clear that coelobites, the inhabitants of cavities, are by no means restricted to the undersides of rubble exposed on the sea floor. Indeed, the more one examines broken surfaces of reef rock, the more astounding is the quantity of life below the interface. A convenient way to describe the coelobitic clan is to list them according to their modes of life.

1. Encrusting — bryozoans, foraminifers, crustose coralline algae, sponges, corals and tunicates.
2. Attached — bivalves, brachiopods, branching coralline algae, sponges, crinoids, ahermatypic corals and tunicates.
3. Boring — sponges, bivalves, worms, barnacles, fungi and algae (blue-greens).
4. Burrowing (within internal sediments) — crustaceans and worms.
5. Vagil — ostracods, decapods; crustaceans (crabs, lobsters).
6. Nektonic–planktonic — fish and crustaceans.

The list of organisms in each category is necessarily generalized and surely incomplete, but it indicates how varied this biota may be. Further evidences of the variety of coelobites are in the report by Buss & Jackson (1979) of 300 species just from the undersides of foliaceous corals on a Jamaican fringing reef, and the finding of skeletons of eight different phyla in a small sample of internal sediment from an algal cup reef in Bermuda (Ginsburg & Schroeder 1973).

Study of the zonation of coelobites is still in the pioneering phase. Garrett *et al.* (1971) related the variety of coelobites in a lagoon reef of Bermuda to variations in light intensity. In open cavities that receive some incident but mostly reflected light (½ to $\frac{1}{16}$ of that incident on the exposed surface), the distinctive organisms are green algae, hermatypic corals and abundant bivalves. Crustose coralline algae and bivalves extend from open cavities into those described as gloomy, but all light-dependent organisms are absent from the completely dark cavities. In these dark cavities, the surfaces are covered by ectoproct bryozoans, serpulid worm tubes, foraminifers and sponges; forms that are also present in both open and gloomy cavities. An expanded version of this phototrophic zonation is expressed on a small scale in the zonation of coelobites on the undersides of individual pieces of coral rubble from the Florida Reef Tract (D. R. Choi, pers. comm. 1979).

Only a small fraction of the cavity system receives enough light for photosynthesis. It is, therefore, not surprising that most coelobites are filter feeders, detritus feeders or predators in one way or another; and much, possibly most, of the particulate and dissolved nourishment for coelobites must come from outside the cavity system. The circulation system or systems that bring food and fresh sea water to the myriads of animals that line the extensive cavity system also flush out dissolved waste products and larvae. Once these exports from the internal cavity system reach open water, they become a potentially significant food resource to the life of the exposed reef.

How far below the surface do living coelobites extend? No systematic observations are available to provide an answer. From my own explorations into reefs in Bermuda with Johannes Schroeder and studies of the reef front in Belize with Noel James, I have the impression that coelobites, particularly the wide-ranging ectoproct bryozanos, serpulid worms and foraminifers may extend some metres into reefs that stand well above the sea floor and have some larger growth cavities or tunnels. Conversely, in reefs with little relief, or those where deposition of fine sediment is so rapid as to clog the cavity system, coelobites are limited to the outer tens of centimetres or less.

INTERNAL SEDIMENTS

The cavity systems of reefs are traps for sediments produced on the living surfaces and within the cavities, as well as for suspended sediment from outside the reefs. On the surfaces of reef and within the cavities, sediment of widely varying grain size is produced by the death and disintegration of fragile and lightly mineralized skeletons and by the bioerosion of more resistant skeletons. Among the principal contributors to these skeletal sediments are segmented green algae, corals, coralline algae, molluscs, foraminifers, sponges, alcyonarians, and echinoderms. Much of this sediment is of sand and silt sizes (50-2000 μm) but grains larger and smaller than these limits are also present. In addition to entire skeletons and skeletal debris, peloids, rounded grains of fine carbonate in the fine-sand and silt size range, are a common constituent of internal sediments (Land & Goreau 1970; Ginsburg & Schroeder 1973; Macintrye 1977; Alexandersson 1978).

The cavity systems of most reefs have so many openings to the surface that grains produced on the surfaces soon accumulate in the cavities. Wave action on the surface promotes movement of grains into the openings and it also shifts sediment progressively downward into the cavity system. Fine-grained sediment that originates on the reef surface or is brought to the reef in suspension from the surrounding area moves into the cavity system where it settles in the smaller or more remote cavities.

From my own examinations of the interiors of reefs in Bermuda and Belize and casual examination of reefs in Australia, I believe that internal sediments are frequently a major component of the geological record of reef growth. One of the more distinctive attributes of internal sediments in reefs, is large variation in grain size and composition between adjacent cavities and even within a single cavity. The grain size of internal sediments is determined by the range of sizes available and by the size, orientation, and configuration of the openings or passages. Cavities with openings of about a centimetre or more are usually, although not invariably, partly to completely filled with sand-sized and coarser sediment; cavities with openings smaller than several millimetres commonly have mud-sized internal sediment. In shelf-margin, algal cup reefs on Bermuda, Ginsburg & Schroeder (1973) found that larger cavities, including the borings of bivalves, are usually partly to completely filled with coarse skeletal sand, and in smaller cavities, like the chambers of vermetid gastropods, the internal sediment is mud-sized.

Another characteristic feature of internal sediments is that they are commonly multigenerational, and the successive generations that have varying grain size and composition may alternate with coelobites growing on the earlier sediment or alternate with rinds or layers of submarine cement. In shelf-margin reefs, it is not uncommon to find three or four generations of internal sediment within a single sample of reef rock and individual cavities of a few centimetres or less can have two or more generations (Ginsburg & Schroeder 1973; Schroeder & Zankl 1974).

Commonly, there is a decrease in the grain size of successive generations of internal sediment and individual cavities may have an upward decrease in grain size. These progressive changes in grain size, as well as the alternation of internal sediment with coelobites or cement, probably reflect changes in the passageway connecting cavities. The growth of ceolobites, the deposition of internal sediment, and the emplacement of submarine cement all tend to constrict the cavity system and therefore lead to the observed decreases in grain size of internal sediment. These restrictions, or even the closure of passages through which sediment moves, are counterbalanced by the creation of new passages by boring animals.

SUBMARINE CEMENTATION

The pervasive, submarine cementation of internal sediments and growth framework has been documented in several different shelf-margin reefs (Ginsburg *et al.* 1967; Land & Goreau 1970; Ginsburg & Schroeder 1973; James *et al.* 1976; Macintyre 1977; James & Ginsburg 1979). The most common cement is microcrystalline magnesium calcite, but fibrous aragonite, often in spherulitic or botryoidal form, is also present in varying amounts. The conversion of soft internal sediment to well-cemented limestone provides new substrate for attached coelobites and borers and it strengthens the reef against erosion by wave action.

The mechanism that triggers this massive cementation is unknown. However, the preferential cementation of reefs in areas of open circulation where the cavity system is maintained, indicates that frequent flushing of the cavities is a necessary condition.

IMPLICATIONS

The formation of cavities, the growth of coelobites within them, and their filling with sediment and cement are three quite distinct and separate processes. However, because these three processes are related synergistically, their individual effects are amplified so as to produce a signifcant part of the geological record of reefs; furthermore, because the cavity systems interact with the living surfaces of reefs, they also appear to be an integral part of the reef ecosystem.

The role of cavities as a refuge for fish and vagil benthos has long been recognized. An additional biological linkage between cavities and the surface can be inferred from the existence of a biomass of coelobites that equals or possibly exceeds that of the surface. Detritus and waste products from the surface biota surely provide most of the nourishment for coelobites; in turn, the excretions of coelobites, their larvae, and the demersal plankton living in the cavities must contribute to the food chain that sustains the surface biota.

The sedimentological interactions between the biological, physical and chemical processes within the cavity systems of reefs are most obvious in shelf-margin reefs, submarine cementation of internal sediments provides new hard substrate for the growth of attached coelobites and for borers. Borers living within the cavity system produce additional cavities that provide new surfaces for coelobites and space for internal sediments. The reiteration of boring, growth of coelobites, deposition of internal sediments and cementation has two significant consequences: first, it can fill much of the cavity system and transform growth framework onto internal sediment and second, it constantly renews the internal surface area available for the growth of coelobites.

It is hoped that this preliminary account of the 'inner life' of coral reefs will encourage further study both of the separate processes as well as their interactions.

LITERATURE CITED

Alexandersson, E. T., 1978. Distribution of submarine cements in a modern Caribbean fringing reef, Galeta Point, Panama. *Journal of Sedimentary Petrology* 48, 665-83.

Buss, L. W. and J. B. C. Jackson, 1979. Competitive networks: nontransitive competitive relationships in cryptic coral reef environments. *The American Naturalist* 113, 223-34.

Garrett, P., D. L. Smith, A. O. Wilson and D. Patriquin, 1971. Physiography, ecology, and sediments of two Bermuda patch reefs. *Journal of Geology* 79, 647-68.

Ginsburg, R. N. and J. H. Schroeder, 1973. Growth and submarine fossilization of algal cup reefs, Bermuda. *Sedimentology* 20, 575-614.

Ginsburg, R. N., E. A. Shinn and J. H. Schroeder, 1967. Submarine cementation and internal sedimentation within Bermuda reefs. *Geological Society of America, Special Paper* 115, 78-9.

Jackson, J. B. C., T. F. Goreau and W. D. Hartman, 1971. Recent brachiopod-coralline sponge communities and their paleoecological significance. *Science* 173, 623-5.

James, N. P., R. N. Ginsburg, D. S. Marszalek and P. W. Choquette, 1976. Facies and fabric specificity of early subsea cements in shallow Belize (British Hondouras) reefs. *Journal of Sedimentary Petrology* 46, 523-44.

James, N. P. and R. N. Ginsburg, 1979. The seaward margin of Belize barrier and atoll reefs. *International Association of Sedimentologists, Special Publications* 3, 191 pp.

Land, L. S. and T. F. Goreau, 1970. Submarine lithification of Jamaican reefs. *Journal of Sedimentary Petrology* 40, 457-62.

Macintyre, I. G., 1977. Distribution of submarine cements in a modern Caribbean fringing reef, Galeta Point, Panama. *Journal of Sedimentary Petrology* 47, 503-516.

Otter, G. W., 1937. Rock-destroying organisms in relation to coral reefs. *Scientific Report of the Great Barrier Reef Expedition, 1928-29, British Museum (Natural History)* 1, 323-52.

Schroeder, J. H. and H. Zankl, 1974. Dynamic reef formation: A sedimentological concept based on studies of Recent Bermuda and Bahama reefs. In, *Proceedings of the Second International Symposium on Coral Reefs*. Brisbane: Great Barrier Reef Committee, vol. 2, pp. 413-28.

Scoffin, T. P., 1972. Fossilization of Bermuda patch reefs. *Science* 178, 1280-82.

Zankl, H. and J. H. Schroeder, 1972. Interaction of genetic processes in Holocene reefs off North Eleuthera Island, Bahamas. *Geologische Rundschau* 61, 520-41.

Zankl, H. and H. G. Multer, 1977. Origin of some internal fabrics in Holocene reef rocks, St. Croix, U.S. Virgin Islands. In D. L. Taylor (ed.), *Proceedings: Third International Coral Reef Symposium*. Miami: University of Miami, vol. 2, pp. 127-34.

9 Seismic Investigations of Late Quaternary Reefal and Inter-reefal Sediments of the Great Barrier Reef Province

N. Harvey
Department of Geography, University of Adelaide, Box 498, Adelaide, S.A. 5001.
and
D. E. Searle
Geological Survey of Queensland, 41 George Street, Brisbane, Qld. 4000.

INTRODUCTION

Seismic methods play an important part in providing subsurface data to enable the geological history and structure of reefs to be determined. The principal techniques used are the reflection and refraction methods. These methods rely upon the reflection or refraction of a seismic pulse (also referred to as a sonic, acoustic, or elastic pulse) at a seismic discontinuity, where there is a physical property contrast in the density and elasticity of the contiguous materials (figs. 1 & 2). In a geologic context, the contrast may represent changes in mineralogy, lithology, porosity, cementation, diagenesis, or lithostatic pressure. Seismic methods can thus be used to map subsurface discontinuities, determine thickness of strata, and also to make stratigraphic inferences.

APPLICATION OF SEISMIC METHODS TO REEF RESEARCH

The use of seismic methods in the Great Barrier Reef province was advocated in 1936 by Spender (in Steers 1937, p. 142) but seismic methods were not used in the Great Barrier Reef until 1973 for inter-reefal subsurface investigations, and until 1975 for subsurface investigations of modern coral reefs. Elsewhere, seismic methods have been used to map fossil reefs as potential hydrocarbon reservoirs but the scale and resolution of these data make them largely irrelevant to the study of Quaternary reefs.

In 1946, 1950 and 1952 seismic refraction programmes were conducted on Pacific atolls and data were obtained on the deep structure beneath Bikini, Kwajalein and Enewetak atolls (Dobrin & Perkins 1954; Raitt 1954, 1957). Near-surface seismic refraction investigations were also carried out at Enewetak Atoll in 1971-72 (Henny *et al.* 1974) and indicated the utility of the seismic refraction method for shallow subsurface investigations of reefs.

Seismic reflection data recorded in 1962 across reefs of the Belize shelf (Purdy 1974a,b) provided evidence for Purdy's exposition of the 'Antecedent Karst Theory' (Purdy 1974a) of modern reef development. This theory postulates that an underlying

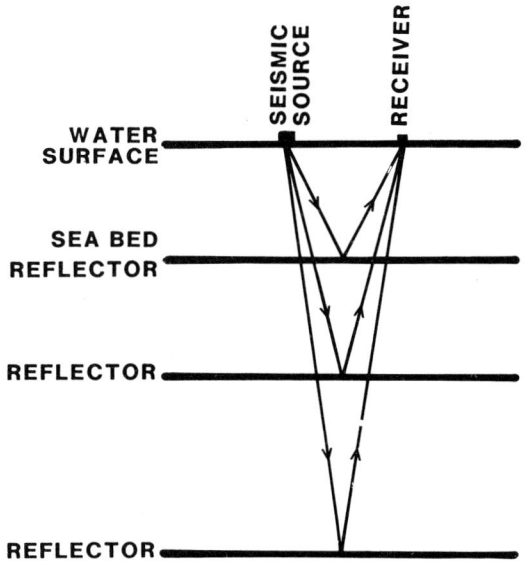

DEPTH TO ANY REFLECTOR=1/2T$\bar{\text{V}}$

WHERE T IS REFLECTION TIME
FOR RAY PATH: SEISMIC
SOURCE-REFLECTOR-RECEIVER

AND $\bar{\text{V}}$ IS AVERAGE VELOCITY
BETWEEN WATER SURFACE
AND REFLECTOR

Figure 1   Seismic reflection methods: raypaths

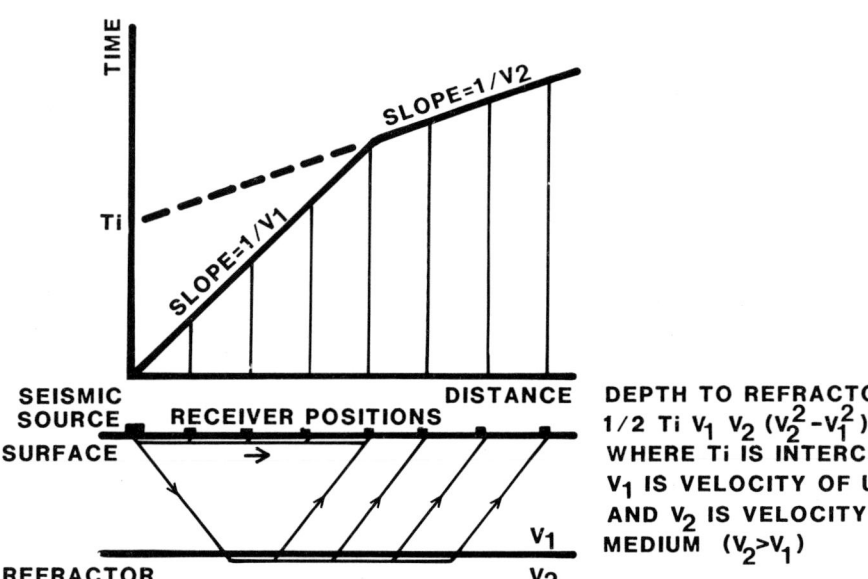

DEPTH TO REFRACTOR =
1/2 Ti V_1 V_2 $(V_2^2 - V_1^2)^{-1/2}$
WHERE Ti IS INTERCEPT TIME
V_1 IS VELOCITY OF UPPER MEDIUM
AND V_2 IS VELOCITY OF LOWER
MEDIUM $(V_2 > V_1)$

Figure 2   Seismic refraction method: raypaths and time-distance plot

karst substrate is a major determinant of the shape attributes of modern coral reefs in general, whereas the growth factor has generally less significance. Thus, the theory highlights the influence of a pre-Holocene substrate on the morphology and distribution of modern reefs.

In the Great Barrier Reef province, high resolution seismic investigations of reefal areas (fig. 3) began in 1973 with the Royal Society (London) and the Universities of Queensland Expedition to the Great Barrier Reef (Orme & Flood 1977; Orme *et al.* 1978a,b). Since that time, the continuous seismic profiling (CSP) method has been used by the Geological Survey of Queensland (Searle, in press) in co-operation with the Bureau of Mineral Resources (Searle *et al.* 1978; Searle 1979) and James Cook University (Searle *et al.*, in press), by the University of Queensland (Jell & Flood 1978), and by the Bureau of Mineral Resources (Marshall 1977).

Shallow seismic refraction investigations by James Cook University (fig. 3) began in 1975 (Harvey 1977a,b, 1978; Backshall *et al.* 1979) and later continued in co-operation with the Bureau of Mineral Resources (Harvey *et al.* 1978, 1979; Davies *et al.* 1977a,b). Details of an integrated refraction and reflection study were presented to the Australian Institute of Marine Science, Coral Reef Workshop in 1979.

EQUIPMENT AND METHODOLOGY

Studies of the inter-reefal areas of the Great Barrier Reef province have mainly involved the use of the high resolution boomer system (Sargent 1969; Searle 1977). This system provides subsurface penetration in excess of 75 m and resolution better than about 0.1 m. However, the CSP method is equipment-intensive and requires a vessel for its operation, thus restricting its use to navigable waters.

Reefal investigations, using the seismic refraction technique, have been conducted with a single channel portable seismograph using either hammer impacts or explosive detonators as the energy source. The penetration in the reefal environment is usually limited to the first seismic discontinuity since a requirement of this method, unlike the reflection method, is an increase of seismic velocity with depth across the refracting interface. Details of equipment and operation have been discussed by Harvey (1977b).

INTERPRETATION OF SEISMIC DATA

Reflection Data

The seismic reflection record represents a somewhat distorted vertical geological cross section with reflection time as the vertical axis (fig. 4). Time is related to depth by the average velocity through the overlying layers. The reflectors are related to bedding, erosional or structural interfaces in the subsurface. The nature of the reflectors, and of the material on either side of the reflecting surface, may sometimes be interpreted from the character and geometry of the reflectors (fig. 4). Thus, features such as solution disconformities, karst surfaces, terraces, bedded sediments, reef-rock, cross bedding, ancient river channels, and fluviatile deposits have been recognized in the Great Barrier Reef province.

Refraction Data

Field data obtained using the seismic refraction method are represented by a time-distance record of the passage of each seismic pulse through the ground (fig. 5a). Results

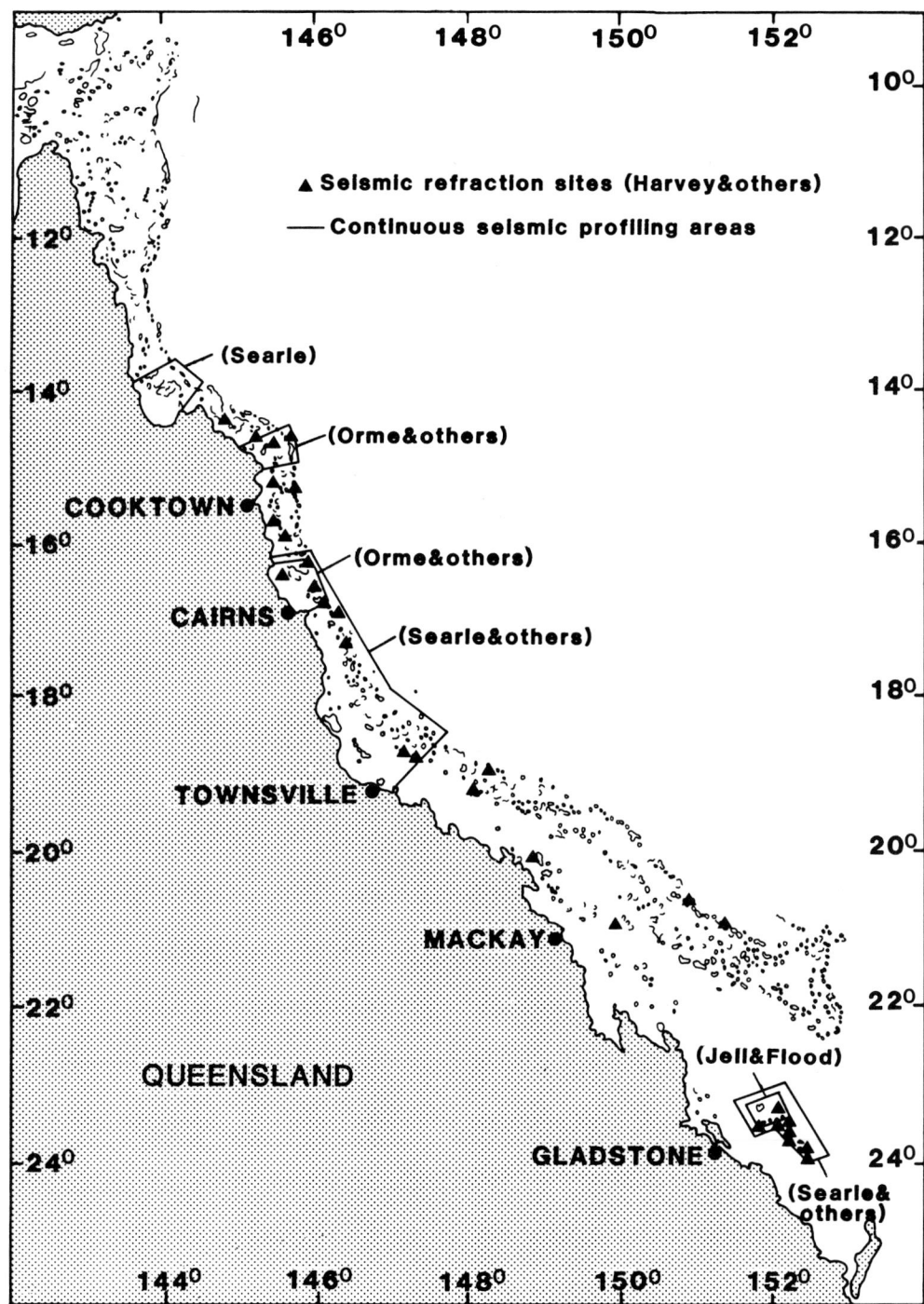

Figure 3   Seismic investigation in the Great Barrier Reef province

(a) Inter-reef channel between Grub and Corbett Reefs, Princess Charlotte Bay. The record shows ancient river channels (C) incising a probable late Pleistocene land surface (E); the channels have been buried beneath bedded sediments (D) following the rise in sea level

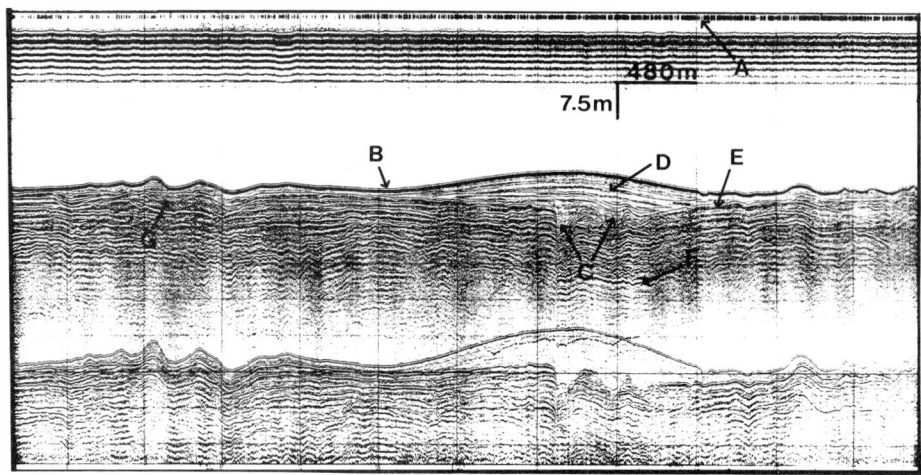

(b) Hervey Shoals, a shelf edge reefal shoal, east-south-east of Cairns. Note loss of subsurface data beneath pinnacles

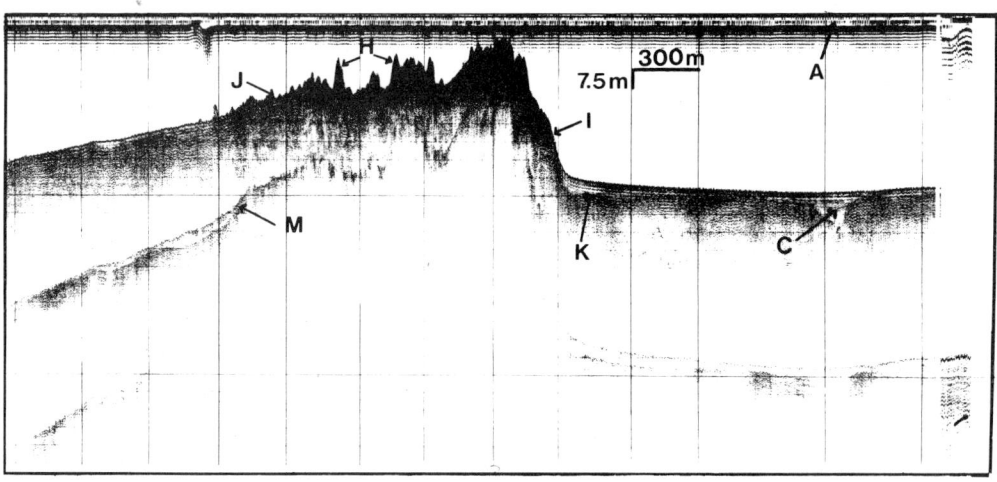

Legend 4a,b
A water surface
B sea bed
C buried channel
D bedded sediments
E major erosional disconformity
F strong reflector
G cross bedding

H coral pinnacles
I fore-reef slope
J back-reef slope
K erosional disconformity
L reef rock
M multiple

Figure 4   Reflection (SP) records from the Great Barrier Reef

(a) Refraction record from Fitzroy Reef using Huntec F53 portable facsimile seismograph in conjunction with detonators.

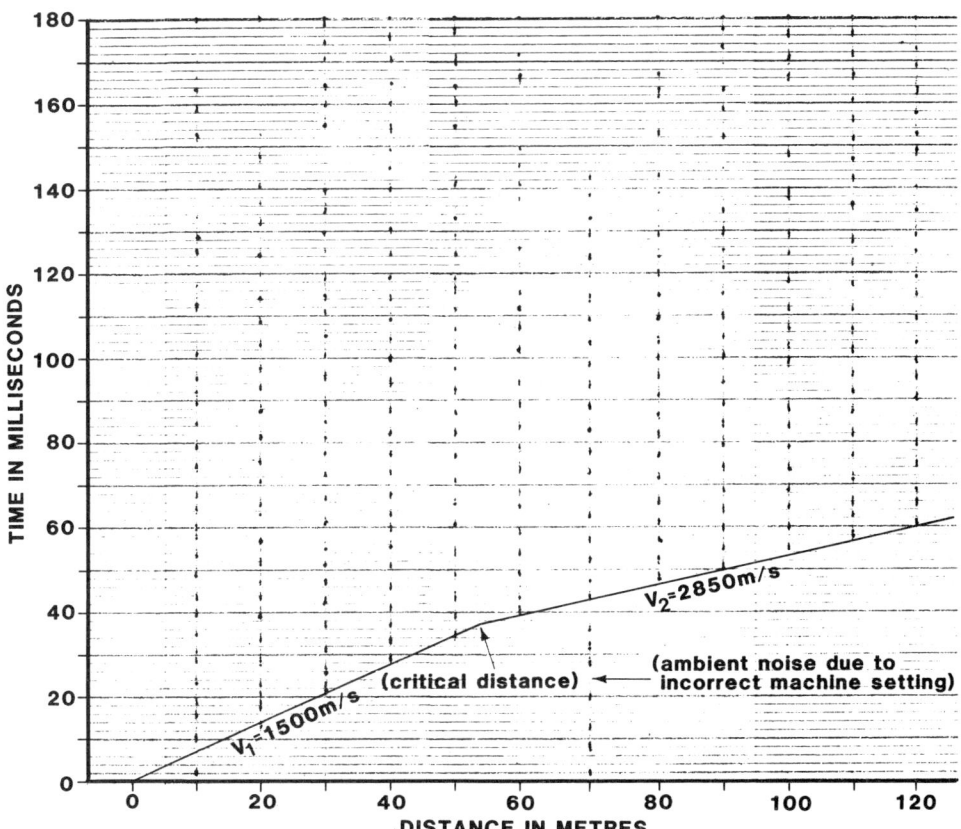

(b) Interpreted seismic refraction section across rim of the blue hole on Cockatoo Reef. Field data consisted of three reversed profiles. Note resemblance between Holocene reef and the morphology of the seismic discontinuity

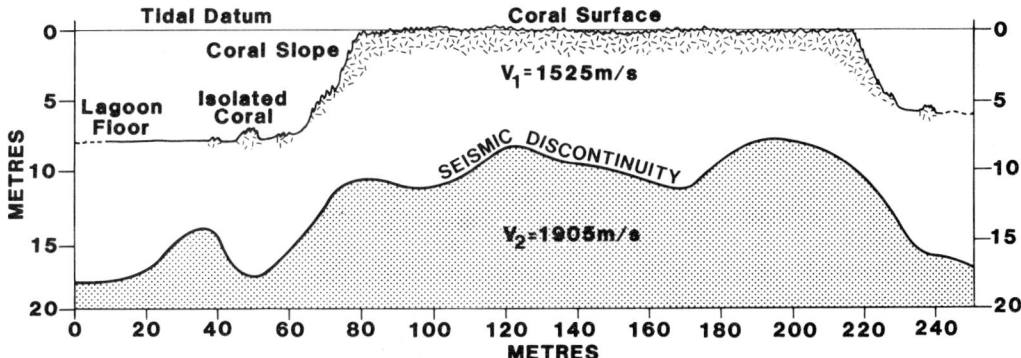

Figure 5   Seismic refraction data from the Great Barrier Reef

are computed from the critically refracted ray path at each interface. Various computational methods are available (see Dobrin 1976), but that devised by Hales (1958) has proved the most satisfactory in handling the relatively high relief of the shallow refractor beneath reefs (Harvey 1977a). Results give a depth profile and seismic velocities of material above and below the refractor (fig. 5b). These velocities often indicate the nature of the material.

CONCLUSIONS FROM SEISMIC INVESTIGATIONS IN THE GREAT BARRIER REEF PROVINCE

Seismic techniques, using both the reflection (CSP) and refraction methods, have been used to map the subsurface in certain areas of the Great Barrier Reef province (fig. 3). Interpretations of seismic data have had little confirmation by drilling, particularly in the inter-reefal areas. However, some provisional conclusions can be drawn.

Although, it is possible that Tertiary sediments are represented on some seismic sections (Marshall 1977; Searle 1979), major thrusts of seismic investigations have been towards understanding the Holocene and pre-Holocene reefal and inter-reefal sediments, and the effect of the morphology and depth of antecedent surfaces upon subsequent reef growth.

The pre-Holocene is represented on the CSP sections by several reflectors. The stronger reflections probably represent disconformities formed by subaerial weathering during Pleistocene glacial low sea levels.

The reflectors coalesce in places, indicating that old surfaces have been exhumed during subsequent low stands of sea level. The antecedent surface may thus be the product of more than one period of subaerial exposure (see Jell & Flood 1978; Orme *et al.* 1978a; Searle, *et al.* 1978).

In other places, the pre-Holocene reflectors have been incised by streams and subsequently infilled. This indicates that fluvial processes were active during emergence of the continental shelf (Orme *et al.* 1978a,b; Searle *et al.,* in press).

Correlation of seismic refraction data with dated borehole data from Bewick, Hayman and Heron Islands (Harvey *et al.* 1979) indicates that there is an empirical relationship between a seismic velocity close to 1500 m s^{-1} for Holocene carbonate material, and greater than 2000 m s^{-1} for pre-Holocene material. Thus, the shallow seismic discontinuity mapped by the refraction method can be equated with the solution unconformity which is recognized in the boreholes and which separates the Holocene reef from its substrate.

Seismic refraction results indicate that the Holocene reef is only a thin veneer (4-24 m thick) over a pre-Holocene reefal substrate. In the northern Great Barrier Reef, a shallow substrate beneath small 'low wooded islands' suggests a relationship between the substrate depth and the degree of development of patch reefs and associated cay and mangrove cover, as envisaged by Thom *et al.* (1978). In the same region, the substrate beneath outer ribbon reefs is shallower than beneath mid-shelf reefs, adding support to Purdy's (1974a) proposition that the Barrier Platform acted as a drainage-divide with minimal breaching, which contrasts with Maxwell's (1968) suggestion of cross-shelf drainage through reef passes during times of emergence.

Refraction results from the central region of the Reef suggest a slightly deeper refractor in this region than elsewhere. This may be related to present reef morphology (see Hopley, this volume). Further south, in the shelf-edge Pompey Reefs, karst analogues (blue holes) in the modern reef resemble the morphology of the underlying substrate (see

Backshall *et al.* 1979). In the southern-most reefs, Holocene reef morphology also resembles pre-Holocene substrate morphology (Harvey *et al.* 1979).

Direct correlation of reefal refraction and inter-reefal reflection results (Searle & Harvey, in prep.) allows the Holocene–pre-Holocene solution unconformity to be identified in inter-reefal areas and traced across the continental shelf into the terrigenous coastal environment. The relief of this surface beneath reefal and inter-reefal areas suggests that karst processes as well as fluvial processes (evidenced by infilled, buried channels) have played a part in determinig the morphology of the substrate. Karst processes are likely to have acted on carbonate terrains throughout the Great Barrier Reef province during successive exposures of the continental shelf associated with Quaternary sea level fluctuations.

CSP records indicate that minor disconformities, terraces and changes of slope have affected Holocene sedimentary accumulations of the inter-reefal areas. This suggests still-stands or minor regressions during the post-glacial transgression (see Orme *et al.* 1978a).

CSP results also indicate that inter-reefal Holocene sedimentation in the Great Barrier Reef province forms a thin but uneven veneer over the pre-Holocene surface, except in proximity to modern reefs and reefal shoals where carbonate sediments, probably derived from the adjacent reef and transported by waves and currents, form a lee-side sediment wedge several metres thick. Where there are significant bottom currents the carbonate material is formed into sand waves, which may have a wavelength of 100–200 m and an amplitude of 3–4 m (see Jell & Flood 1978; Searle *et al.* 1978).

The Holocene inter-reefal sedimentary pattern in the southern Great Barrier Reef contrasts strongly with that interpreted from seismic data in the central and northern provinces. In the northern region (unlike the southern region), the pre-Holocene surface is rarely exposed, commonly incised by channels which are now buried, and usually covered by several metres of Holocene sediment. This is in contrast to Purdy's data from Belize (Purdy 1974a,b) where it is apparent that ancient channels are seldom infilled and masked by later sedimentation.

FURTHER RESEARCH

Most refraction surveys have been restricted to the reefal environment where it is possible to correlate seismic data with available borehole data. These surveys should be extended to provide direct correlation with the CSP data in the inter-reefal areas, which would permit calibration of the CSP data from the velocity of the seismic refracting layers and thus allow a more reliable interpretation of the stratigraphy. In addition, the use of multichannel signal enhancement equipment would make the field process of refraction surveying more efficient.

Reefal refraction results are generally limited to the first seismic discontinuity since the method requires an increase of seismic velocity with depth across the refracting interface. The same limitations do not apply to reflection techniques which could be used to investigate the pre-Holocene structure of the reefs.

Investigations of the deeper structure in the Great Barrier Reef province should be attempted with seismic reflection equipment using a higher input power and multi-channel recording techniques.

Interpretation of seismic data needs to be verified by drilling, which could provide further information if the boreholes were cased and subsequently subjected to geophysical logging techniques. In particular, sonic logging should be conducted to determine vertical and interval velocities, together with cross-hole surveys.

It is apparent from seismic data that effects of late Quaternary sea level fluctuations can be detected in the stratigraphic record of reefal and inter-reefal sediments. In particular, the subaerial exposure of the Queensland Continental Shelf during the last glacial has produced a surface which can be mapped using seismic methods. The accuracy in identifying this seismic discontinuity from seismic data will improve when additional work, particularly confirmatory drilling of seismic profiles, is carried out. It will then be possible to determine the morphology of the surface upon which modern reefs are growing, to assess the drainage pattern in both terrigenous and biogenic terrains during late Pleistocene emergence of the shelf and to assess the volume of sediment produced in carbonate and non-carbonate environments during the Holocene.

ACKNOWLEDGEMENTS

D. E. Searle published with the permission of the Under-Secretary, Department of Mines, Queensland.

LITERATURE CITED

Backshall, D. G., J. Barnett, P. J. Davies, D. C. Duncan, N. Harvey, D. Hopley, P. J. Isdale, J. N. Jennings and R. Moss, 1979. Drowned dolines — the blue holes of the Pompey Reefs, Great Barrier Reef. *BMR Journal of Australian Geology and Geophysics* 4, 99-109.

Davies, P. J., J. F. Marshall, D. Foulstone, B. G. Thom, N. Harvey, A. D. Short and K. Martin, 1977a. Reef growth, southern Great Barrier Reef — preliminary results. *BMR Journal of Australian Geology and Geophysics* 2, 69-72.

Davies, P. J., J. F. Marshall, B. G. Thom, N. Harvey, A. D. Short and K. Martin, 1977b. Reef development — Great Barrier Reef. In D. L. Taylor (ed.), *Proceedings: Third International Coral Reef Symposium*. Miami: University of Miami, vol. 2, pp. 331-7.

Dobrin, M. B., 1976. Introduction to Geophysical Prospecting. New York: McGraw Hill, 630 pp.

Dobrin, M. B. and B. Perkins, 1954. Seismic studies of Bikini Atoll. *United States Geological Survey Professional Paper* 260-J, 487-504.

Hales, F. W., 1958. An accurate graphical method of interpreting seismic refraction lines. *Geophysical Prospecting* 3, 285-94.

Harvey, N., 1977a. The identification of subsurface solution disconformities on the Great Barrier Reef, Australia between 14°S and 17°S, using shallow seismic refraction techniques. In D. L. Taylor (ed.), *Proceedings: Third International Coral Reef Symposium*. Miami: University of Miami, vol. 2, pp. 45-51.

Harvey, N., 1977b. The application of shallow seismic refraction techniques to coastal geomorphology: A coral reef example. *Catena* 4, 333-9.

Harvey, N., 1978. Wheeler reef: morphology and shallow reef structure. In D. Hopley (ed.), *Geographical Studies of the Townsville Area*. Townsville: James Cook University, Geography Department, Monograph Series, Occasional paper 2, pp. 51-3.

Harvey, N., P. J. Davies and J. F. Marshall, 1978. Shallow reef structure: southern Great Barrier Reef. *Bureau of Mineral Resources, Geology and Geophysics Record* 96.

Harvey, N., P. J. Davies and J. F. Marshall, 1979. Seismic refraction — a tool for studying coral reef growth. *BMR Journal of Australian Geology and Geophysics* 4, 141-7.

Henny, R. W., J. W. Mercer and R. T. Zbur, 1974. Near surface geologic investigations at Eniwetok Atoll. In, *Proceedings of the Second International Symposium on Coral Reefs*. Brisbane: Great Barrier Reef Committee, vol. 2, pp. 615-26.

Jell, J. S. and P. G. Flood, 1978. Guide to the geology of reefs of the Capricorn and Bunker Groups, Great Barrier Reef Province, with special reference to Heron Reef. Brisbane: *University of Queensland, Department of Geology, Papers* 8, 3.

Marshall, J. F., 1977. Marine geology of the Capricorn Channel area. *Bureau of Mineral Resources, Geology and Geophysics Bulletin* 163.

Maxwell, W. G. H., 1968. Atlas of the Great Barrier Reef. Amsterdam: Elsevier Scientific Publishing Company, 258 pp.

Orme, G. R. and P. G. Flood, 1977. The geological history of the Great Barrier Reef: a reappraisal of some aspects in the light of new evidence. In D. L. Taylor (ed.), *Proceedings: Third International Coral Reef Symposium.* Miami: University of Miami, vol. 2, pp. 37-43.

Orme, G. R., P. G. Flood and G. E. G. Sargent, 1978a. Sedimentation trends in the lee of outer (ribbon) reefs, Northern Region of the Great Barrier Reef province. *Philosophical Transactions of the Royal Society of London, Series A* 291, 85-99.

Orme, G. R., J. D. Webb, N. C. Kelland and G. E. G. Sargent, 1978b. Aspects of the geological history and structure of the northern Great Barrier Reef. *Philosophical Transactions of the Royal Society of London, Series A* 291, 23-35.

Purdy, E. G., 1974a. Reef configurations: cause and effect. *Society of Economic Paleontologists and Mineralogists. Special Publication* 18, 9-76.

Purdy, E. G., 1974b. Karst determined facies patterns in British Honduras: Holocene carbonate sedimentation model. *American Association of Petroleum Geologists Bulletin* 58, 825-55.

Raitt, R. W., 1954. Seismic refraction studies of Bikini and Kwajalein atolls. *United States Geological Survey Professional Paper* 260-K, 507-26.

Raitt, R. W., 1957. Seismic refraction studies of Eniwetok Atoll. *United States Geological Survey Professional Paper* 260-S, 685-98.

Sargent, G. E. G., 1969. Further notes on the application of sonic techniques to submarine geological investigation, with special reference to continuous reflection seismic profiling. *Ninth Commonwealth Mining and Metallurgical Congress.* London: Institution of Mining and Metallurgy, Paper 28.

Searle, D. E., 1977. The application of continuous seismic profiling to marine geology. *Geological Survey of Queensland Record* 19.

Searle, D. E., 1979. Results of a continuous seismic profiling survey in the Capricorn/Bunker Group, southern Great Barrier Reef. *Geological Survey of Queensland Record* 19.

Searle, D. E., in press. Results of a continuous seismic profiling survey in the Princess Charlotte Bay area. *Geological Survey of Queensland Record.*

Searle, D. E., D. J. Davies, H. Hekel, J. Kennard, J. F. Marshall and B. G. Thom, 1978. Preliminary results of a continuous seismic profiling survey in the Capricorn Group, southern Great Barrier Reef. *Geological Survey of Queensland Record* 46.

Searle, D. E., N. Harvey and D. Hopley, in press. Preliminary results of continuous seismic profiling in the Great Barrier Reef province between 16°10'S and 19°20'S. *Geological Survey of Queensland Record.*

Steers, J. A., 1937. The coral islands and associated features of the Great Barrier Reefs: a discussion. *Geographical Journal* 89, 140-6.

Thom, B. G., G. R. Orme and H. A. Polach, 1978. Drilling investigation of Bewick and Stapleton Islands. *Philosophical Transactions of the Royal Society of London, Series A* 291, 37-54.

10 Grazing in Coral Reef Ecosystems

Bruce G. Hatcher
Botany Department,
Sydney University,
Sydney, N.S.W. 2006*

INTRODUCTION

The term grazing has several meanings. In common usage it refers to herbivory. In its broadest ecological sense it refers to ingestion of discrete units of living animal or plant material. Scientific use of the terms grazing and grazer has been extremely variable in its specificity. Dietary classification is further complicated because animals are frequently less particular than the terms used to describe their feeding habits. For instance, marine herbivores inadvertently ingest varying amounts of detritus and living animal material. Thus, any classification is an attempt to partition a continuum.

In practice, dietary classification is based on the relative proportions of food items in gut samples, and observations of feeding behaviour. Numerous authors have defined terms that describe various feeding specializations of coral reef grazers (e.g. Hiatt & Strasburg 1960; Bakus 1966, 1976; Jones 1968; Smith & Tyler 1973; Ogden 1977). The classification shown in figure 1 summarizes and clarifies these terms. The grazers considered here are those that obtain the majority of their sustenance from plants living in or on coral substrates; scrapers, browsers, croppers and suckers.

Studies of marine grazing benefited from earlier work on feeding activities of terrestrial animals (see Crisp 1964). Initial aquatic studies concentrated on grazing by zooplankton (see Raymont 1963). The term grazer was subsequently used to describe invertebrates that feed on benthic plants. Studies of temperate and tropical echinoids comprise much of the recent literature on grazing (see Lawrence 1975).

Grazing has often been proposed as the explanation for the paucity of benthic algae on coral reefs (Bakus 1972; Dahl 1973; Marsh 1976). Early work dealing with intertidal habitats was reviewed by Bakus (1964), who included carnivores feeding on sessile invertebrates among the grazers. Stephenson & Searles (1960) carried out the first experimental study of the effects of grazing on a coral reef. In the past decade, research has centered on grazing fish and echinoids and their effects on coral reef community structure. This work, mostly carried out in the Caribbean, has been reviewed by Ogden & Lobel (1978).

Several generalizations about grazing processes and the role of grazing organisms in coral reef ecosystems may be extracted from the literature:

* Present address: CSIRO Marine Laboratories, P.O. Box 20, North Beach, W.A. 6020.

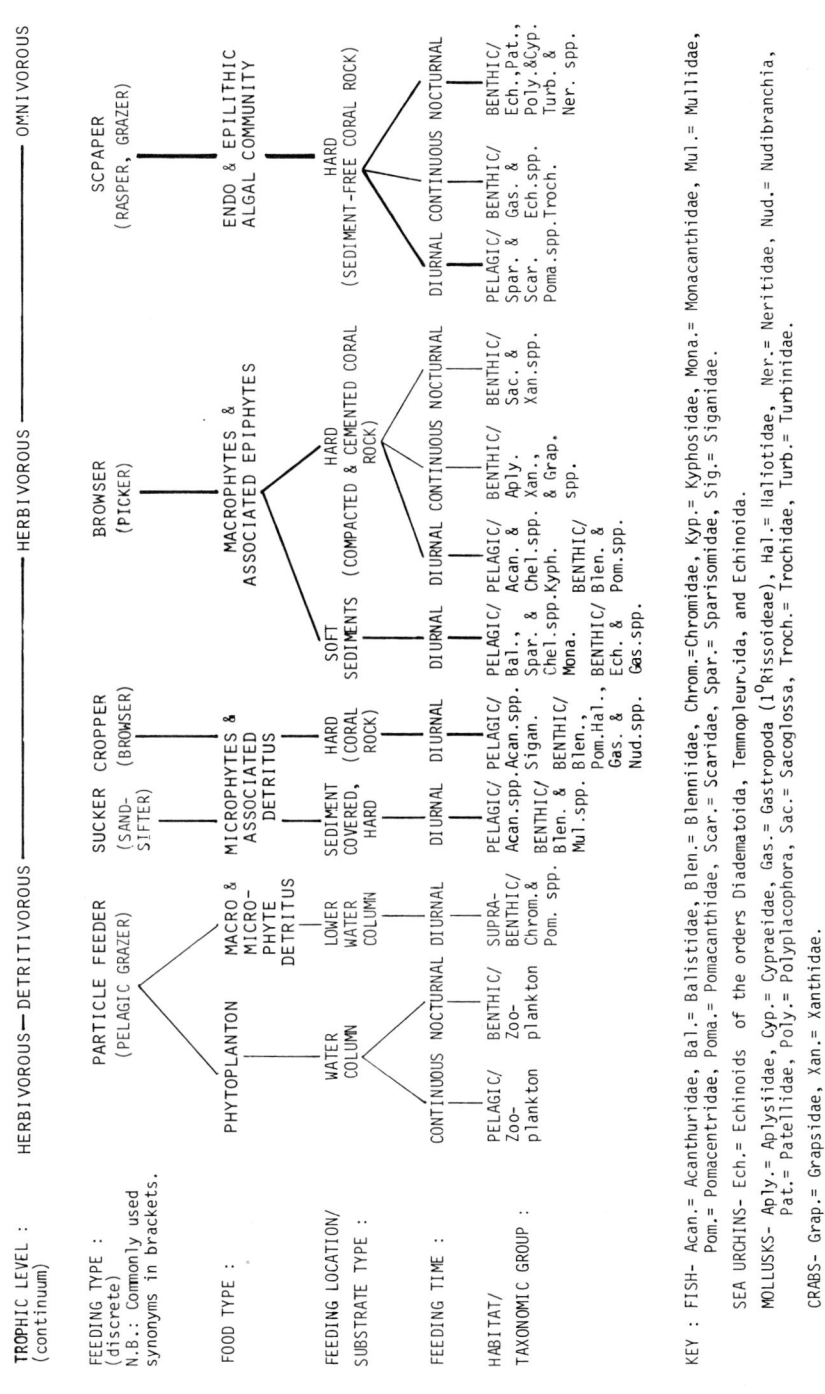

Figure 1 A functional classification of grazing organisms. Feeding types thought to be of major importance are indicated by thicker lines. It should be noted that grazing organisms often show a high degree of opportunistic feeding

1. trophic flow in reef systems, unlike that in most benthic marine systems studied to date, is primarily through grazing, rather than detritus-based pathways;
2. fish (namely scarids and acanthurids) are the predominant grazing organisms on Pacific reefs, but echinoids dominate in the Caribbean;
3. grazing is the major factor controlling various aspects of the structure of benthic communities, notably the standing crop and diversity of primary producers;
4. the grazer-primary producer interaction is highly evolved.

These generalizations are based on observational and experimental evidence which is often circumstantial and of variable quality. They should be viewed as inferences rather than assertions. In this review I shall examine these inferences in the context of trophodynamics, community structure and interactions, and evolution and adaptation.

SIGNIFICANCE OF GRAZING AT THE ECOSYSTEM LEVEL

A number of gut content studies indicate that plant components of the epilithic algal community (EAC) dominate the diet of the majority of herbivorous, and many omnivorous, coral reef fish (Hiatt & Strasburg 1960; Randall 1967; Tsuda & Randall 1971; Hobson 1974). The EAC is the so-called 'algal mat' or 'turf' occurring on dead coral rock. It is the major contributor to the very high productivity of reefs (Wanders 1976a; Marsh 1976; Rogers & Salesky 1981). It is important to note that the EAC often includes large animal and microbial components which contribute significantly to its nutritional value (Menzel 1959; Lobel 1980).

Most studies of the diets of tropical echinoids have been conducted in the Caribbean, where seagrasses growing in beds adjacent to reefs may be an important dietary component as well as the EAC (McPherson 1965, 1968; Ogden *et al.* 1973a; Abbott *et al.* 1974). The fish on reefs feed diurnally and often migrate from deep resting areas to shallow feeding areas (Bardach 1958; Hobson 1973; Ogden & Buckman 1973; Smith & Paulson 1974). In shallow areas, the EAC is likely to have the greatest productivity (Vine 1974; Wanders 1976a). Between-habitat variation in the standing-crop of benthic plants has often been related to qualitative estimates of grazer abundance or activity (Randall 1961a; Bakus 1966, 1967; Mathieson *et al.* 1971; John & Pople 1973; Connor & Adey 1977; Benayahu & Loya 1977; van den Hoek *et al.* 1978; Montgomery *et al.* 1982), but within-habitat variations in EAC biomass have often been attributed to the protective activities of territorial fish (Foster 1972; Vine 1974; Belk 1975; Kaufman 1977; Brawley & Adey 1977; Robertson *et al.* 1979; Montgomery 1980a). A number of experimental studies have demonstrated that the result of exclusion of grazers is an increase in the standing-crop of the EAC (Stephenson & Searles 1960; Randall 1961a; Mathieson *et al.* 1971; Earle 1972; John & Pople 1973; Tsuda & Kami 1973; Sammarco 1975, 1977a; Sammarco *et al.* 1974; Ogden *et al.* 1973b; Wanders 1977; Lassuy 1980; Montgomery 1980a). Where segments of substrate supporting a high EAC biomass were transplanted into areas where grazers are common the material was rapidly consumed (Mathieson *et al.* 1971; Randall 1961a, 1965; Bakus 1967; Vine 1974; Lobel 1980). On a larger scale, reduced algal biomass in tropical benthic communities (as compared to temperate counterparts) has often been attributed to the activities of grazing organisms (Randall 1961a; Bakus 1966, 1972; Dahl 1972). Overgrazing, or complete denudation may occur in temperate echinoid grazing systems (Kitching & Ebling 1961; Breen & Mann 1976). In the tropics, overgrazing by echinoderms has been observed in the form of 'halos' or bare patches in seagrass beds proximate to reefs (Randall 1965; Camp *et al.* 1973; Ogden *et al.*

1973c; Ogden & Zieman 1977). Destructive overgrazing appears to be unique to echinoids feeding on macrophytes and releases large quantities of plant material to detritus food chains (Leighton *et al.* 1966; Mann 1977). Denudation of the EAC may occur in the presence of extremely high echinoid densities (Sammarco 1977b; Benayahu & Loya 1977).

Thus, there is a large body of evidence that suggests that grazing plays an important role in controlling the standing-crop of primary producers on coral reefs. However, grazing may not be the only controlling factor, or indeed, the major controlling factor. High standing crops have often been noted in areas accessible to grazers (Doty 1971; Wanders 1976b; Adey *et al.* 1977; Hatcher, unpublished data) and may be related to factors, such as inorganic nutrient levels, that cause production to exceed losses due to grazing.

Estimates of the relative and absolute abundances of reef fish indicate that grazing species constitute a major feeding group in terms of biomass and that they are the dominant group in shallow (<10 m) water (Bardach 1959; Randall 1963; Talbot 1965; Bakus 1966, 1967; Hobson 1974; Goldman & Talbot 1975; Jones & Chase 1975; Nagelkerken 1975; Frydl & Stearn 1978; Robertson *et al.* 1979). Biomass values range from 14-70 g fresh weight m^{-2} or about 25% of total fish biomass. Scarids, Acanthurids and Blennies are usually the most abundant families. The estimates suffer from a number of sampling errors, including a bias introduced because itinerant species form a significant proportion of the samples and these do not obtain all of their nutritional requirements from the reef (Bardach 1959). The biomass of omnivorous and carnivorous fish feeding primarily on benthic invertebrates may be equal to or greater than that of herbivorous fish. This suggests that a large proportion of the total energy flow passes through these invertebrates, many of which are herbivorous gastropods, decapods and echinoids.

The scarcity of intertidal invertebrates in the tropics has been noted qualitatively by Endean *et al.* (1956) and Newman (1960). Estimates of the abundance of invertebrate grazers on reefs are rare except for echinoids in the Caribbean, which range from 8.5-100 individuals m^{-2} (Hunt 1969; Sammarco 1972; Bak & van Eys 1975; Ogden 1977; Sammarco 1977b). The only figures for molluscs indicate that herbivorous gastropods account for 50-70% of all molluscs (Taylor 1971; Kay 1978).

There is a greater abundance and diversity of grazing fish in the tropics than in temperate regions (Bakus 1966), although echinoids may be important in both temperate and tropical coastal ecosystems (Ogden 1976; Mann 1977). Direct comparisons of the biomass of fish and other vertebrate or invertebrate grazers are not available at the reefal or the geographic scale. Furthermore, there have been no tests of the correlation between the feeding rates of grazers in various habitats and standing-crop of primary producers in those habitats. Hence, quantitative assessment of either the relative or absolute importance of various grazer groups in terms of tropho-dynamics is not possible. Qualitatively, it seems likely that there is a reversal of the relative importance of invertebrate versus vertebrate grazing between Atlantic and Indo-Pacific reefs (Ogden 1976). It is possible that the predominance of echinoids on Caribbean reefs is a recent development related to heavy fishing (Ogden *et al.* 1973b).

The grazing of macrophytes by sea turtles is another potentially important pathway for which there are very few data. Recent work by Bjorndal (1980) indicates that green turtles specialise on either seagrasses or algae, and maintain distinct grazing plots. The relative importance of turtles as grazers will depend upon their abundance, which prior to the advent of the turtle fishery was probably so great as to make them the dominant grazers in Caribbean seagrass beds.

No direct measurements of the trophic exchange between benthic primary producers and grazers on coral reefs have been published. Greenway (1976) estimated the amount of seagrass removed by echinoids in Kingston Harbour, Jamaica, to be 2.0 g organic weight m^{-2} d^{-1} (1.4% standing-crop d^{-1}). The amount of seagrass grazed by fish in this study was insignificant. Montgomery (1980b) estimated the amount of EAC removed by a territorial pomacentric fish *Eupomacentrus rectifraenum)* from granite boulder reefs in the Gulf of Mexico to be about 8.0 g dry weight m^{-2} d^{-1} (1.4% standing-crop d^{-1}).

There have been some measurements of the amount of inorganic material removed by grazers. Such work has been primarily concerned with reef bioerosion (e.g. Gygi 1975; Trudgill 1976; Frydl & Stearn 1978) and was reviewed by Ogden (1977). These measurements may be roughly converted to biomass turnover using the percentage organic figures given by Ogden (1977). Values are in the range 0.09-0.49 g organic weight m^{-2} d^{-1} for fish and 0.85-1.55 g organic weight m^{-2} d^{-1} for echinoids.

The yield from the EAC to grazing organisms greater than 1 cm in size has been measured at One Tree Reef, southern Great Barrier Reef. Values obtained vary with habitat and time from 0.13 to 3.8 g organic weight m^{-2} d^{-d}, which represent up to 6.0% of the standing-crop d^{-1} (Hatcher 1981). These data indicate that grazing organisms remove 20- 90% of the net daily production by the EAC, and demonstrate quantitatively the primary importance of grazing pathways in coral reef tropho-dynamics.

The relationship between grazing and benthic primary productivity is not well understood. Increased productivity is sometimes equated with increased standing-crop (e.g. Brawley & Adey 1977) although there is not necessarily a positive correlation (Kohn & Helfrich 1957; Odum 1971). In fact, by maintaining a low standing crop, grazing can actually increase turnover and hence productivity (Hayne & Ball 1956; Greenway 1974; Matson & Addy 1975; Dyer & Bakhari 1976; Porter 1976; Wanders 1977; Fowler & Robson 1978; McNaughton 1979; Montgomery 1980a) Possible mechanisms for this effect include: (a) reduction of over-growth and self-shading; (b) enchancement of nutrient exchange with the water mass; (c) maintenance of plants in the exponential growth phase; (d) rapid recycling of nutrients and (e) high rates of disturbance that interrupt succession and favour rapidly growing species. If the production enhancement effect occurs in coral reef communities, grazing is probably the single most important factor maintaining their high productivity.

Subjective data and circumstantial evidence has produced some general hypotheses. There is now a need for quantitative measurements of ecosystem parameters including the yield from primary producers to grazers, production to biomass ratios of both producers and consumers, and the variation in these parameters through time and space. Such studies should be reinforced by measurements of the nutritional value of foods, the nutrient specific assimilation efficiencies of grazers, and the energy and nutrient feedbacks between the two. Such data will allow the assessment of the major pathways of trophic exchange and valid comparisons with other ecosystems.

SIGNIFICANCE OF GRAZING AT THE COMMUNITY LEVEL

Grazing can affect benthic community structure in a variety of ways. Best documented are controls on EAC biomass, succession and diversity inferred from grazer exclusion experiments. Figures for biomass increase inside artificial or 'natural' cages range from

1.5 to 15 times those in control areas (Sammarco *et al.* 1974; Vine 1974; Sammarco 1977b; Wanders 1977; Birkeland 1977; Brawley & Adey 1977; Lassuy 1980). In short term caging experiments, biomass and species numbers generally increase with duration of protection. However, within 15-60 days there is a shift in species composition favouring a few prominent macroalgae (Mathieson *et al.* 1971; John & Pople 1973; Sammarco *et al.* 1974; Day 1977; Wanders 1977; Montgomery 1980a; Lassuy 1980). Only two studies have been of sufficient duration to determine whether the observed changes are stable community features. Stephenson & Searles (1960) placed cages over intertidal beach rock at Heron Island, southern Great Barrier Reef. Enclosed substrates showed no significant change in species composition after 15 months. At regular intervals, the community became unstable and sloughed off the substrate. Ogden (1976) reported persistence of an initial increase in biomass on a patch reef in the Virgin Islands maintained free of echinoids for eight months.

The increase in algal diversity resulting from the short-term exclusion of grazers is in contrast to the decrease in diversity in benthic invertebrate communities when predators are exluded (Paine 1966, 1969; Menge 1976; Russ 1980). Day (1977) suggested that this occurs because herbivores are less selective feeders than carnivores. It is likely that, despite a copious literature, insufficient data are available for derivation of the principles of tropical algal–herbivore relationships in terms of species diversity. Certainly, where hypotheses have been tested in terrestrial and temperate subtidal environments, the results are conflicting, showing both increases and decreases in diversity (Harper 1969; Paine & Vadas 1969a; Vance 1979). A number of factors have to be considered when such studies are made in coral reef communities: (a) autotrophs are more closely coupled to the physical environment than heterotrophs, thus presence and success in a given habitat may often be controlled by physical rather than biotic factors; (b) micro-habitat may be the major factor controlling the small scale distribution of the minute forms which comprise the bulk of the EAC; (c) components of the EAC possess a broad and often complex range of reproductive adaptations which in turn affect the availability of colonists; (d) there is a strong morphological division between the encrusting, micro-filamentous and macro-algae to which grazers respond with differing degrees of selectivity; (e) these differences in scale complicate diversity measures so that simple indices like species number are often not suitable (furthermore, the taxonomy of all but the macro-algae is difficult), and (f) the inherent complexity and lack of clear dominants in the EAC makes for complicated experimental design. A more fruitful approach to understanding the effect of grazing on benthic community composition is through disturbance-dependent models of plant succession (see Connell & Slatyer 1977) with grazing invoked as the predominant form of disturbance. Montgomery (1980a) presents such a model and uses algal assemblages ranked accordingly to their productivity as the successional unit. Such an approach avoids some of the problems outlined above.

Exclusion experiments have demonstrated the importance of a number of secondary effects of algal grazing. It has been shown that filamentous algae and sediment trapped by the EAC interfere with the settlement and growth of encrusting coralline algae (Littler & Doty 1975; Vine 1974; Wanders 1977; van den Hoek *et al.* 1978) and sessile invertebrates, particularly corals (Taylor 1968; Bakus 1969; Dart 1972; Vine 1974; Pearson 1974; Neudecker 1977; Potts 1977; Birkeland 1977; Sammarco 1980). From these studies it may be deduced that in natural reef communities grazing serves to restrict the thickness of the algal cover, thus enhancing settlement and survival of certain benthic organisms. Recently, Brock (1979) has demonstrated this effect experimentally. As both coral and coralline algae are vital to reef construction, their protection from algal competition may be the single most important effect of grazing.

Conversely, it has been suggested that the activities of grazers have a detrimental effect on reef growth through the destruction of living coral (Bakus 1964, 1969; Glynn *et al.* 1972; Bak & Van Eys 1975; Neudecker 1979; Frydl 1979; Sammarco 1980). Brock (1979) emphasized the importance of physical refuges in protecting coral recruits from inadvertent removal. However, Randall (1974) noted that few grazers eat living coral, and Birkeland (1977) found that grazers did not remove corals from settlement plates. Thus, the detrimental effect of grazers on reefs remains uncertain and their net effect on reef growth cannot be assessed at present.

The questions of community structure and interactions outlined above offer the most fertile areas for the generation of hypotheses. Testing these hypotheses requires manipulative experimentation that should meet the following criteria where possible: (a) the duration must be great enough to allow assessment of the persistence of any effect observed; (b) artificial substrates introduce significant biases in colonization and succession (Tsuda & Kami 1973; Wanders 1977; Day 1977) and should be avoided; (c) confounding factors due to caging effects must be kept to an absolute minimum through careful design and frequent cleaning, and be assessed with controls (in a tropical, subtidal environment a cage can become heavily fouled in two weeks, thereby decreasing light and water movement and increasing sedimentation) and (d) experiments should be designed in accordance with a valid statistical model including sufficient replication and controls.

SIGNIFICANCE OF GRAZING IN THE ADAPTATION AND EVOLUTION OF CORAL REEF ORGANISMS

The effect of grazing on community structure is, in part, dependent on the degree of selectivity exhibited by the grazing organisms. This has been demonstrated in the temperate intertidal by Lubchenco (1978); on tropical granite and sandstone reefs by Montgomery (1980a,b) and in a coral reef lagoon by Day (1977). It begs the question: how specialized are grazers in their feeding adaptations? Ogden & Lobel (1978) reviewed this topic for the two major groups of grazers: fish and sea urchins. Lines of evidence include the morphology of the digestive tract, jaws and dentition, analysis of gut contents, field observations of feeding behaviour and experimental determinations of food preferences. Considerable data is available for echinoids (see Lawrence 1975). Most tropical sea urchins appear to be generalist feeders, consuming algae, seagrasses and even invertebrates in rough proportion to availability. However, certain species of macro-algae are favoured and others avoided (Sammarco 1972; Ogden 1976). A higher degree of selectivity is exhibited by some temperate sea urchins (Vadas 1977; Vance 1979).

Unlike the echinoids, grazing fish embrace a range of morphological adaptations to herbivory. These are a comparatively recent development in evolutionary time involving adaptations to removal and digestion of vegetable matter such as fused dentition and elongated guts (Hobson 1974). Nutritional analysis and biochemical considerations imply that certain components of the EAC should be better food than others (Paine & Vadas 1969b; Lobel 1980; Montgomery & Gerking 1980). There is some evidence for the preferential selection of food species, which generally supports predictions based on nutritional value and digestability (Randall 1961b; Tsuda & Bryan 1975; Montgomery 1977, 1980b; Lassuy 1980; Montgomery & Gerking 1980). As with the echinoids, the case is stronger for avoidance rather than the selection of certain species. The criteria for distinction are broad morphological and biochemical groupings rather than species specific characteristics.

Comparisons between gut contents and the relative abundance of food types *in situ* have rarely been made. They are complicated by the highly variable nature of EAC species composition through space and time (Biggs & Eminson 1977; Borowitzka *et al.* 1978). The experimental evidence is of limited value because experiments usually involve the presentation of discrete pieces of macro-algae in an enclosed environment. In natural feeding situations, a highly mixed assemblage of food species and sizes is available to the fish (see for example species lists in Price *et al.* 1976). On exposed reef faces, where grazing is most intense, macro-algae are virtually absent (Bakus 1969; John & Pople 1973). Micro-filamentous forms dominate (Randall 1961a; van den Hoek 1969) and species are intermixed on a scale which may be an order of magnitude below that of a fish's jaws (Montgomery *et al.*, 1982). Selection or avoidance of individual food species would be precluded in such habitats.

Herbivorous fish and echinoids do not include examples of highly specialized feeding adaptations whereas certain gastropods (e.g. Sacoglossids) and carnivorous fish (e.g. Chaetodontids) often have such adaptations. Grazing fish and echinoids are adapted on a coarser scale, responding to different EAC and habitat types (Fig. 1). Although coral reef grazers prefer or dislike certain food species, they may rarely get the opportunity to express these tendencies. More work is required, but on present evidence it seems unlikely that these grazers have evolved highly selective feeding strategies. From the standpoint of their effect on community structure, they should be viewed as non-selective, opportunistic feeders which do, however, have the potential to modify the productivity and relative abundances of components of the EAC. As Montgomery (1980a) has pointed out, the realization of this potential does not necessarily require specialized feeding behaviours.

Grazers also exhibit adaptations in their use of space in terms of vagility and distribution. Territoriality and schooling of fish may be explained as foraging strategies (Barlow 1974a,b; Thresher 1976; Robertson *et al.* 1976, 1979). However, they can often be explained with equal parsimony as predator avoidance or reproductive strategies (Hobson 1969; Buckman & Ogden 1973; Ehrlich & Ehrlich 1973; Barlow 1975; Alvezion 1976; Ebersole 1977). In the case of tropical echinoids and at least one species of grazing gastropod (Tsuda & Randall 1971), it is best to avoid the term strategy because they probably feed when and where they can within the physical and biotic confines of their habitat (Kier & Grant 1965; Ogden *et al.* 1973, 1976; Lawrence & Kafri 1979; Vance & Schmitt 1979). Although there is evidence that certain behaviours confer a nutritional advantage, these will be overridden by other behaviours having a greater survival value, particularly when food is abundant. There is as yet little reason to believe that the movements and distribution of grazers on coral reefs are controlled primarily by behavioural adaptations aimed at optimizing nutrition.

The effect of grazing on the evolution of benthic invertebrates in the tropics has been discussed by Bakus (1964). The predominance of forms exhibiting cryptic, chemical or morphological defence adaptations can be taken as evidence for strong selective pressure exerted by predators. Some predators have evolved counter-defence mechanisms, probably as a result of a long history of close interaction. Ogden & Lobel (1978) suggested that benthic algae have also evolved defense mechanisms in response to grazing. However, such defense mechanisms may simply be the by-products of biochemical and morphological adaptations to the physical environment. Certainly, the postulated enhancement effect of grazing on primary productivity (discussed above) does not necessarily imply a highly evolved interaction. The same effect can be obtained from a lawn with a lawn mower. Very little evidence exists for a 'rich fabric of co-evolution' between grazers and algae. The unpredictable and multi-scaled patchy distribution of reef algae through space and time, their low degree of endemism, and the relative youth of the herbivorous branch of Teleost evolution suggests that co-evolution in algae-herbivore interactions on reefs in unlikely to be highly developed. Finally, hypotheses regarding evolutionary adaptations are essentially untestable, and thus will always be open to speculation.

Most experimental investigations of the fish-benthos interaction involve the territorial damselfishes. Hence models describing the interaction are based primarily on this easily studied group. Until their relative importance in comparison with other grazers is assessed, caution should be exercised when extrapolating from such studies to reef grazing interactions in general.

CONCLUSION

Grazing can be a major factor affecting coral reef ecosystems at all levels of organization. Present understanding of the interactions is poor. A holistic understanding will only be attained if future research concentrates on the quantification of trophic exchange between functional groups at the ecosystem level. Experimental testing of hypotheses concerning both first and second order interactions at the community level is essential if predictive models are to be developed. Studies concerning diversity and evolution-adaptation in response to grazing are based primarily at the species level; they often generate untestable hypotheses and are unlikely to be of predictive value given the complexity of coral reefs. Considerations of scale, physical factors and the basic differences between plants and animals must not be lost sight of by researchers investigating grazing effects.

The interaction between primary producers and their consumers is a universal 'key' interaction which determines the structure and dynamics of the system that depends upon it. As such, it transcends the often incomprehensible differences between ecosystems; thereby allowing derivation of general principles from comparisons of different ecosystems.

ACKNOWLEDGEMENTS

The author is grateful to Tony Ayling, Dave Barnes, Annamarie Hatcher, Ross Robertson and Paul Sammarco for helpful discussions, and constructive criticism of this manuscript.

LITERATURE CITED

Abbot, D. P., J. C. Ogden and I. A. Abbot, 1974. Studies on the activity pattern, behaviour and food of the echinoid *Echinometra lucunter* (Linnaeus) on beach rock and algal reefs at St. Croix, U.S. Virgin Is. *West Indies Laboratory Special Publication* 4, 1-111.

Adey, W. H., P. J. Adey, R. B. Burke and L. Kaufman, 1977. The Holoscene reef systems of eastern Martinique, French West Indies. *Atoll Research Bulletin* 218, 1-40.

Alvezion, W. S., 1976. Mixed schooling and its possible significance in a tropical Western Atlantic parrot fish and surgeon fish. *Copeia*, 796-8.

Bak, R. P. M. and G. van Eys, 1975. Predation of the sea urchin *Diadema antillarum* Philippi on living coral. *Oecologia* 20, 111-5.

Bakus, G. J., 1964. The effects of fish grazing on invertebrate evolution in shallow tropical waters. *Occasional Paper, Allan Handcock Foundation* 27, 1-29.

Bakus, G. J., 1966. Some relationships of fishes to benthic organisms on coral reefs. *Nature* 210, 280-4.

Bakus, G. J., 1967. The feeding habits of fishes and primary production at Eniwetok, Marshall Islands. *Micronesica* 3, 135-49.

Bakus, G. J., 1969. Energetics and feeding in shallow marine waters. *International Review of General and Experimental Zoology* 4, 275-369.

Bakus, G. J., 1972. Effects of the feeding habits of coral reef fishes on the benthic biota. In C. Mukundan and C. S. Gopinadha Pillai (eds.), *Proceedings of the Symposium on Coral and Coral Reefs.* Cochin: The Marine Biological Association of India, pp. 445-8.

Bardach, J. E., 1958. On the movements of certain Bermuda reef fishes. *Ecology* 39, 139-45.

Bardach, J. E., 1959. The summer standing crop of fish on a shallow Bermuda reef. *Limnology and Oceanography* 4, 77-85.

Barlow, G. W., 1974a. Extraspecific imposition of social grouping among surgeon fishes (Pisces: Acanthuridae). *Journal of Zoology* 174, 333-40.

Barlow, G. W., 1974b. Contrasts in social behaviour between Central American cichlid fishes and coral reef surgeon fishes. *American Zoologist* 14, 9-34.

Barlow, G. W., 1975. On the sociobiology of four Puerto Rican parrot fishes (Scaridae). *Marine Biology* 33, 281-94.

Belk, M. S., 1975. Habitat partitioning in tropical reef fishes *Pomacentrus lividus* and *P. albofasciatus. Copeia* 603-17.

Benayahu, Y. and Y. Loya, 1977. Seasonal occurrence of benthic-algae communities and grazing regulation by sea urchins at the coral reefs of Eilat, Red Sea. In D. L. Taylor (ed.), *Proceedings: Third International Coral Reef Symposium.* Miami: University of Miami, vol, 1, pp. 383-9.

Biggs, P. and D. E. Eminson, 1977. Studies on algal recolonization of coral predated by the crown of thorns starfish *Acanthaster planci* in the Sudanese Red Sea. *Biological Conservation* 11, 41-7.

Birkeland, C., 1977. The importance of rate of biomass accumulation in early successional stages of benthic communities to the survival of coral recruits. In D. L. Taylor (ed.), *Proceedings: Third International Coral Reef Symposium.* Miami: University of Miami, Vol. 1, pp. 15-21.

Bjorndal, K. A., 1980. Nutrition and grazing behaviour of the green turtle *Chelonia mydas. Marine Biology* 56, 147-54.

Borowitzka, M. A., A. W. D. Larkum and L. J. Borowitzka, 1978. A preliminary study of algal turf communities of a shallow coral reel lagoon using an artificial substratum. *Aquatic Botany* 5, 365-81.

Brawley, S. H. and W. H. Adey, 1977. Territorial behaviour of threespot damselfish *(Eupomacentrus planifrons)* increases reef algal biomass and productivity. *Environmental Biology of Fishes* 2, 45-51.

Breen, P. A. and K. H. Mann, 1976. Destructive grazing of kelp by sea urchins in Eastern Canada. *Journal of the Fisheries Research Board of Canada* 33, 1278-83.

Brock, R. E., 1979. An experimental study on the effects of grazing by parrotfishes and role of refuges in benthic community structure. *Marine Biology* 51, 381-8.

Bryan, F. G., 1975. Food habits, functional digestive morphology and assimilation efficiency of the rabbit fish, *Siganus spinus* (Pisces: Siganidae) on Guam. *Pacific Science* 29, 269-77.

Buckman, N. S. and J. S. Ogden, 1973. Territorial behaviour of the striped parrotfish *Scarus croicensis* Bloch (Scaridae). *Ecology* 54, 1377-81.

Camp, K., S. P., Copp and J. F. van Breedveld, 1973. Overgrazing of seagrasses by the regular urchin *Cytechinus varregatus. Bioscience* 23, 37-8.

Connell, J. H. and R. O. Slatyer, 1977. Mechanisms of succession in natural communities and their role in community stability and organisation. *American Naturalist* 111, 1119-44.

Connor, J. L. and W. H. Adey, 1977. The benthic algal composition, standing crop, and productivity of a Caribbean algal ridge. *Atoll Research Bulletin* 211, 1-40.

Crisp, D. J., 1964. Grazing in Terrestrial and Marine Environments. Oxford: Blackwell, 362 pp.

Dahl, A. L., 1972. Ecology and community structure of some tropical reef algae in Samoa. In K. Nisizawa (ed.) *Proceedings, Seventh International Seaweed Symposium*. New York: John Wiley & Sons, vol. 1, pp. 36-9.

Dahl, A. L., 1973. Benthic algal ecology in a deep reef and sand habitat off Puerto Rico. *Botanica Marina* 16, 171-5.

Dart, J. A. G., 1972. Echinoids, algal lawn and coral colonization. *Nature* 239, 50-1.

Day, R., 1977. Two contrasting effects of predation on species richness in coral reef habitats. *Marine Biology* 44, 1-6.

Doty, M., 1971. The productivity of benthic frondose algae at Waikiki Beach, 1967-8. *Hawaii Botanical Science Paper* 22, 1-119.

Dyer, M. I. and U. G. Bakhari, 1976. Plant-animal interactions: Studies on the effects of grasshopper grazing on blue grama grass. *Ecology* 57, 762-72.

Earle, S. A., 1972. The influence of herbivores on the marine plants of Great Lameshur Bay. In B. B. Collette (ed.), *Results of the Tektite Program: Ecology of Coral Reef Fishes*. Natural History Museum Los Angeles County, Science Bulletin 14, pp. 17-44.

Ebersole, J. P., 1977. The adaptive significance of interspecific territoriality in the reef fish *Eupomacentrus leucostictus*. *Ecology* 58, 914-20.

Ehrlich, P. R. and A. H. Ehrlich, 1973. Coevolution: heterotypic schooling in Caribbean reef fishes. *American Naturalist* 107, 157-80.

Endean, R., W. Stephenson and R. Kenny, 1956. The ecology and distribution of intertidal organisms on certain islands off the Queensland coast. *Australian Journal of Marine and Freshwater Research* 7, 317-42.

Foster, M. S., 1972. The algal turf community in the nest of the ocean goldfish; *Hypsypops rubicunda*. In K. Nisizawa (ed.), *Proceedings, Seventh International Seaweed Symposium*. New York: John Wiley & Sons, vol. 1, pp. 55-60.

Fowler, M. C. and T. O. Robson, 1978. The effects of the food preferences and stocking rates of grass carp *(Ctenophoryagodon idella,* Val.) on mixed plant communities. *Aquatic Botany* 5, 261-76.

Frydl, P., 1979. The effect of parrotfish (Scaridae) on coral in Barbados, West Indies. *International Revue der gesamten Hydrobiologie* 64, 737-48.

Frydl, P. and C. W. Stearn, 1978. Rate of bioerosion by Parrotfish in Barbados reef environments. *Journal of Sedimentary Petrology* 48, 1149-57.

Glynn, P. W., G. M. Wellington and C. Birkeland, 1979. Coral reef growth in the Galapagos: Limitation by sea urchins. *Science* 203, 47-9.

Goldman, B. and F. H. Talbot, 1975. Aspects of the ecology of coral reef fishes. In I. A. Jones and R. Endean (Editors), *Biology and Geology of Coral Reefs*. Academic Press, New York, vol. 3, Biology 2, pp. 125-54.

Greenway, M., 1974. The effects of cropping on the growth of *Thalassia testudinum* (Konig) in Jamaica. *Aquaculture* 4, 199-206.

Greenway, M., 1976. The grazing of *Thalassia testudinum* in Kingston harbour, Jamaica. *Aquatic Botany* 2, 117-26.

Gygi, R. A., 1975. *Sparisoma viride* (Bonnaterre), the Stoplight Parrotfish, a major sediment producer on coral reefs of Bermuda? *Eclogae Geologicae Helvetiae* 68, 327-59.

Harper, J. C., 1969. The role of predation in vegetational diversity. *Brookhaven Symposia in Biology* 22, 48-62.

Hatcher, G. B., 1981. The interaction between algae and grazers on a coral reef. Ph.D. dissertation, University of Sydney, 161pp.

Hayne, D. W. and R. C. Ball, 1956. Benthic productivity as influenced by fish predation. *Limnology and Oceanography* 1, 162-75.

Hiatt, R. W. and D. W. Strasburg, 1960. Ecological relationships of the fish fauna on coral reefs of the Marshall Islands. *Ecological Monographs* 30, 65-127.

Hobson, E. S., 1969. Possible advantages to the blenny *Runula azalea* in aggregating with the wrasse *Thallosoma lucasanum* in the tropical eastern Pacific. *Copeia* 191-3.

Hobson, E. S., 1973. Diel feeding migrations of tropical reef fishes. *Helgoländer wissenshaftliche Meeresuntersuchungen* 24, 361-70.

Hobson, E. S., 1974. Feeding relationships of teleostean fishes on coral reefs in Kona, Hawaii. *Fishery Bulletin* 72, 915-1031.

Hunt, M., 1969. A preliminary investigation of the habits and habitat of the rock-boring urchin *Echinometra lucunter* near Devonshire Bay, Bermuda. In R. N. Ginsburg and P. Garrett (eds.), *Seminar on Organism — Sediment Interrelationships. Bermuda Biological Station Special Publication.* 2, pp. 31-40.

John, D. M. and W. Pople, 1973. The fish grazing of rocky shore algae in the Gulf of Guinea. *Journal of Experimental Marine Biology and Ecology* 11, 81-90.

Jones, R. S., 1968. Ecological relationships in Hawaiian and Johnston Island Acanthuridae (surgeon fishes). *Micronesica* 4, 309-61.

Jones, R. S. and J. A. Chase, 1975. Community structure and distribution of fishes in an enclosed high island lagoon in Guam. *Micronesica* 11, 127-48.

Kaufman, L., 1977. The three spot damselfish: effects on benthic biota of Caribbean coral reefs. In D. L. Taylor (Editor), *Proceedings: Third International Coral Reef Symposium.* Miami: University of Miami, vol. 1, pp. 559-64.

Kay, E. A., 1978. Molluscan distribution patterns at Canton Atoll. *Atoll Research Bulletin* 221, 161-9.

Kier, P. M. and R. E. Grant, 1965. Echinoid distribution and habits, Key Largo Coral Reef Preserve, Florida. *Smithsonian Miscellaneous Collection* 149, 1-68.

Kitching, J. A. and F. J. Ebling, 1961. The ecology of Lough Ine. XI. The control of algae by *Paracentrotus lividae* (Echinoidea). *Journal of Animal Ecology* 30, 373-83.

Kohn, A. J. and P. Helfrich, 1957. Primary organic productivity of a Hawaiian coral reef. *Limnology and Oceanography* 2, 241-51.

Lawrence, J. M., 1975. On the relationship between marine plants and sea urchins. *Oceanography and Marine Biology an Annual Review* 13, 213-86.

Lawrence, J. M. and J. Kafri, 1979. Numbers, biomass and calorific content of the echinoderm fauna of the rocky shores of Barbados. *Marine Biology* 52, 87-92.

Leighton, D. C., L. G. Jones and W. J. North, 1966. Ecological relationships between giant kelp and sea urchins in southern California. In E. Gordon-Young and T. L. McLachlan (eds.), *Proceedings, Fifth International Seaweed Symposium.* Oxford: Permagon Press, pp. 141-53.

Littler, M. M. and M. S. Doty, 1975. Ecological components structuring the seaward edges of tropical Pacific reefs: The distribution, communities and productivity of *Porolithon. Journal of Ecology* 63, 117-29.

Lubchenco, J. M., 1978. Plant species diversity in a marine intertidal community: Importance of herbivore food preference and algal competitive abilities. *American Naturalist* 112, 23-9.

McNaughton, S. J., 1979. Grazing as an optimization process: grass-ungulate relationships in the Seregeti. *American Naturalist* 113, 691-703.

McPherson, B. F., 1965. Contributions to the biology of the sea urchin *Tripheustes ventricosus. Bulletin of Marine Science* 15, 228-44.

McPherson, B. F., 1968. Contributions to the biology of the urchin *Eucidaris tribuloides* (Lammarck). *Bulletin of Marine Science* 18, 400-33.

Mann, K. H., 1977. Destruction of kelp beds by sea urchins. A cyclical phenomenon, or irreversible degradation? *Helgoländer wissenshaftliche Meeresuntersuchungen* 30, 455-67.

Marsh, J. A., Jr., 1976. Energetic role of algae in reef ecosystems. *Micronesica* 12, 13-21.

Mathieson, A. C., R. A. Fralick, R. Burns and W. Flahive, 1971. Comparative studies of subtidal vegetation in the Virgin Islands and the New England coastlines. In P. Miller (ed.), *Scientists in the Sea.* Washington: United States Government Printing Office.

Mattson, W. J. and N. D. Addy, 1975. Phytophagous insects as regulators of forest primary production. *Science* 190, 515–22.

Menge, B. A., 1976. Organization of the New England rocky intertidal community: role of predation, competition, and environmental heterogeneity. *Ecological Monographs* 46, 355–69.

Menzel, D. W., 1959. Utilization of algae for growth by the angelfish, *Holacanthus bermudensis*. *Journal du Conseil. Conseil Permanent International pour l'Exploration de la Mer*. 24, 308–13.

Montgomery, W. L., 1977. Diet and gut morphology in fishes with special reference to the monkey face prickle back, *Cebidichthys violaeceous* (Stichaeidae: Blennioidei). *Copeia* 178–82.

Montgomery, W. L., 1980a. The impact of non-selective grazing by the giant blue damselfish, *Microspathodon dorsalis*, on algal communities in the Gulf of California, Mexico. *Bulletin of Marine Science* 30, 290–303.

Montgomery, W. L., 1980b. Comparative feeding ecology of two herbivorous damselfishes (Pomacentridae: Teleostei) from the Gulf of California, Mexico. *Journal of Experimental Biology and Ecology* 47, 9–24.

Montgomery, W. L., and S. D. Gerking, 1980. Marine macroalgae as foods for fishes: an evaluation of potential food quality. *Environmental Biology of Fishes* 5, 143–53.

Montgomery, W. L., T. Gerrodette and C. D. Marshall, 1982. Effects of grazing by the yellow tail surgeon fish, *Prionurus punctatus* on algal communities in the Gulf of California, Mexico. *Bulletin of Marine Science* 30, 901–8.

Nagelkerken, W. P., 1975. On the occurrence of fishes in relation to corals in Curacao. *Studies of the Fauna of Curacao and other Caribbean Islands* 147, 118–40.

Neudecker, S., 1977. Transplant experiments to test the effect of fish grazing on coral distribution. In D. L. Taylor (ed), *Proceedings: Third International Coral Reef Symposium*. Miami: University of Miami, vol 1, pp. 317–23.

Neudecker, S., 1979. Effects of grazing and browsing fishes on the zonation of corals in Guam. *Ecology* 60, 666–72.

Newman, W. A., 1960. The paucity of intertidal barnacles in the tropical western Pacific. *Veliger* 2, 89–94.

Odum, E. P., 1971. Fundamentals of Ecology. Philadelphia: W. B. Saunders. 574pp.

Ogden, J. C., 1976. Some aspects of herbivore-plant relationships on Caribbean reefs and seagrass beds. *Aquatic Botany* 2, 103–16.

Ogden, J. C., 1977. Carbonate-sediment production by parrot fish and sea urchins on Caribbean reefs. *American Association of Petroleum Geologists. Studies in Geology* 4, 281–8.

Ogden, J. C., D. P. Abbott and I. A. Abbott, 1973a. Studies on the activity and food of the echinoid *Diadema antillarum* Philippi on a West Indian patch reef. *West Indies Laboratory Special Publication* 2, 1–96.

Ogden, J. C., D. P. Abbott, I. A. Abbott and C. Kitting (eds.), 1976. Population; movement, food and winter spawing in the sea urchin *Tripneustes ventricosus* (Lamarck) at St. Croix, United States, Virgin Islands. *West Indies Laboratory Special Publication* No. 9.

Ogden, J. C., R. Brown and N. Salesky, 1973b. Grazing by the echinoid *Diadema antillarum* Philippi: Formation of halos around West Indian patch reefs. *Science* 1982, 715–17.

Ogden, J. C. and N. S. Buckman, 1973. Movements, foraging groups and diurnal migrations of the striped parrot fish *Scarus croicensis* Block (Scaridae). *Ecology* 54. 589–96.

Ogden, J. C. and P. S. Lobel, 1978. The role of herbivorous fishes and urchins in coral reef communities. *Environmental Biology of Fishes* 3, 49–63.

Ogden, J. C., P. W. Sammarco and I. A. Abbott, 1973c. *Diadema antillarum* and plants: A study of a shallow water marine food chain. Abstract, Association of the Island Marine Laboratory of the Caribbean, Mayaguez, Puerto Rico, 1973.

Ogden, J. C. and J. C. Zieman, 1977. Ecological aspects of coral reef-seagrass bed contacts in the Caribbean. In D. L. Taylor (ed.), *Proceedings: Third International Coral Reef Symposium.* Miami: University of Miami, vol. 1, pp. 377–82.

Paine, R. T., 1966. Food web complexity and species diversity. *American Naturalist* 100, 65–75.

Paine, R. T., 1969. The *Pisaster-Tegular* interaction: prey patches, predator food preference, and, intertidal community structure. *Ecology* 50, 950–61.

Paine, R. T. and R. L. Vadas, 1969a. The effects of grazing by sea urchins, *Strongylocentrotus* spp., on benthic algal populations. *Limnology and Oceanography* 14, 710–19.

Paine, R. T. and R. L. Vadas, 1969b. Calorific values of benthic marine algae and their postulated relation to invertebrate food preferences. *Marine Biology* 4, 79–86.

Pearson, R. G., 1974. Recolonization by hermatypic corals of reefs damaged by *Acanthaster*. In, *Proceedings of the Second International Symposium on Coral Reefs.* Brisbane: Great Barrier Reef Committee, vol. 2, pp. 207–15.

Porter, K. G., 1976. Enhancement of algal growth and productivity by grazing zooplankton. *Science* 192. 1332–4.

Potts, D. C., 1977. Suppression of coral populations by filamentous algae within damselfish territories. *Journal of Experimental Marine Biology and Ecology* 28, 207–16.

Price, I. R., A. W. D. Larkum and A. Bailey, 1976. Check list of marine benthic plants collected in the Lizard Island Area. Appendix to: N. E. Tolbert and C. B. Osmond, The Great Barrier Reef Photorespiration Expedition: Introduction. *Australian Journal of Plant Physiology* 3, 3 –8.

Randall, J. E., 1961a. Overgrazing of algae by herbivorous marine fishes. *Ecology* 42, 812–14.

Randall, J. E., 1961b. A contribution to the biology of the convict surgeon fish of the Hawaiian Islands, *Acanthurus triostegus sandvicensis. Pacific Science* 15, 215–72.

Randall, J. E., 1963. An analysis on the fish populations of artificial and natural reefs in the Virgin Islands. Caribbean Journal of Science 3, 31–47.

Randall, J. E., 1965. Grazing effect on sea grasses by herbivorous reef fishes in the West Indies. *Ecology* 46, 255–60.

Randall, J. E. 1967. Food habits of reef fishes of the West Indies. *Studies in Tropical Oceanography* 5, 665–847.

Randall, J. E., 1974. The effect of fishes on coral reefs. In, *Proceedings of the Second International Symposium on Coral Reefs.* Great Barrier Reef Committee, Brisbane, vol, 1, pp. 159–66.

Raymont, J. E. G., 1963. Plankton and the Productivity in the Oceans. New York: Pergamon, Chapter 10, 660pp.

Robertson, D. R. and H. P. A. Sweatman, E. A. Fletcher and M. G. Cleland, 1976. Schooling as a mechanism for circumventing the territoriality of competitors. *Ecology* 57, 1208–20.

Robertson, D. R., N. V. C. Polunin and K. Leighton, 1979. The behavioural ecology of three Indian Ocean surgeon fishes *(Acanthurus lineatus, A. leucosternon,* and *Zebrasoma scopas):* their feeding strategies, and social mating systems. *Environmental Biology of Fishes* 4, 125–70.

Rogers, C. S. and N. H. Salesky, 1981. Productivity of *Acropora palmata* (Lamarck), macroscopic algae, and algal turf from Tague Bay Reef, St. Croix, U.S. Virgin Islands. *Journal of Experimental Marine Biology and Ecology* 49, 179–87.

Russ, G. R., 1980. Effects of predation by fishes, competition and structural complexity of the substratum on the establishment of a marine epifaunal community. *Journal of Experimental Marine Biology and Ecology* 42, 55–69.

Sammarco, P. W., 1972. Some aspects of the ecology of *Diadema antillarum* (Philippi): Food preference and effect of grazing. In J. C. Ogden (ed.), *Special Problems Report.* West Indies Laboratory, Fairleigh Dickinson University, Christiansted, St. Croix, United States, Virgin Islands, 47pp.

Sammarco, P. W. 1975. Grazing by *Diadema antillarum* Philippi (Echinodermata: Echinoidea): Density-dependent effects on coral and algal community structure. In, *Proceedings of the Association of Island Marine Laboratories of the Caribbean.* 11th. Meeting, Christiansted, St. Croix, United States, Virgin Islands, Abstract, pp. 19.

Sammarco, P. W. 1977a. Regulation of competition and disturbance in a reef community by *Diadema antillarum.* In, *Fourth Simposium Internactional de Ecologia Tropical Panama.* Abstract. Sammarco, P. W., 1977b. The effects of grazing by *Diadema antillarum* Philippi on a shallow-water coral reef community. Ph.D. dissertation, Department of Ecology and Evolution, State University of New York at Stony Brook, 371pp.

Sammarco, P. W., 1980. *Diadema* and its relationship to coral spat mortality: grazing, competition, and biological disturbance. *Journal of Experimental Marine Biology and Ecology* 45, 245–72.

Sammarco, P. W., J. S. Levinton and J. C. Ogden, 1974. Grazing and control of coral reef community structure by *Diadema antillarum* Philippi (Echinodermata: Echinoidea): A preliminary study. *Journal of Marine Research* 32, 47–53.

Smith, C. L. and J. C. Tyler, 1973. Direct observations of resource sharing in coral reef fish. *Helgol änder wissenschaftliche Meeresuntersuchungen* 24, 264–75.

Smith, R. L. and A. C. Paulson, 1974. Food transit times and gut pH in two Pacific parrot fishes. *Copeia* 796–99.

Stephenson, W. and R. B. Searle, 1960. Experimental studies on the ecology of intertidal environments at Heron Island. 1. Exclusion of fish from beach rock. *Australian Journal of Marine and Freshwater Research* 11, 241–67.

Talbot, F. H., 1965. A description of the coral structure of Tutia Reef (Tanganyika Territory, East Africa), and its fish fauna. *Proceedings of the Zoological Society of London* 145, 431–70.

Taylor, J. D., 1968. Coral reef and associated invertebrate communities (mainly molluscan) around Mahe, Seychelles. *Philosophical Transactions of the Royal Society of London* 13, 129–206.

Taylor, J. D., 1971. Reef associated molluscan assemblages in the western Indian Ocean. *Symposia of the Zoological Society of London* 28, 501–34.

Thresher, R. E., 1976. Field analysis of the territoriality of the threespot damselfish, *Eupomacentrus planifrons* (Pomacentridae). *Copeia* 266–76.

Trudgill, S. T., 1976. The marine erosion of limestone on Aldabra Atoll, Indian Ocean. *Zeitschrift für Geomorphologie,* Supplementband, 26, 164–200.

Tsuda, R. T. and J. E. Randall, 1971. Food habits of the gastropods *Turbo argyrostoma* and *T. setosus* reported as toxic from the tropical Pacific. *Micronesica* 7, 153–62.

Tsuda, R. T. and H. T. Bryan, 1973. Food preferences of the juvenile *Siganus rostratus* and *S. spinus* in Guam. *Copeia* 604–6.

Tsuda, R. T. and H. T. Kami, 1973. Algal succession on artificial reefs in a marine lagoon environment on Guam. *Journal of Phycology* 9, 260–4.

Vadas, R. L., 1977. Preferential feeding: An optimization strategy in sea urchins. *Ecological Monographs* 47, 337–71.

Van den Hoek, C., 1969. Algal vegetation — types along the open coast of Curacao, Netherlands Antilles. *Proceedings, Konference Netherlands Academie Wetenshappelijke, Serial C* 72, 537–77.

Van den Hoek, C., A. M. Breeman, R. P. M. Bak and G. van Buurt, 1978. The distribution of algae, corals and gorgonians in relation to depth, light attentuation, water movement and grazing pressure in the fringing coral reef of Curacao, Netherland Antilles. *Aquatic Botany* 5, 1–46.

Vance, R. R., 1979. Effects of grazing by the sea urchin, *Centrostephanus coronatas* on prey community composition. *Ecology* 60, 537–46.

Vance, R. R. and R. J. Schmitt, 1979. The effect of the predator-avoidance behaviour of the sea urchin, *Centrostephanus coronatas,* on the breadth of its diet. *Oecologia* 44, 21–5.

Vine, P. J., 1974. Effects of algal grazing and aggressive behaviour of the fishes *Pomacentrus lividus* and *Acanthurus sohal* on coral reef ecology. *Marine Biology* 24, 131–6.

von Westernhagen, H., 1974. Food preference in cultured rabbitfishes (Siganidue). *Aquaculture* 3, 109–17.

Wanders, J. B. W., 1976a. The role of benthic algae in the shallow reef of Curacao (Netherlands Antilles). I: Primary productivity in the coral reef. *Aquatic Botany* 2, 235–70.

Wanders, J. B. W., 1976b. The role of benthic algae in the shallow reef of Curacao (Netherlands Antilles). II: Primary productivity of the *Sargassum* beds on the north-east coast submarine plateau. *Aquatic Botany* 2, 327–35.

Wanders, J. B. W., 1977. The role of benthic algae in the shallow reef of Curacao (Netherlands Antilles). III: The significance of grazing. *Aquatic Botany* 3, 357–90.

11 Morphological Classifications of Shelf Reefs: A Critique with Special Reference to the Great Barrier Reef

David Hopley

Department of Geography, James Cook University of North Queensland, Qld. 4811.

INTRODUCTION

Any landform classification should acknowledge the processes involved in producing morphological variation if it is to be a useful tool in recognizing regional variation which, in turn, may lead to recognition of spatial variations in parameters controlling such processes. The most commonly quoted classification of reef forms is still that of Darwin (1842). Within each class of reef: fringing, barrier or atoll, which are presumed to have a sequential relationship, there is much morphological diversity. This is particularly true of the reefs of continental shelves which generally fall within Darwin's barrier reef class, and nowhere is the diversity as great as in the Great Barrier Reef province. Two significant attempts have been made to produce a genetic classification of these reefs. A simple classification by Fairbridge (1950, 1967) was elaborated and extended by Maxwell (1968). Since the late 1960s the amount of research carried out on coral reefs has escalated and the results suggest that the three most important parameters in the determination of Holocene reef morphology are:

1. the morphology and depth of the pre-Holocene reefal surface,
2. the nature of the Holocene sea level curve, which for hydro isostatic and tectonic reasons may vary geographically and
3. the net rate of reef accretion.

Data on these factors are discussed by Davies in this volume. Such information was not available to either Fairbridge or Maxwell, and its importance to reef classification would appear to necessitate a re-evaluation of their approaches which are so widely quoted.

INADEQUACIES OF FAIRBRIDGE'S AND MAXWELL'S CLASSIFICATIONS

Both classifications are based on organic and sedimentary growth of reefs in response to prevailing wind and wave conditions during a single period of relatively stable eustatic sea level. Both authors, either directly or by implication, presume that the reefs have developed exclusively during the Holocene; it is presumed that during each glacial low sea level stage, pre-existing reefs were removed by erosion. Fairbridge (1950, p. 359) presumes a 'relative degree of stability' throughout the period of reef development and:

> the developments in surficial reef morphology during the last few thousands years

cannot be related to major changes in level of land or sea. The evolution of reef morphology can thus be related to the molding action of wind, waves and currents.

His classification (1950, fig. 7; 1967, fig. 17.1) thus envisages reefs developing downwind detrital horns, which, on larger reefs at least, will form open lagoons. With time, lagoons become enclosed and finally infilled by sedimentation and organic growth.

Similarly, the more elaborate classification of Maxwell (1968, fig. 65) is based around the premise that:

> organic reefs, once initiated, will expand in directions controlled by the hydrologic-bathymetric-biological balance (Maxwell 1968, p. 99).

Eustatic changes of sea level were seen as modifying these basic controls but Maxwell considered the whole reef column to be Holocene in age (e.g. Maxwell 1973, p. 267). Such a concept is no longer tenable in the light of drilling results and radiometric dating (Hopley *et al.* 1978; Thom *et al.* 1978; Davies & Marshall 1979) which demonstrate a thin veneer (5 to 25 m) of Holocene material over older reefal foundations. Seismic surveys (Davies *et al.* 1977 a,b; Harvey 1977, 1978; Harvey *et al.* 1978, 1979; Backshall *et al.* 1979) suggest that Holocene veneers of similar thickness have wide distribution in the Great Barrier Reef province. In detail, Maxwell's genetic classification differentiates between reefs where there is a near equal balance of the controlling parameters from all directions (resulting in compact platform reefs) and reefs where there is a strong control from a single direction (and where development is much along the lines suggested by Fairbridge). Maxwell further suggests that where strong tidal currents carry turbulent aerated oceanic water into the back-reef zone, prongs normal to the reef-front may form, and 'plug' reefs evolve in passes between the larger reefs. The ultimate fate of the shelf reef is seen as being 'resorbtion' — an enigmatic term not fully explained by Maxwell but suggesting retarded reef growth or even:

> the extinction of reef organisms and the progressive resorbtion of the reef mass (Maxwell 1968, p. 107).

This has been interpreted by Bloom (1974) as 'geochemical resorption' *(sic)* and apparently refers to the erosion by solution of the older reef mass. Such a process, reminiscent of the ideas of Semper (1863, 1873) and Murray (1880, 1889) is also suggested as forming the shallow lagoons of platform reefs by 'central decay' (Maxwell 1968, fig. 65). The supersaturation of tropical sea surface waters in calcium carbonate make such processes highly dubious (Revelle & Emery 1957; Trudgill 1976).

It is not disputed that, once reefs have reached modern sea level, their growth is strongly influenced by winds, waves and tides. However, to give these ambient environmental factors prime place in a genetic classification too greatly subordinates substrate control, sea level change and reef growth rate. More specifically, Maxwell's classification needs re-examination because drilling and seismic surveys on the Great Barrier Reef (see Harvey, this volume) and elsewhere show that few reefs have developed solely during the Holocene transgression. Multi-cyclic development with the addition of thin veneers of younger coralline caps during transgressive periods appears to be the most common model of reef development. Growth from embryonic colonies to mature platform, ring or mesh reefs requires more time than is available during the Holocene alone, given what is presently known about reef accretion rates and carbonate productivity (e.g. Smith & Kinsey 1976; Kinsey & Davies 1979; Davies, this volume). This again demonstrates that the present morphology of reefs derives from the morphology of earlier reefs, and that the earlier morphology was modified by exposure during glacially lowered sea level. Maxwell's (1976) claim that the growth on the Great

Barrier Reef is more rapid than elsewhere, especially than in the Caribbean, is not borne out by either stratigraphic or chemical methods (Adey 1978; Davies & Marshall 1979; Davies, this volume), which suggest average accretion rates of between 2 and 6 mm yr^{-1} (Davies & Marshall 1979). The closest that Maxwell approaches the recognition of the influence of pre-existing platforms is to suggest that 'resorbed' reefs rise from larger submerged platforms which:

> represent the planed surfaces of earlier reefs, and . . . with rising sea level, new hydrologic conditions prevailed that prevented the same degree of growth (Maxwell 1968, p. 197).

He did not recognise similar platforms must exist beneath most reefs and exert considerate control on modern day morphology.

If Maxwell's scheme represents an evolutionary sequence then resorbed reefs should be the oldest reef type in the Great Barrier Reef: small platform reefs and wall or cuspate reefs should be the youngest. This is not born out by radiocarbon dating of windward reef cap (upper 3 m) materials, where it may be presumed that the reef first reached modern sea level. Table 1 is derived from the evolutionary scheme of Maxwell with maximum age for the various reef types indicated, where available. Little or no pattern related to the Maxwell scheme can be seen in these ages. Reefs of most types can have near surface ages of between 4000 and 6000 years BP. Of particular importance is the group of ages obtained for 'resorbed' type reefs. Only one, Long Reef north of Cooktown, is older than 4500 years, the remainder are all younger than 2700 years. Far from being the ultimate stage of reef development, these figures suggest that the morphology described as 'resorbed' is a young reef form.

FRAMEWORK FOR A GENETIC CLASSIFICATION

Recent Advances in Understanding Reef Growth

Both Purdy (1974) and Davies (this volume) suggest that the morphology of modern reefs is derived from the morphology of the antecedent platforms from which they grow. The extent to which this antecedent morphology has been initially mimicked and later masked by Holocene growth is the basic framework for a classification of shelf reefs. Such a framework has been approached by Flood (1977), Jell & Flood (1978) and Orme & Flood (1977) who incorporate tectonic, isostatic and biological growth as variables in an explanation of the succession of reef types (see Flood & Orme 1977, fig. 2). The rise and fall of sea level relative to an upward growing reef is seen as producing a succession of reef types which may be reversed by a change in direction of the rise or fall of relative sea level. Over most of the Great Barrier Reef, the course of evolution during the Holocene has been one of drowning of pre-Holocene reefal platform in response to a eustatic sea level rise aided by local isostatic subsidence. Initially, the reef morphology closely mimics that of the underlying platform. Flood (1977), Jell & Flood (1978) and Orme & Flood (1977) suggest that planar surfaces produce platform reefs, antecedent platforms with shallow depressions produce lagoonal platform reefs, and broad surfaces with very deep centrally located depressions produce closed ring reefs (all terms referring to morphological types of Maxwell). If the rate of vertical reef growth equals the rate of relative sea level rise, the reef type will persist. If reef growth is greater than this rise then the sequence through which the morphology progresses is (1) closed ring, (2) lagoonal platform, (3) platform and (4) platform with sediment veneer, starting at whatever reef type was in existence when the reef first reached sea level. Where reef growth is overtaken by the relative sea level rise, then the sequence may be reversed.

Major Reef Type	Evolutionary Sequence	Example	Latitude (S)	Age	Reference or, if unpublished, C14 identification
1 Embryonic colonies		n.a.			
2 Compact reefs of radial growth	Platform	Fisher	12°15'	6310±90	Polach et al. (1978)
		Stainer	13°57'	4980±80	"
		Turtle Is.	14°44'	4910±90	"
		Three Is.	15°06'	4520±110	GaK–7667
		Wheeler	18°48'	4390±100	GaK–7841
		Lady Elliot	24°07'	5300±90	Flood et al. (1979)
	Lagoonal or Elongate Platform	Redbill	20°58'	6550±150	GaK–7228
		One Tree	23°30'	c. 5000	Davies (pers. comm.)
3 Reefs with elongation	Wall Reef	Michaelmas	16°35'	5100±130	GaK–7675
		Lark Pass	15°05'	5910±110	GaK–6683
		Bowl	18°30'	2720±100	GaK–7842
	Cuspate Reef (with or without prongs)	Carter	14°33'	5800±100	Hopley (1977)
		Opal	16°15'	4780±120	GaK–7669
		Darley	19°13'	6210±140	Hopley (1978b)
	Open Ring Reef (with or without mesh)	Cairns	15°41'	4110±130	GaK–6686
	Closed Ring Reef (with or without mesh)	Cockatoo	20°45'	4100±100	Backshall et al. (1979)
4 Final Reef Form	Resorbed Reef	Long	15°03'	4740±120	Hopley (1977)
		Thetford	16°49'	2680±120	GaK–7681
		Channel	16°56'	1930±110	GaK–7683
		Hedley	17°14'	2160±130	GaK–7687
		Keeper	18°45'	2610±90	GaK–7275
		Viper	18°52'	2660±90	Hopley (1978b)

Table 1   Maximum radiocarbon age from upper 3 m of reef tops arranged according to Maxwell's evolutionary classification

This simplistic model does much to explain the progression of morphology developed over antecedent platforms of moderate size and generally oval shape. Not only does it use the terms of Maxwell, it also parallels his sequence for the development of closed ring and closed mesh reefs. Radiocarbon dating of reefs of this type suggests that the sequence may be correct. However, a major criticism of the scheme of Flood and co-authors relates to their basic presumption that growth occurred equally over the whole antecedant platform after it was drowned, thus leading to a maintenance of form after drowning. Recent evidence, for example, Smith & Kinsey (1976) and Davies (this volume) indicates that maximum growth can be expected on the windward reef margins, and slower rates will operate elsewhere. Thus, whatever the initial form of a drowned platform, a crescentic reef would develop first and, after growing up to a stable sea level, slowly change into an open ring, then closed ring as growth on less exposed margins of the platform reaches sea level.

Additional Considerations in Building a Genetic Framework

The morphology of modern reefs is particularly affected by the size of their antecedent platform. Large modern reefs have necessarily grown over large pre-Holocene platforms, although the antithesis is not necessarily true; small reefs may have grown from larger foundations. For instance, the Arlington complex near Cairns comprises several reefs — Arlington, Michaelmas, Oyster and Upolu, which have grown from a single pre-Holocene platform. The degree to which lagoons are developed is a major diagnostic feature in reef classification. Modern lagoon development may be related to the size of the antecedent platform. The largest pre-Holocene reef platforms can develop complex relief consisting of multiple enclosed depressions, whilst smaller platforms may produce simpler saucer-like morphology consisting of a single depression. Smallest reef platforms may not have any central depression. The question of origin of these depressions, whether growth induced or the result of karst erosion, is immaterial (Hopley, 1982).

To determine the size of reef platform required to produce single and multiple depressions, 150 reefs were randomly selected between latitudes 15°S and 24°S. None had reached a planar stage of development (i.e. had extensive reef flats), but all had considerable areas of reef-top close to present sea level. Fifty had multiple lagoon cells, 50 had single lagoon cells, and 50 had no lagoon whatsoever. Results are seen in table 2. The groups are clearly differentiated on a size basis, although some overlap occurs. Generally, multiple cells are developed where reef tops are wider than 3.25 km; single cells are developed where reef width is between 1.75 and 3.25 km, and no lagoons are present on reefs less than 1.75 km wide. These measurements are used in the following classification when morphological data cannot be used; for instance, where completely planar surfaces are now present, suggesting infill of lagoons. Pre-existing platform width thus produces a tripartite classification. It is possible that some Holocene reefs have no antecedent platform. These will be small and will rise from shallow water. Primary reefs, lacking antecedent platforms, form a fourth basic group.

The secondary set of criteria which need to be introduced into the classification are those which determine the degree to which the pre-existing substrate is masked by recent growth, and the extent to which the modern reef has grown up to present sea level. A complex series of interrelated factors is involved and it may not be possible to separate the individual factors contributing to the morphology of a particular reef without intensive investigation.

	Sample Size	Mean Width (km)	σ	Minimum Width (km)	Maximum Width (km)
No lagoon	50	1.05	0.58	0.25	3.25
Single lagoon	50	2.53	0.73	1.25	4.25
Multiple lagoons	50	4.93	2.00	2.25	12.75

Table 2   Width of reef platforms without lagoons, with single lagoons and multiple lagoons

The depth of the pre-existing platform (or, in the case of reefs without an antecedent platform, the depth of the sea floor) is of paramount importance in determining the time at which a reef reached modern sea level. During much of the Holocene transgression, sea level rise outstripped the capability of reefs for upward growth. Thus, reefs growing from the shallowest platforms will, assuming no temporal differences related to the establishment of a living veneer, have reached sea level earliest and are more likely to have the greatest lateral growth (as opposed to vertical growth) masking the underlying morphology.

Variation in carbonate productivity will also be a factor. At the extremes, Holocene growth varies from almost nothing to several centimetres per year (based on the most rapid growth rates quoted for Caribbean fringing reefs, e.g. Macintyre & Glynn 1976). Again, the more productive reefs are most likely to have masked their pre-Holocene foundations.

The final major factor is the variation which may have occurred in the apparent sea level curve. Subsidence would have emphasized vertical growth from topographic highs in the antecedent platform. Despite more rapid growth on windward margins, the gross morphology of the antecedent platform may have been retained. In contrast, where sea level has been relatively stable (with respect to the antecedent platform and for over 5000 years) lateral growth will have occurred after the reef grew up to sea level and masking of pre-Holocene relief by lateral growth will be significant.

In combination, these factors will produce a continuum from no modification of the antecedent platform (where Holocene colonization has not taken place), through a stage where Holocene growth mimics the relief of the platform (essentially during upward growth), to a final stage where the pre-existing relief is masked by Holocene growth, largely through lateral reef accretion including transport of sediments from the most productive reefal margins.

A CLASSIFICATION OF HOLOCENE SHELF REEFS

A classification of Holocene shelf reefs is presented in table 3 and figure 1a, b, c, d. The classification presumes that most reefs are located over shallow pre-Holocene foundations and it is set in the eustatic history of the Holocene, that is, a period of transgression over these antecedent platforms. The basic three fold division is based on the size criteria previously discussed which influences the presence and nature of lagoonal depressions. In the antecedent platform these depressions may be either karstic or be constructional features of Pleistocene age. A karst origin suggests that there is a minimum size of limestone platform required for the development of single or multiple depressions, possibly as solution dolines (see Backshall *et al.* 1979). Alternatively, carbonate budget studies suggest that reef growth will result in living reef flats up to a maximum width of 0.5 km, beyond which negative feedback mechanisms restrict further widening to detrital infilling of former lagoons (Kinsey & Davies 1979). Thus, it is unlikely that a significant central depression will form by growth mechanisms on reef platforms less than 1 km in width. Even on platforms of slightly greater diameter, the high ratio of productive reef margin to central lagoon may result in rapid infilling of any initial depression. The diameter of 1.75 km derived from measurement of reefs lacking lagoons (table 2) supports this suggestion. On larger reefs, multiple lagoons produced by growth of anastomosing reef ridges may be dependent on a minimal lagoon size to retain circulation. However, it is probable that the development of enclosed depressions in the pre-Holocene substrate is the product of a combination of pre-Holocene morphology and karst solution.

Linear reefs commonly grow from narrow antecedent platforms where lack of substrate of suitable width precludes the development of lagoons of any size. Distinctive linearity places these reefs in a separate sub-group. Again, alternative explanations for linear antecedent platforms may be found. Fairbridge (1950, 1967) suggested they represented upgrowth of reefs fringing former low sea level shorelines. On the northern Great Barrier Reef in particular, when sea level was below -50 m, the shallow shelf and steep continental slope probably resulted in rocky shorelines along which narrow linear reefs would have developed. However, it has also been suggested (Harvey 1977) that linear or ribbon reefs may have developed over solution rims on larger karst eroded platforms. Purdy (1974) has suggested that solution would have caused rims to develop.

Secondary criteria of the classification relate to growth of reefs from their antecedent platforms and the degree to which the morphology of the pre-Holocene foundation has been masked. Initial growth on eminences in the foundations, as suggested by Purdy (1974) and apparently supported by drilling on the Belize barrier by Halley *et al.* (1977), may initially reinforce the relief of the underlying morphology. It is likely that the rate of upward growth of the Holocene reefs has been outstripped by the rate of sea level rise (Smith & Kinsey 1976). Thus, for a period, the Holocene reefs may have been largely submerged features initially reaching modern sea level as an irregularly shaped reef grossly reflecting the antecedent platform's shape, the morphology being the equivalent

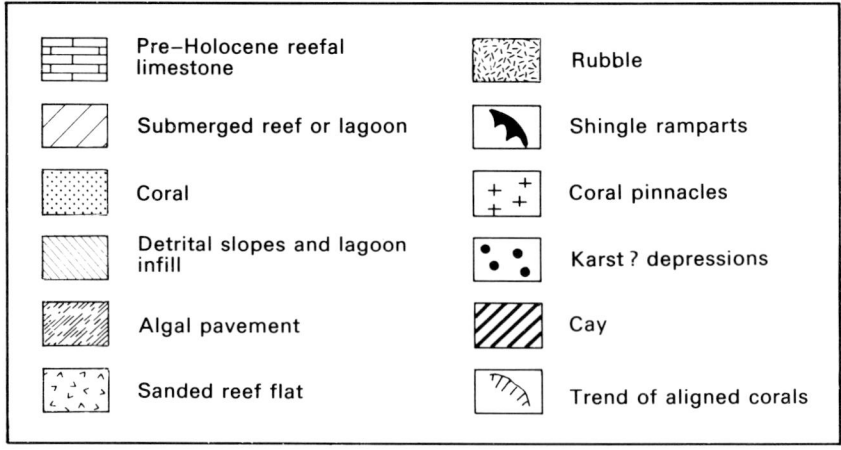

Figure 1(a)   Evolutionary classification of reefs. Key to fig. 1(b), (c), (d)

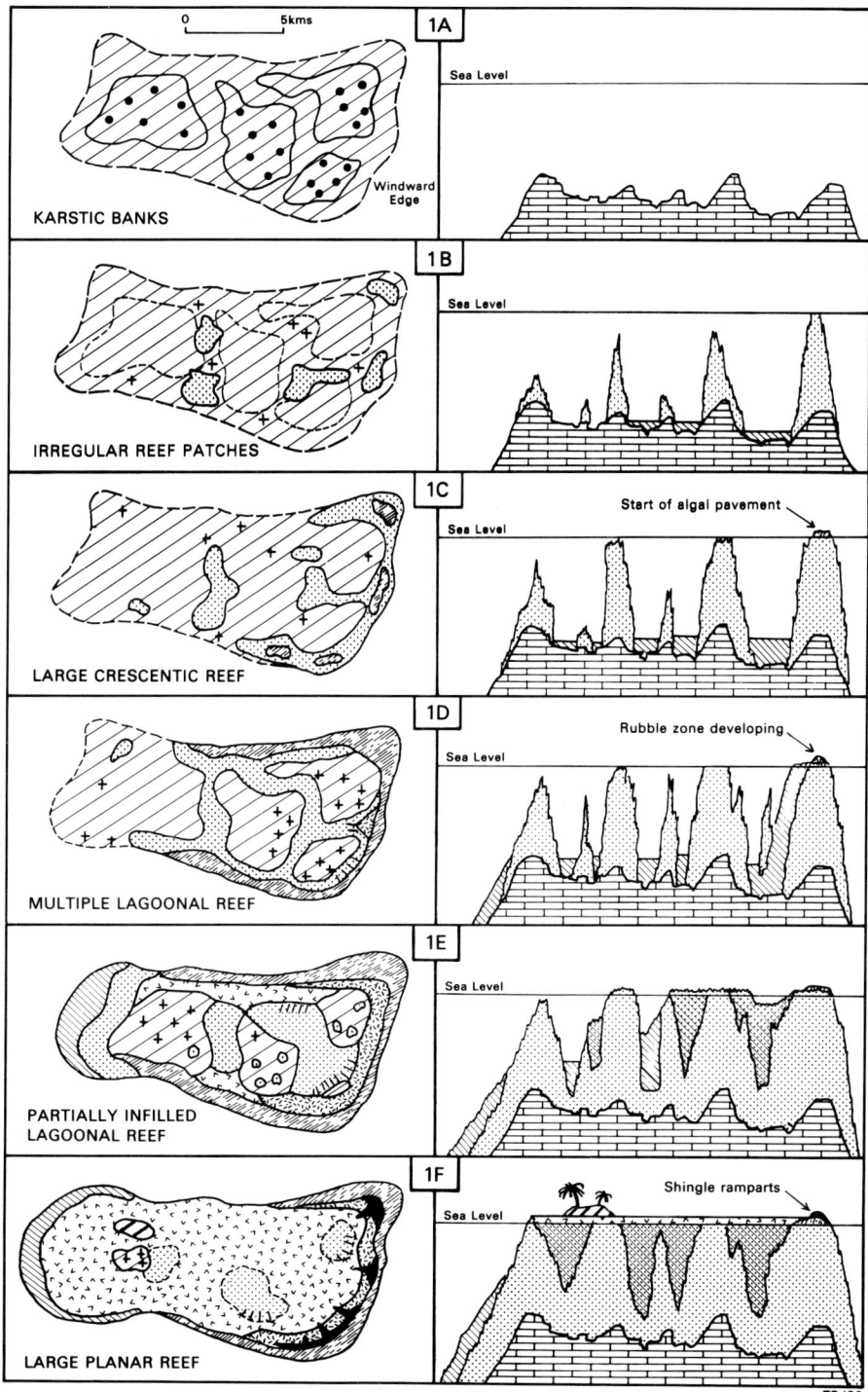

Figure 1(b)   Evolutionary classification of larger reefs (class 1 reefs)

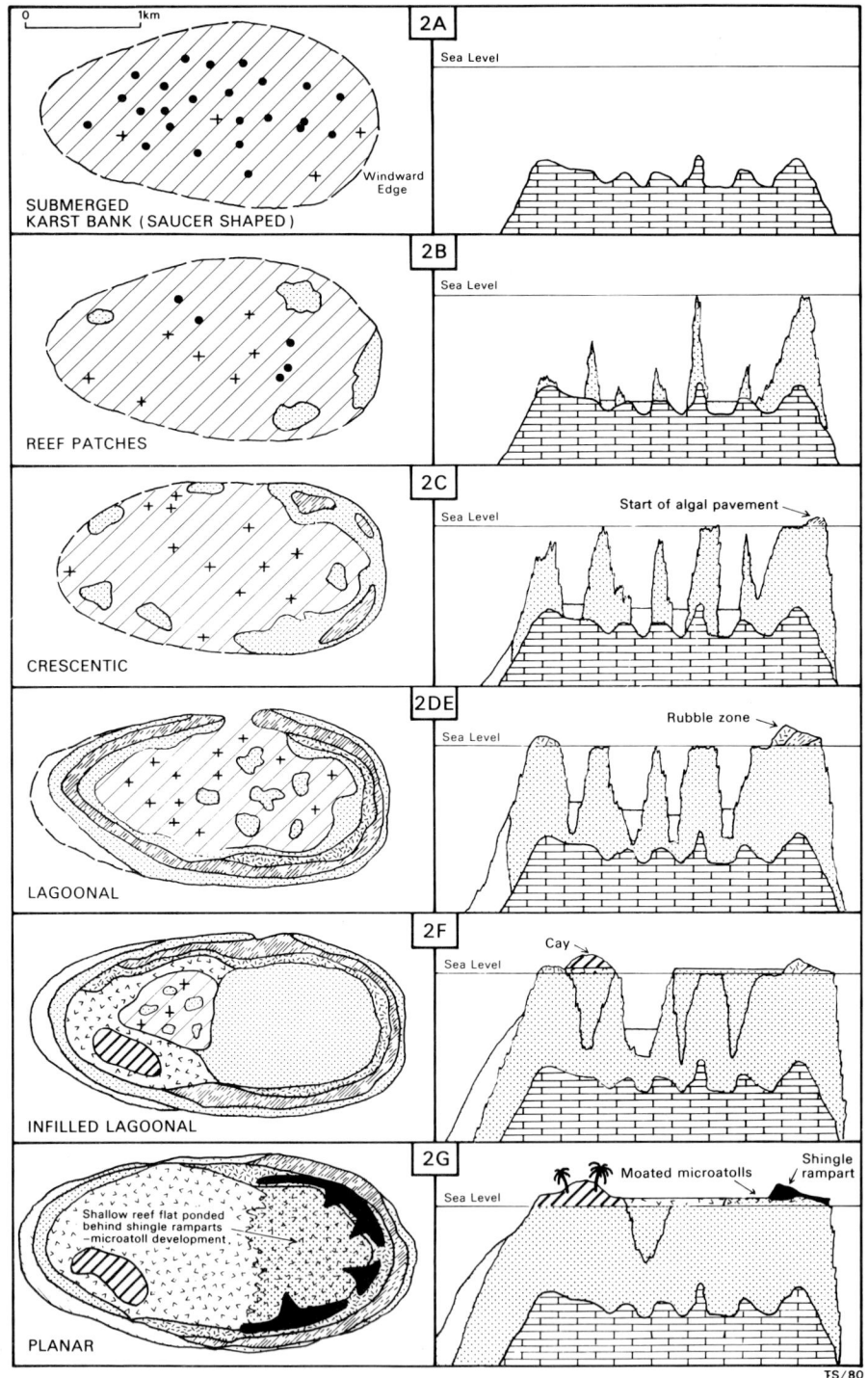

Figure 1(c)   Evolutionary classification of medium size reefs (class 2 reefs)

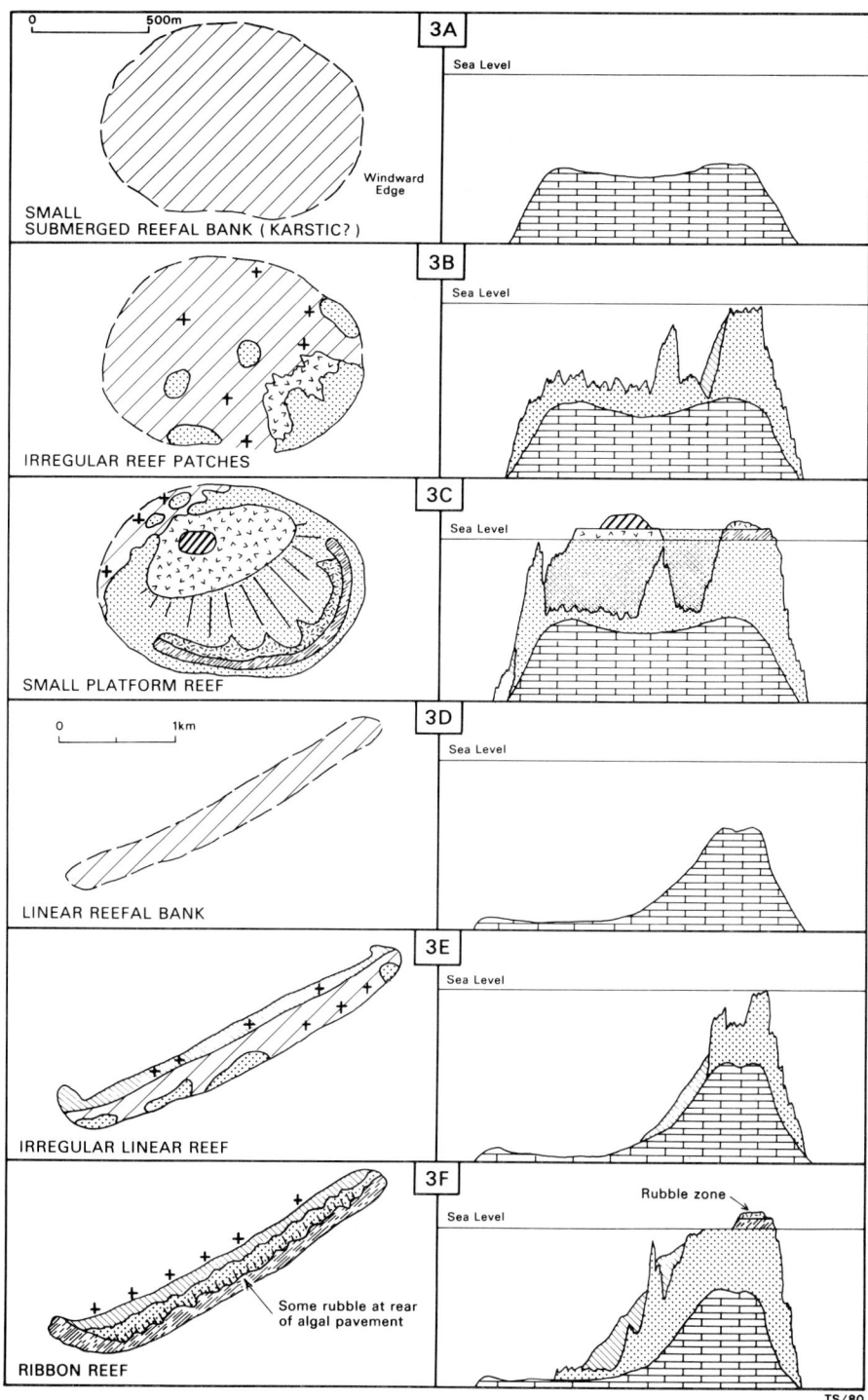

Figure 1(d)   Evolutionary classification of small and linear reefs (class 3 reefs)

Primary Criteria Diameter of Antecedent Platform	Secondary Criteria	
	Largely below present S.L. — Mainly upward growth — Retaining antecedent morphology	Largely above present S.L. — Mainly lateral growth — Masking antecedent morphology
1. Large (>3.25 km width) (Multiple enclosed depressions)	1$_A$ Karstic banks 1$_B$ Irregular reef patches (less than 50% hardline) 1$_D$ Large reefs with multiple lagoons (more than 50% hardline) 1$_E$ Large reefs with partially infilled lagoons 1$_F$ Large planar reefs with or without islands	
2. Medium (1.75–3.25 km width) (Single enclosed depressions)	2$_A$ Submerged lagoonal reef 2$_B$ Irregular reef patches 2$_C$ Crescentic reef — less than 50% hardline 2$_D$ Open lagoonal reef — more than 50% hardline 2$_E$ Lagoonal reef 2$_F$ Partially infilled lagoonal reef 2$_G$ Medium size planar reefs with or without islands	
3. Small (<1.75 km width) (no enclosed depression)	Compact 3$_A$ Small submerged reef patch 3$_B$ Partially submerged reef patch (less than 50% hardline) 3$_C$ Small planar reef (more than 50% hardline)	Linear (<0.85 km width) 3$_D$ Linear submerged 3$_E$ Linear irregular 3$_F$ Ribbon reef
4. (Primary prior platform)	Probably a similar course to type 3 Reefs	

Table 3   A classification of shelf reefs of the Great Barrier Reef

of the 'resorbed' reefs of Maxwell. Subsequently, the greater potential of the well aerated windward margin for growth (Smith & Kinsey 1976; Kinsey & Davies 1979) may result in hardline crescentic reefs developing on the windward rim and slowly extending around larger platforms to eventually form a lagoonal reef. Largest reefs may develop ridges between lagoonal cells. Reef flats grow to about 500 m width (Smith *et al.* 1978; Kinsey & Davies 1979), after which sediment produced on the productive margins is transported inwards over the reef to accumulate in lagoons. Lagoons may be eventually infilled, resulting in planar reef tops with extensive sediment veneer. The latter stages of reef development involve a reduction and final obliteration of the relative relief inherited from the antecedent platform.

The exact stage reached by any reef at the present time is the result of a number of closely related factors. A combination of the depth of the antecedent platform, and the rate of upward reef accretion, will initially have determined the exact time at which the reef first reached modern sea level and changed from a process of enhancement of the inherited relief to one of masking of that relief. Subsidence of the shelf, producing an apparently later attainment of modern sea level, may delay this important change of direction in reef growth. However, it is not necessary for all reefs to go through the full cycle. Acceleration of the earlier phases will take place if the antecedent platform is very shallow. In such a situation, there may be a rapid change from irregular reef patches to planar reef stages. Development of a planar reef may also be accelerated by a relative fall in sea level exposing more reef flat and hastening the production of sediment for infilling of lagoons. As the rate of infilling is largely dependent on the ratio of productive reef margin to the volume of the central depression, it is probable that larger reefs will develop through the later stages at a slower rate than their smaller counterparts.

DISCUSSION AND APPLICATION OF THE CLASSIFICATION

The relationships between present reef morphology and the influences of pre-Holocene platform morphology and depth, reef accretion rates and sea level variations are demonstrated by existing data. Davies *et al.* (1977a,b), Harvey (1977), Harvey *et al.* (1978, 1979) and Backshall *et al.* (1979) have demonstrated that pre-Holocene substrate highs have been responsible for the development of morphological highs in the modern reefs such as reefal rims and lagoonal reef patches, and low points on the platform can determine location of lagoons and blue holes. Harvey (1977) has shown that there is considerable antecedent platform relief beneath some planar reefs, and it may be assumed that this relief had considerable influence on the early development of these planar reefs. For example, 10 m of relief is seen in the pre-Holocene substrate of Three Isles Reef, a low wooded island (see Stoddart *et al.* 1978 for description).

Similarly, the depth to the pre-Holocene can be clearly shown to influence the age of the reef surface, on the scale of an individual reef or at a regional scale. The greater the depth of the pre-Holocene surface, the younger the surface of the overlying reef. Darley Reef (fig. 2) and Redbill Reef (fig. 3) provide excellent examples. Even more convincing is the evidence from the area between Michaelmas and Bewick Reefs. Within this 250 km stretch of the Great Barrier Reef, data are available on nine reefs for both age of the reef caps (upper 3 m) and depth of the Holocene-Pleistocene interface (Harvey 1977; Harvey, pers. comm.). Depth of recent growth and age of reef cap have a correlation coefficient of 0.90, significant at the 99% level. It is tempting to extrapolate from these data and suggest unconformity depths of 8.0 m and 9.2 m for Long Reef and Cairns Reef

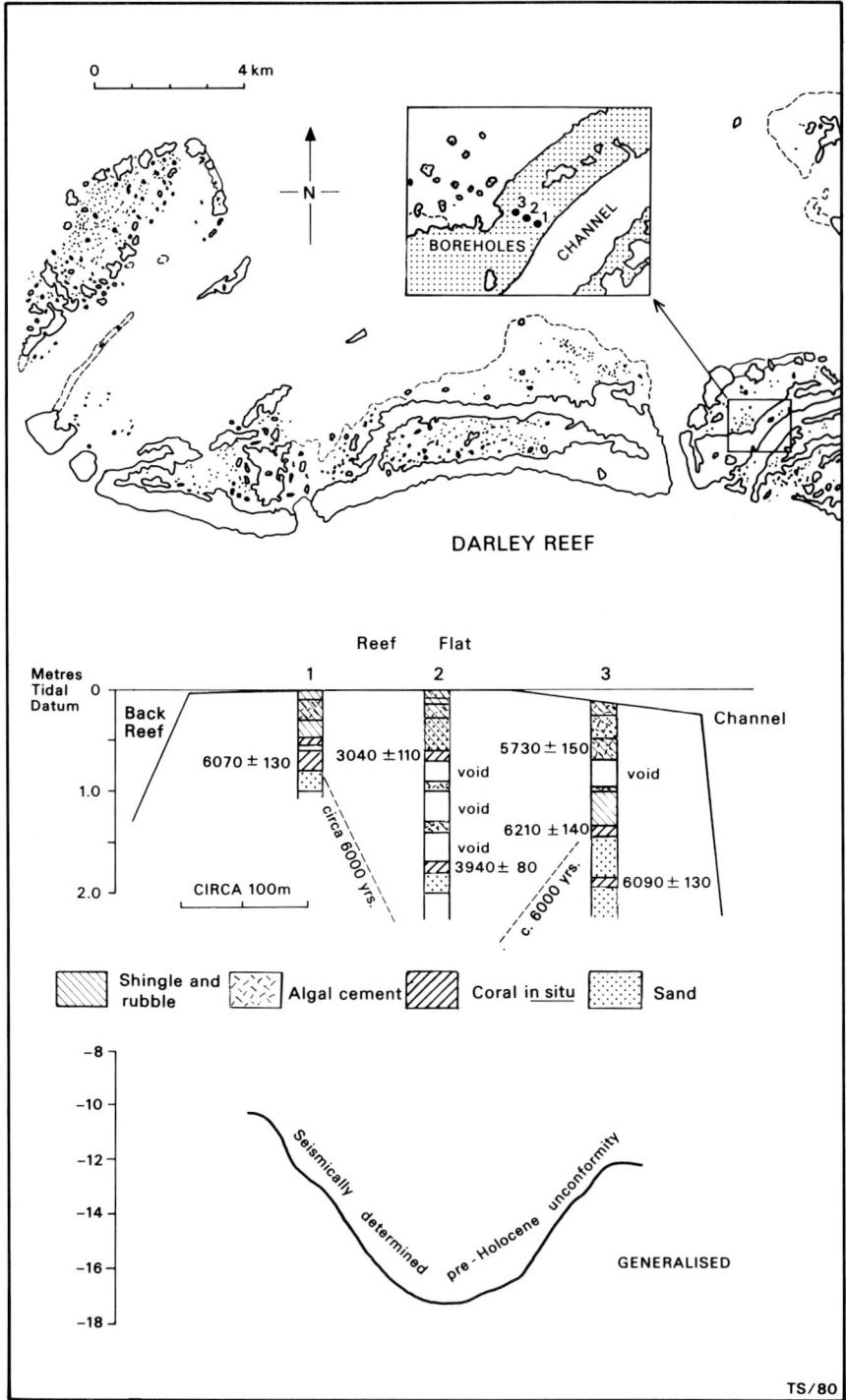

Figure 2 Darley Reef. Radiocarbon ages and pre-Holocene morphology

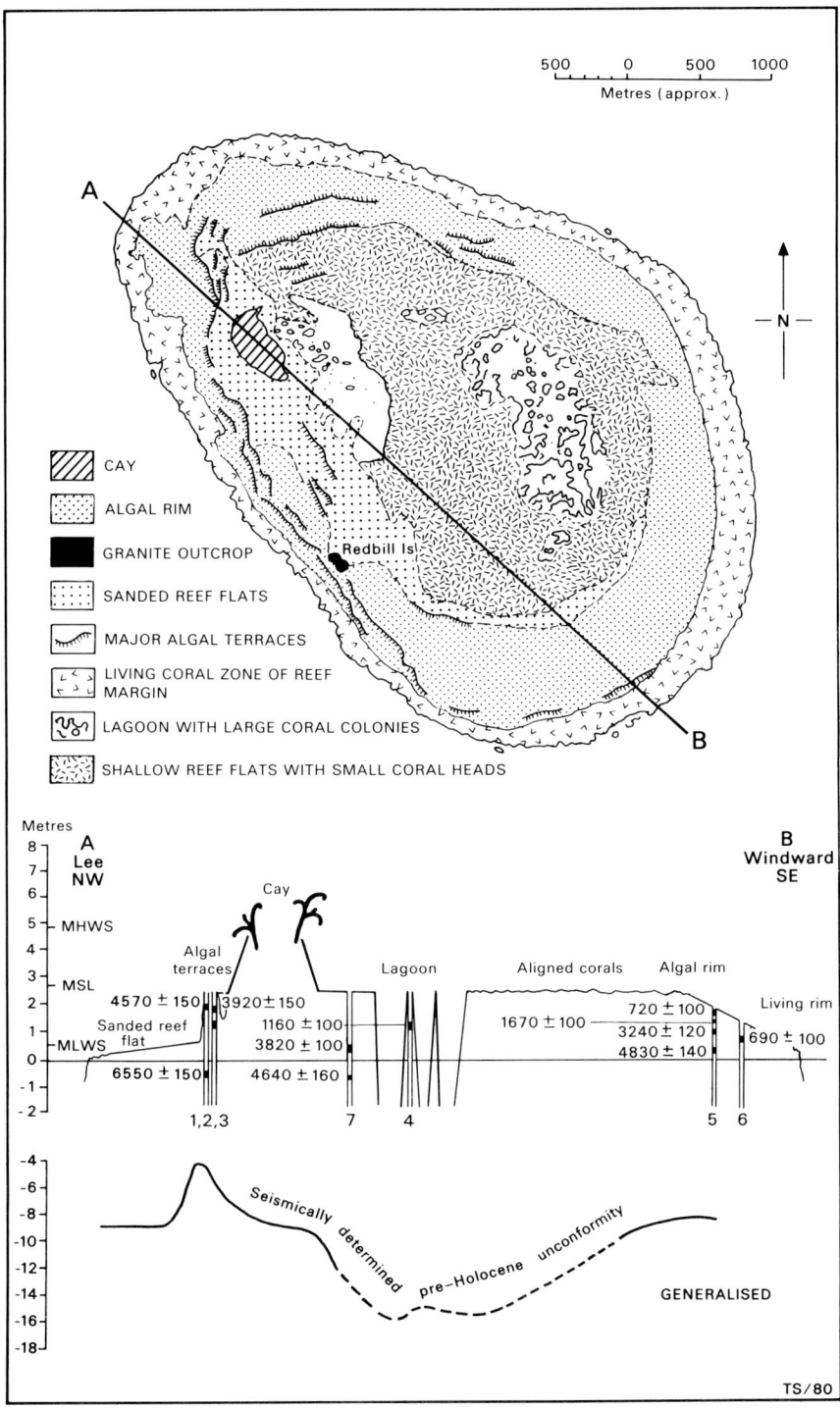

Figure 3   Redbill Reef. Radiocarbon ages and pre-Holocene morphology

respectively. Both reefs lie within this area; their oldest radiocarbon dates are 4740 ± 120 yr BP (GaK-6483) and 4110 ± 130 yr BP (GaK-6686) respectively, however, there is a lack of seismic information. These data imply that it took 5200 years to add 10 m of recent cover. Such a difference is resolved by an upward reef accretion rate of 1.92 mm yr^{-1}. Unfortunately, the Holocene veneer on Bewick Reef is too thin to provide a stratigraphic check on growth rate. However: a rate of 2.0 mm yr^{-1} is provided by radiocarbon ages from the nearby Stapleton Islands borehole which was put down through largely leeward detrital materials (see Thom *et al.* 1978).

This growth rate contrasts with that obtained for the south central section of the Great Barrier Reef. Boreholes on Hayman Island fringing reef (Hopley *et al.* 1978) gave a mean accretion rate of between 4 and 5 mm yr^{-1}. Within this area of the Great Barrier Reef, only five reefs have been dated and had their pre-Holocene substrate determined by seismic survey (although more data is being acquired). The available data produces a correlation coefficient of 0.81, the regression for which is resolved by a vertical accretion rate of 5.3 mm yr^{-1}. This is remarkably close to the stratigraphic result from Hayman Island. Davies & Marshall (1979) have suggested growth rates as rapid as 6 mm yr^{-1} based largely on data from the southernmost Great Barrier Reef. It is evident that, as suggested in the original premise behind the classification, reef accretion rates vary in different parts of the Great Barrier Reef.

The final tenet behind the classification is that, due to tectonic and/or isostatic warping of the continental shelf, different relative sea levels may apply to different areas of the Great Barrier Reef. Such variations have been noted along the inner shelf (Hopley 1974, 1975, 1978a). Most inner shelf areas display evidence for modern sea level being achieved prior to 6000 years BP, and some areas experienced higher than present sea levels shortly afterwards. In contrast, no such evidence exists on the outer shelf (e.g. Hopley 1977). It seems unlikely that all former evidence could have been subsequently removed by erosion. Thus, there is a strong suggestion of both latitudinal and cross shelf warping of a nature similar to that suggested by the hydro-isostatic models of Walcott (1972), Chappell (1974) and Clark *et al.* (1978).

Of particular relevance to the question of shelf warping and relative sea levels are the radiocarbon dates acquired for the central Great Barrier Reef between Townsville and Cairns. Here, little relationship exists between the maximum ^{14}C date from the upper reef cap and the depth to the pre-Holocene antecedent platform. Instead, the ages from the six reefs dated so far are all within the narrow time span of 2720 ± 100 to 1930 ± 110 yr BP. Arguments are made elsewhere (Hopley, 1982) for the outer shelf between Townsville and Cairns being an area of down-warping where the apparent Holocene sea level curve shows present sea level being achieved only by about 2000 yr BP (i.e. about 4000 years later than on the shelf to north and south, and on the adjacent inner shelf). Thus the Holocene sea level curve here is similar to that of the Caribbean. It seems likely that the upward growth of many reefs has caught up with a sea level which was still rising towards its present position and, in their later stages, the reefs have grown upward at the rate of rise in sea level, thus explaining the close clustering of reef cap ages. It is significant that the surfaces of these younger reefs caps are lower than those for which dates older than 5000 years BP have been obtained. The younger reef tops are generally 0.2 to 0.4 m below MLWS, but the older reefs are usually at or slightly higher than MLWS. Perhaps, because of their height and age, much of top the younger reefs is covered in live coral, and algal pavement is poorly developed. In contrast, the older reefs to north and south have extensive algal pavement which appears to overlie older reef-top corals. On windward margins of reefs north of Cairns, the algal pavement has a thickness of about 1 m. The nature of the reef cap is reflected in the proportion of unconsolidated material found in the upper metre of reef caps which have been drilled. A correlation coefficient of -0.81 ($n = 20$) was obtained between the proportion of unconsolidated material and the (oldest) reef cap age, significant at the 99.9% level. A regression line based on this data suggests that very young reefs are completely unconsolidated, and that reefs of about 6000 years age are about 50% unconsolidated in their upper metre.

This analysis of the progression of reefs from crescentic forms to planar paltforms is similar to that of both Fairbridge and Maxwell. However, this analysis takes account of the important role played by substrate morphology. More complete understanding of the multiple influences on the rate of reef development helps to explain the apparent lack of continuity shown by radiometric ages of different reef types (table 1). Reefs growing from shallow pre-Holocene platforms or reefs with rapid accretion rates may be expected to have reached later stages in the classification and to provide older radiometric ages. In areas of relative subsidence, reefs reaching sea level may be maintained at a particular stage for a considerable period. Lateral growth and lagoon infilling can only occur where the reef accretion rate exceeds the rate of rise in relative sea level. This advances the classification. Thus, in areas of susidence, relatively young dates would be expected for reef cap materials regardless of the stage reached.

Application of the classification to the Great Barrier Reef, bearing in mind the factors which influence the stage reached, produces some insight into the evolution of this area of Australia's continental shelf. Distribution of the different reef types and discussion of the significance is made in Hopley (1982) but a summary is seen in table 4. The distribution of reef types is plotted by latitude at 30-minute intervals, starting at Cape Grenville (12°00'S). This is the northern most limit of comprehensive aerial photograph cover. Features of the distribution which appear to be significant include the following:

1. Large reefs (Class 1) have a very limited distribution. They are common at the southern end of the Great Barrier Reef between 19° and 21°S. As accumulation of such large amounts of calcium carbonate can be presumed to have taken place over many glacial-interglacial cycles, these latitudes may represent areas of relative stability or only slow shelf subsidence.

2. The number of reefs and area of shelf covered by reefs is relatively constant as far south as 19°S. Maximum numbers of reefs occur between 19°00' and 22°30'S. Shelf stability may again be a factor in this distribution.

3. Small planar reefs (class 3C) have a wide distribution, being rare only between 17°00' and 10°30'S. Up to 50% of reefs of some areas (particularly around 15°00'S) are of this category. Small reefs of this type develop quickly once they reach sea level, but their size suggests they may grow from relatively young reef foundations. However, it seems likely that some grow from much larger antecedent platforms which cannot be detected on aerial photographs.

4. Irregular and crescentic reefs (classes 1B, 1C, 2B, 2C, 3B) have a distinctive distribution, making up over 50% of all reefs between latitudes 15°00'-13°30'S and 15°30'-19°30'S. The form of these reefs probably reflects slow accretion rates; deep pre-Holocene foundations, or late Quaternary shelf subsidence, or a combination of these factors. These are the 'resorbed' reefs of Maxwell.

5. Planar or partially infilled lagoonal reefs of classes 1E, 1F and 2F, 2G possibly result from rapid accretion rates, shallow foundations and stable or even rising shelf tectonics. A feature of their distribution which is not seen from table 4 is that the majority of these

Latitude	1A	1B	1C	1D	1E	1F	Reef Type (number) 2A	2B	2C	2D	2E	2F	2G	3A	3B	3C	3D	3E	3F
12.00-12.30	Major	1	4				3			1	2	2	7		7	17			7
12.30-13.00	Major	1					5	2			1	4			9	17			8
13.00-13.30	Major	3				1	1	2			1		6	2	21	7			5
13.30-14.00	Minor	1	4			6	1	1			4		6		6	11			4
14.00-14.30	Minor		7					4			1	1	11	1	3	6			5
14.30-15.00	Minor						3	2	1				4	2	5	23			7
15.00-15.30			2				3	7	2			2		2	3	16			6
15.30-16.00		1	2				4	8	1						3	5			3
16.00-16.30		2	3	1			1	1	4					5	3	2	2		1
16.30-17.00			3					2	4				1	6	1	3	1		
17.00-17.30		3	3	1			4		6						3		5		
17.30-18.00		2	6					2	6					4					1
18.00-18.30		3	4				5	7	2					2	2				
18.30-19.00		2	5		1		2	2	14	2	1	1		3	4	3			1
19.00-19.30	Minor	4	7	4			9	22	12	1			1	5	6				
19.30-20.00	Minor	10	7	7	2		3	9	6	1	2	1	4	6	6	11			
20.00-20.30		4	5	12	3			1	6		3	3	2	20	8	6			
20.30-21.00		2	8	19	6	1	3	4	8	2	3	5	13	12	5	13			1
21.00-21.30		3	22	19	11		1	15	17	15	18	23	4	4	8	72		1	2
21.30-22.00		1	3	7	3		2	4	17	7	12	24	3	4	6	46			2
22.00-22.30							3	3	4	3	3	6	2	7	18	15		1	2
22.30-23.00							3							2					
23.00-23.30							4				1	3	2			4			
23.30-24.00											5	2	2			1			
Total		43	95	70	26	8	35	90	137	38	57	78	67	87	127	278	8	2	55

Many reefs of types 1A, 2A and 3A cannot be detected on aerial photographs
An indication only of the known area of distribution of reef type 1A is given

Table 4   Distribution of reef types, Great Barrier Reef, by latitude

reefs have inner shelf locations. Latitudinally they are concentrated between 13°00' and 14°30'S, 20°30' and 22°00'S and in the Bunker-Capricorn Group between 23°00' and 24°00'S. They are very rare in areas where irregular or crescentic reefs are found.

6.   The most distinctive distribution is for class 3F reefs, the linear or ribbon reefs. Most occur north of 16°00'S. Many of those which are south of this latitude are not parallel to the continental shelf edge, as are their counterparts to the north, and may thus have different origins. Tectonic stability, a shallow shelf and steep continental slope, may be the influential factors in this distribution. From 16°S to at least Townsville (19°15'S) there are a series of drowned ribbon reefs on the shelf margin. However, their full distribution cannot be determined. These deep features and the distribution of other reef types mentioned, are suggestive of outer shelf subsidence in the late Quaternary, south of 16°00'S.

CONCLUSIONS

The much quoted classifications of Fairbridge (1950-1967) and Maxwell (1968) are no longer tenable because of the false premises upon which they are based, although at least parts of their schemes of changing reef morphology appear to be correct. The most misleading reef type is the 'resorbed' reef of Maxwell (1968) and it is strongly recommended that this term not be used in describing reefs. Not only can these reefs be shown to be young forms rather than the final reef form as suggested by Maxwell, but it seems highly unlikely that processes which are suggested by the term are possible. Major influences on reef morphology are identified as: size, depth and morphology of pre-Holocene antecedent reefal platforms; rates of vertical reef accretion; and shelf tectonics and isostasy affecting local sea level curves. The classification presented here is based on a scheme of initial upward growth from pre-Holocene foundations during which the morphology of the foundations is maintained until reefs reach modern sea level, followed by a period of largely horizontal growth and sediment movement during which the original forms are lost. Application of the system to the Great Barrier Reef suggests that some reef types may indicate geographical variation in some of the parameters influencing Holocene reef growth which may deserve more detailed research.

ACKNOWLEDGMENTS

Much of the original research included in this paper was financed by the Australian Research Grants Committee. Seismic data used was provided by Dr N. Harvey.

LITERATURE CITED

Adey, W. H., 1978. Coral reef morphogenesis: A multidimensional model. *Science* 202, 831-6.

Backshall, D. G., J. Barnett, P. J. Davies, D. C. Duncan, N. Harvey, D. Hopley, P. J. Isdale, J. N. Jennings and R. Moss, 1979. Drowned dolines — the blue holes of the Pompey Reefs, Great Barrier Reef. *BMR Journal of Australian Geology and Geophysics* 4, 99-109.

Bloom, A. L., 1974. Geomorphology of reef complexes. *Society of Economic Paleontologists and Mineralogists. Special publication* 18, 1-8.

Chappell, J., 1974. Late Quaternary glacio- and hydro-isostasy on a layered earth. *Quaternary Research* 4, 405-28.

Clark, J. A., W. E. Farrell and W. R. Peltier, 1978. Global changes in postglacial sea level: A numerical calculation. *Quaternary Research* 9, 265-87.

Darwin, C. R., 1842. The Structure and Distribution of Coral Reefs. London: Smith, Elder and Co., 214pp.

Davies, P. J. and J. F. Marshall, 1979. Aspects of Holocene reef growth — substrate age and accretion rate. *Search* 10, 276-9.

Davies, P. J., J. F. Marshall, D. Foulstone, B. G. Thom, N. Harvey, A. D. Short and K. Martin, 1977a. Reef growth, southern Great Barrier Reef — preliminary results. *BMR Journal of Australian Geology and Geophysics* 2, 69-72.

Davies, P. J., J. F. Marshall, B. G. Thom, N. Harvey, A. D. Short and K. Martin, 1977b. Reef development, Great Barrier Reef, In D. L. Taylor (ed.), *Proceedings: Third International Coral Reef Symposium*. Miami: University of Miami, vol, 2, pp. 331-7.

Fairbridge, R. W., 1950. Recent and Pleistocene coral reefs of Australia. *Journal of Geology* 58, 330-401.

Fairbridge, R. W., 1967. Coral reefs of the Australian region. In J. N. Jennings and J. A. Mabbutt (eds.), *Landform Studies from Australia and New Guinea*. Canberra; Australian National University Press, pp. 386-417.

Flood, P. G., 1977. The three southernmost reefs of the Great Barrier Reef Province — an illustration of the sequential/evolutionary nature of reef type development. In R. W. Day (ed.), *Geological Society of Australia. Queensland Division. Field Conference, Lady Elliot Island, Fraser Island, Gayndah, Biggenden*. Brisbane: Geological Society of Australia, pp. 37-45.

Flood, P. G., S. Harjanto and G. R. Orme, 1979. Carbon-14 dates, Lady Elliot Reef, Great Barrier Reef. *Queensland Government Mining Journal* 444-7.

Flood P. G. and G. R. Orme, 1977. A sedimentation model for platform reefs of the Great Barrier Reef, Australia. In D. L. Taylor (ed.), *Proceedings: Third International Coral Reef Symposium*. Miami: University of Miami, vol, 2, pp. 111-7.

Goreau, T. F. and L. S. Land, 1974. Fore-reef morphology and depositional processes, North Jamaica. *Society of Economics Palentologists and Mineralogists. Special Publication* 18, 77–89.

Halley, R. B., E. A. Shinn, J. H. Hudson and B. Lidz, 1977. Recent and relict topography of BooBee Patch reef, Belize. In D. L. Taylor (ed.), *Proceedings: Third International Coral Reef Symposium*. Miami: University of Miami, vol 2, pp. 29–35.

Harvey, N., 1977. The identification of subsurface solution disconformities on the Great Barrier Reef, Australia, between 14°S and 17°S, using shallow seismic refraction techniques. In D. L. Taylor (ed.), *Proceedings: Third International Coral Reef Symposium*. Miami: University of Miami, vol. 2, pp. 45–51.

Harvey, N., 1978. Wheeler reef: morphology and shallow reef structure. In D. Hopley (ed.), *Geographical Studies of the Townsville Area*. James Cook University Geography Department, Monograph Series Occasional Paper 2, pp. 51–3.

Harvey, N., P. J. Davies and J. F. Marshall, 1978. Shallow reef structure: southern Great Barrier Reef. *Bureau of Mineral Resources, Geology and Geophysics Record* 96.

Harvey, N., P. J. Davies and J. F. Marshall, 1979. Seismic refraction — a tool for studying coral reef growth. *BMR Journal of Australian Geology and Geophysics* 4, 141–7.

Hopley, D., 1974. Investigations of sea level changes along the coast of the Great Barrier Reef. In, *Proceedings of the Second International Symposium on Coral Reefs*. Brisbane: Great Barrier Reef Committee, vol. 2, pp. 551–62.

Hopley, D., 1975. Contrasting evidence for Holocene sea levels with special reference to the Bowen-Whitsunday area of Queensland. In I. Douglas, J. E. Hobbs and J. J. Pigram (eds.), *Geographical Essays in Honour of Gilbert J. Butland*. Armidale: University of New England, pp. 51–84.

Hopley, D., 1977. The age of the outer ribbon reef surface, Great Barrier Reef, Australia: implications for hydro-isostatic models. In D. L. Taylor (ed.), *Proceedings: Third International Coral Reef Symposium*. Miami: University of Miami, vol. 2, pp. 45–51.

Hopley, D., 1978a. Sea level change on the Great Barrier Reef: an introduction. *Philosophical Transactions of the Royal Society of London, Series A* 291, 159–66.

Hopley, D., 1978b. The Great Barrier Reef in the Townsville region. In D. Hopley (ed.), *Geographical Studies of the Townsville Area*. James Cook University, Geography Department, Monograph Series Occasional Paper 2, pp. 45–50.

Hopley, D., 1982. Geomorphology of the Great Barrier Reef. New York: Wiley–Interscience, 453 pp.

Hopley, D., R. F. McLean, J. Marshall and A. S. Smith, 1978. Holocene-Pleistocene boundary in a fringing reef: Hayman Island, North Queensland. *Search* 9, 323–5.

Jell, J. S. and P. G. Flood, 1978. Guide to the geology of reefs of the Capricorn and Bunker Groups, Great Barrier Reef Province, with special reference to Heron Reef. *University of Queensland, Department of Geology, Papers* 8, 3.

Kinsey, D. W. and P. J. Davies, 1979. Carbon turnover, calcification and growth in coral reefs. In P. A. Trudinger and D. J. Swain (eds.), *Biogeochemical Cycling of Mineral-forming Elements*. Amsterdam: Elsevier Scientific, pp. 131–62.

Macintyre, I. G. and P. W. Glynn, 1976. Evolution of modern Caribbean fringing reef, Galeta Point, Panama. *American Association of Petroleum Geologists Bulletin* 60, 1054–72.

Maxwell, W. G. H., 1968. Atlas of the Great Barrier Reef. Amsterdam: Elsevier Scientific, 258pp.

Maxwell, W. G. H., 1973. Geomorphology of eastern Queensland in relation to the Great Barrier Reef. In O. A. Jones and R. Endean (eds.), *Biology and Geology of Coral Reefs*. New York: Academic Press, vol. 1, Geology 1, pp. 233–72.

Murray, J., 1880. On the structure and origin of coral reefs and islands. *Proceedings of the Royal Society of Edinburgh* 10, 505–18.

Murray, J., 1889. Structure, origin and distribution of coral reefs and islands. *Nature* 39, 424–8.

Polach, H. A., R. F. McLean, J. R. Caldwell and B. G. Thom, 1978. Radio-carbon ages from the northern Great Barrier Reef. *Philosophical Transactions of the Royal Society of London, Series A* 291, 139-58.

Purdy, E. G., 1974. Reef configurations: cause and effect. *Society of Economic Paleontologists and Mineralogists. Special Publication* 18, 9-76.

Revelle, R. and K. O. Emery, 1957. Chemical erosion of beach rock and exposed reef rock: Bikini and nearby atolls, Marshall Islands. *United States Geological Survey Professional Paper* 260-T, 699-709.

Semper, C., 1863. Reisebericht (Palau-Inseln). *Zeitschrift fur wissenschaftliche Zoologie* 13, 558-70.

Semper, C., 1873. Die Palua-Inseln im Stillen Ocean. Leipzig.

Smith, S. V., and D. W. Kinsey, 1976. Calcium carbonate production, coral reef growth, and sea level change. *Science* 194, 937-9.

Smith, S. V., P. L. Jokiel and G. S. Key, 1978. Biogeochemical budgets in coral reef systems. *Atoll Research Bulletin* 220, 1-11.

Stoddart, D. R., R. F. McLean, T. P. Scoffin and P. E. Gibbs, 1978. Forty-five years of change on low wooded islands, Great Barrier Reef. *Philosophical Transactions of the Royal Society of London, Series B* 284, 63-80.

Thom, B. G., G. R. Orme and H. A. Polach, 1978. Drilling investigation of Bewick and Stapleton Islands. *Philosphical Transactions of the Royal Society of London, Series A* 291, 37-54.

Trudgill, S. T., 1976. The marine erosion of limestones on Aldabra Atoll, Indian Ocean. *Zeitschrift fur Geomorphologie,* Supplementband, 26, 164-200.

Walcott, R. I., 1972. Past sea levels, eustasy and deformation of the earth. *Quaternary Research* 2, 1-14.

12 Cryptofaunal Communities of Coral Reefs

Pat Hutchings
The Australian Museum, P.O. Box A285,
Sydney South, N.S.W. 2000.

INTRODUCTION

Cryptofauna refers to the fauna that lives in coral substrates and certain of the fauna living on the surface of the substrates (Peyrot-Clausade 1974, 1979; Hutchings & Weate 1977). Terms such as infauna and endo-cryptolithic fauna (Hutchings & Weate 1978) have also been used. Cryptofauna consists of two components: the 'true borers' and the 'opportunistic' species. The 'opportunistic species' cannot bore. They utilize the cracks, crevices and holes created by boring species. The abundance of boring species is well documented in the geological record (Warme 1975). Other terminology has been widely used in the literature; with 'boring species' referred to as endolithic species, marine bioeroders (Golubic *et al.* 1975; Warme 1975), or lithophagic species (McLean 1972) and 'opportunistic species' referred to as nestling species (MacGeachy & Stearn 1976), cryptolithic species (Golubic *et al.* 1975) or coelobites (Ginsburg & Schroeder 1973).

Jackson (Jackson 1977; Jackson *et al.* 1971) used the term cryptofauna to describe the cryptic fauna living on the undersurface of foliaceous corals and overhangs, and on walls of crevices and caves. In the discussion that follows, Jackson's epifaunal community and the subsurface fauna are both considered as cryptofauna. Only recent cryptofauna of hard coral substrate are discussed. Such a definition will therefore exclude the community associated with sponges which Uebelacher (1977) called cryptofaunal.

The majority of the cryptofauna is found in reef rock and coral debris, although a few species, such as the polychaete *Spirobranchus giganteus* and some barnacles, are abundant in live coral (Patton 1976). The boring cryptofauna includes sponges (Wilkinson, this volume), bivalve molluscs, sipunculans and some species of polychaetes. The opportunistic cryptofauna belong to a wide variety of groups: molluscs, crustaceans, polychaetes, echinoderms, nematodes, turbellarians, nemerteans, and colonial groups such as bryozoans, zooanthids, hydroids, ascidians and sponges. In addition, a diverse, micro-cryptofauna is certainly present but is as yet undescribed. Of the macro-faunal groups, the polychaetes and crustaceans are represented by very large numbers of species and often by numerous inviduals at any given locality. An example of the diversity and number of polychaetes is given by Grassle (1973). A 4.7 kg (of which 1.6 kg was living material) head of *Pocillopora damicornis* from the reef flat at Heron Island, Great Barrier Reef, contained 1441 polychaetes representing 103 species. Polychaetes constituted approximately two-thirds of the macro-faunal animals in the sample. Other major groups were, in order of abundance, tanaids, amphipods and isopods. Similar high densities of polychaetes have been obtained by other workers (table 1). Direct comparisons between the data of the various authors should be made cautiously because several different extraction techniques were employed. Comparative data for the other cryptofaunal groups is not available.

A rich and diverse cryptoflora of coralline and endolithic algae is also associated with dead coral skeleton habitats (Kobluk & Risk 1977; for review see Borowitzka 1981).

METHODOLOGY

Before 1970, crytofaunal studies were entirely qualitative (Hartman 1954; Yonge 1963; Ebbs 1966; Reish 1968; Kirsteuer 1969). Several quantitative studies have since been undertaken. These have followed two main sampling strategies. In the first, sampling is carried out in particular zones of the reef with no consideration of the physical characteristics of the substrate. For example, Peyrot-Clausade (1974) collected coral samples (with various undefined physical characteristics) in different zones across a reef and Brander *et al.* (1971) collected their samples by removing all the substrate to a uniform depth within quadrants at set intervals along a transect. Brander *et al.* admit that their method is unworkable as it produces such a mixed sample that interpretation of replicate samples is impossible. Peyrot-Clausade (1974) made no attempt to collect replicate samples within any particualr reef zone, instead she has compared the fauna of one sample, say collected from the reef flat, with another, say collected from the lagoon. Such sampling gives no indication of the variation in cryptofauna occurring within a reef zone. The second strategy, which recognizes the importance of the physical characteristics of the substrate, involves collecting particular substrates (Kohn & Lloyd 1973a, b), or coral species (McCloskey 1970), either within a reef zone (Kohn & White 1977), or from and within various parts of the reef (Hutchings 1974). The importance of the physical and biological characteristics of the substrate in determining the biomass and diversity of cryptofauna has only recently been recognized (Brander *et al.* 1971; Hutchings 1974; Kirsteuer 1978). These workers have shown the importance of certain characteristics of the substrate, including surface area, volume, percentage epifaunal and floral cover, ratio of live to dead coral, and porosity (dependent on the coral species). The determination of these physical characteristics is often difficult (see Hutchings 1978) and, as all these factors closely interact with each other, the exact relationships between them are unclear. For example, surface area and percentage of epifaunal and floral cover are important during initial settlement and penetration but subsequently the volume and porosity of the habitat become important in determining the diversity and biomass of the cryptofaunal community.

Problems arise in choosing replicate samples of substrate and in determining the minimum size of sample which is representative of the cryptofaunal community. These problems have not been satisfactorily resolved. Clausade (1970) suggests that 1 dm³ is the minimum volume of each habitat necessary to obtain a representative sample of the fauna but this has not been confirmed by other workers, and may well vary according to the substrate. Various techniques have been used in attempts to collect replicate samples. Kirsteuer (1978) standardized samples by collecting a fixed volume. Kohn & White (1977) chiselled out a rectangular block from the reef limestone platform at Tanguisson, Guam and standardized between samples by using the surface area of the habitat. However, this is not feasible for most coral habitats which have irregular surfaces. Brander *et al.* (1971) and Hutchings (1974) subjectively selected uniform-sized blocks for

Habitat type	Locality	Dominant Polychaete Genera	No. of Polychaete Species	Density	Reference
Predominantly growing corals	Palau, Caroline Is.	*Eunice, Leocrates*	8–15	45–230 per m²	Takahashi 1939
Coral heads (*Oculina arbuscula*)	Beaufort, North Carolina	*Brania, Syllis, Polydora*	26–57	1.49–7.52 species per 100 g of coral	McCloskey 1970
Boulder ridge	Aldabra, Indian Ocean		13	7.45 ± 5.25 per litre	Brander *et al.* 1971
Coral area	Watamu, Kenya		50	9.49 ± 4.62 per litre	Brander *et al.* 1971
Truncated coral reef rock	Sanding Is., Indonesia	*Syllis, Phyllochaetopterus*	22–30	17 000–35 300 per m²	Kohn & Lloyd 1973a
Truncated coral reef rock	Banjak Is., Indonesia	*Syllis, Exogone, Polydora*	32	72 300 per m²	Kohn & Lloyd 1973a
Truncated coral reef rock	Palau Boenda, Indonesia	*Syllis, Nematonereis*	22–25	83 600–91 800 per m²	Kohn & Lloyd 1973a
Truncated coral reef rock	Goh Huyong, Thailand	*Syllis, Palola*	16	16 400	Kohn & Lloyd 1973a
Basalt boulder	Easter Is.	*Syllis, Fabricia, Spirorbis*	33 (16 families)	26 200	Kohn & Lloyd 1973b
Coral rock samples	Kaneohe Bay, Hawaii			50 100–127 900 per m²	Brock & Brock 1977
Coral substrates	Touléar, Madagascar		178	164–479 per dm³	Peyrot-Clausade 1979
Sand rubble under dead coral rocks	Heron Island, Great Barrier Reef	*Pseudoeurythoe, Nereis, Eunice, Nematonereis, Syllis, Micromaldane*	16	28 per litre	Reichelt 1979
Solid limestone	Heron Island, Great Barrier Reef	*Syllis, Brania, Nereis Ceratonereis, Eunice, Lysidice*	41	24 100 per m²	Reichelt 1979
Blocks of recently killed *Porites*	Lizard Island, Great Barrier Reef	*Syllis, Nereis, Eunice, Dorvillea, Polydora, Hydroides* sp.	40–50	70.1–182.7 per 100 g of coral after 12 months	Hutchings & Weate, in press; Hutchings, in prep.

Table 1   Density of cryptofaunal polychaetes

replicate samples. If replicate samples are chiselled off a large massive habitat, samples taken should have similar ratios of exposed surfaces to previously unexposed surfaces and should be cut to a uniform depth into the habitat. Crytofauna is necessarily concentrated around the periphery of the habitat because each animal must maintain an exit to the outside for respiration, feeding and reproduction.

There is an urgent need to establish standard techniques for sampling and measuring the physical parameters of the habitats.

Several techniques have been employed to extract fauna from substrates. Initially, workers pulverized the samples. This often damaged the animals and increased taxonomic problems. Kirsteuer (1978) extracted cryptofaunal nemerteans by allowing the samples to stand in seawater in the shade for several hours; as the oxygen level fell, nemerteans and other errant groups migrated out of the sample. The sessile animals, however, remained in the sample. A more satisfactory method of sampling the solitary cryptofauna is to dissolve the substrate in an acid solution (Brock & Brock 1977; Hutchings & Weate 1978). Colonial cryptofauna are damaged by this method and adequate sampling is difficult. MacGeachy (1977) used radiography to determine the distribution of sponges and their boring rates.

Even where replicates are carefully chosen, considerable variation often exists in the development of the cryptofaunal community because there is little development before a coral dies. A suite of replicate samples may include corals that died at different times. Thus, the cryptofaunal communities may be at different stages of development. Recently, Hudson (1977) has shown that skeletons of *Montastrea annularis* can be aged by stress bands, which develop in response to severe cold. Such bands provide a time marker in the coral skeleton and potentially allow cryptofaunal communities to be aged, and erosion rates to be calculated.

One way of ensuring that all the cryptofaunal communities are of the same age is to prepare blocks of substrate from recently killed coral that is completely unbored. These blocks are then attached to the reef and sampled at intervals to study recruitment and subsequent development of cryptofaunal communities (Hutchings & Weate 1978; Hutchings 1981). Similar techniques have been employed by Neumann (1966) and Rü tzler (1978) for measuring sponge boring rates.

RECRUITMENT

Pelagic larvae are probably responsible for most of the recruitments to cryptofaunal communities (McCloskey 1970). Among the cryptofaunal polychaetes, two families, Syllidae and Cirratulidae *(Dodecaceria),* potentially can reproduce asexually. Rice (1969) has shown that adult sipunculans cannot burrow into hard substrates. Larvae are also responsible for substrate penetration in the boring bivalves. Once a larva has entered or penetrated the substrate, the newly settled individual either gradually expands the burrow by boring or, in the case of opportunistic species, finds a suitable unoccupied burrow and gradually grows to fill it. Adult cryptofauna are often encapsulated in the habitat with just a small communicating passage to the exterior.

Information on recruitment of cryptofauna is almost non-existent, although recently polychaete recruitment has been studied at Lizard Island on the northern part of the Great Barrier Reef (14°40'S, 145°38'E)(Hutchings 1981). Recruitment appears to be strongly seasonal with a maximum in mid-summer. The numbers recruited show considerable variation from year to year, probably reflecting fluctuations in the availability of larval recruits. Information on larval populations in plankton around Lizard Island is not available. However, at Heron Island on the southern part of the Great Barrier Reef, Sale *et al.* (1978) found seasonal variations in the number of polychaetes (epitokes, larvae and pelagic species) in the plankton. McWilliams, one of Sales' co-workers, noted abundant epitokous nereids and syllids carrying eggs in a November-December sample, suggesting potential seasonality of larval polychaete recruits (pers. comm.).

SUCCESSION

Changes in cryptofaunal communities with time are not well documented. McCloskey (1970) studied the communities in coral heads of *Oculina arbuscula* of varying sizes and hence ages. The larger (presumably older) coral heads were proportionally more bored than smaller coral heads and there was a difference in the dominant cryptofaunal species between small and large heads. Initially, the dying coral was bored by algae followed by sponges, such as *Cliona,* which constructed a subsurface labyrinth with numerous openings to the outside. This greatly increased the volume of the habitat and the new voids were colonized by additional species such as bivalves. The mollusc burrows, when vacated, were rapidly colonized by opportunistic polychaetes, shrimps and amphipods (fig. 1). The development of a diverse cryptofaunal community was therefore dependent upon the borers. McCloskey found that the percentage of borers in the community increased with size and age of the coral. A similar increase in the number of sipunculans (all of which were borers) recruited to habitats with increasing time has been found at Lizard Island (Hutchings, unpub.); and by MacGeachy & Stearn (1976) working on sponge boring in *Montastrea.*

STRUCTURE OF COMMUNITY

Little is known about the structure of cryptofaunal communities. McCloskey (1970) has divided the cryptofauna of *Oculina* into feeding categories and described three main types: primarily deposit feeders (65%), suspended detritus feeders (13%) and carnivores (10%). The deposit feeders utilize detritus, silt and organic material which collects in the coral head. McCloskey suggests that mucus, produced in copious amounts by the coral, traps silt and detritus. This is moved by ciliary action away from the living portions of the colony and dropped either into the dead base or off the margins of the coral colony where it is eaten by the deposit feeders. Unpublished data (Hutchings) from Lizard Island also indicates that cryptofaunal communities are dominated by deposit feeders.

No data exist on the rates of turnover of the cryptofaunal communities but it seems likely to be rapid. For instance, some sipunculans and terebellid polychaetes are gravid within three months of settlement (Hutchings, in prep.). Faster growth rates probably occur among the smaller polychaetes such as syllids and spionids which are abundant in all communities. Interactions between species of cryptofauna are poorly documented. Some amphipods are commonly associated with boring sponges and often have the same pigmentation as the sponge, whereas other amphipods seem to avoid sponges. Sipunculans' burrows rarely merge, suggesting an avoidance mechanism.

Although colonial cryptofauna such as sponges, ascidians and bryozoans, tend to dominate the surface fauna of the substrate, probably because of their ability to reproduce asexually and because they are less susceptible to 'fouling and overgrowth' (Jackson 1977), they do not dominate the infaunal community. The reasons for this are unknown.

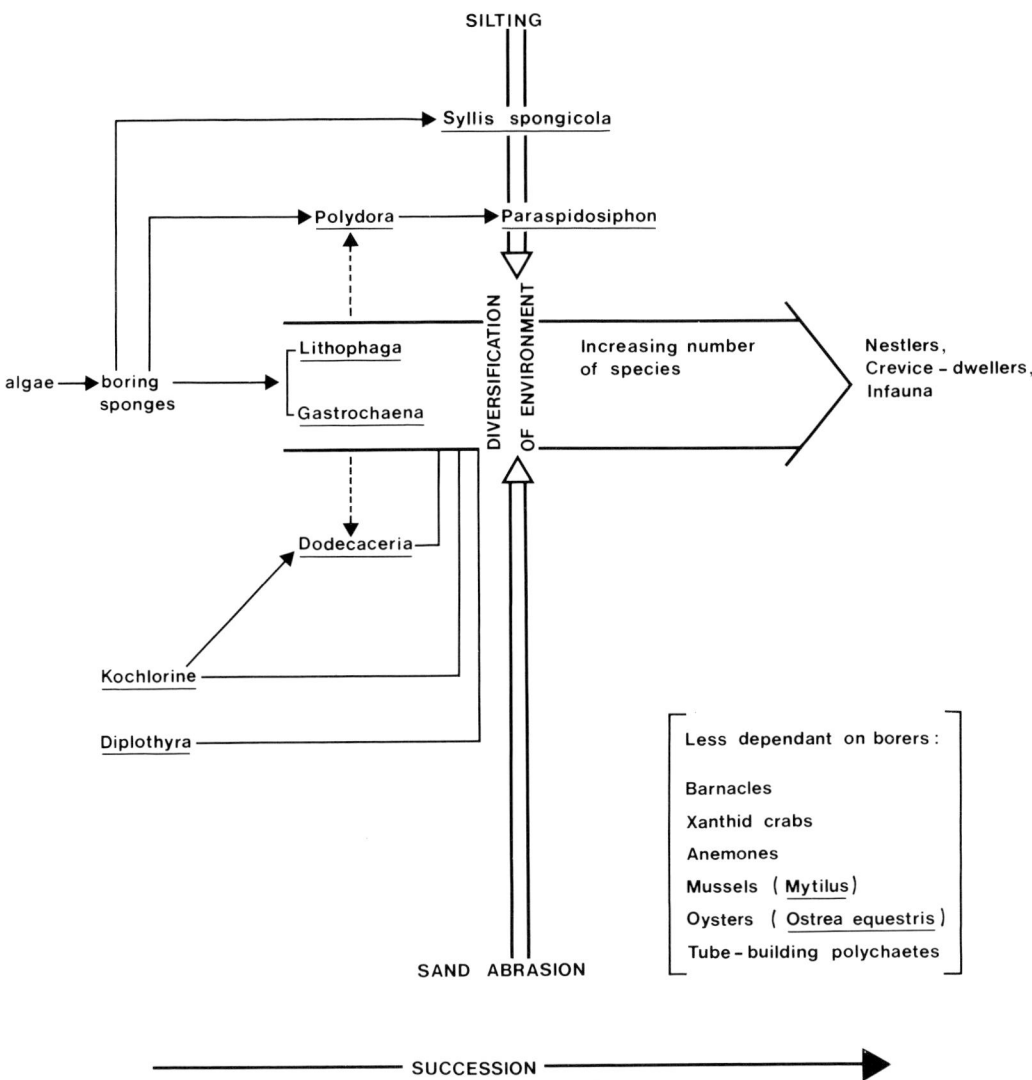

Figure 1 Diagram of the borer-dominated successional sequence in the *Oculina arbuscula* community. (Redrawn from McCloskey 1970 and reproduced by permission of Academie-Verlag.)

ROLE OF THE CRYPTOFAUNA

Cryptofauna are an important food source for holocentrid fish (Vivien & Peyrot-Clausade 1974) and for many predator gastropods (Kohn & Nybakken 1975; Taylor 1976).

Cryptofauna are important in bioerosion (see Wilkinson, this volume). However, only a few measurements have been made of the rate at which cryptofauna erode their substrate and few of these are from the Great Barrier Reef (Davies & Hutchings, in press). Cryptofauna also play an important role in the formation of algal cup reefs (Ginsberg & Schroeder 1973), and in the fossilization of patch reefs (Scoffin 1972). In both cases, boring organisms formed extensive cavities beneath the surface of the reefs. Initially, these cavities had openings to the surface which provided protective niches for the opportunistic species, including many encrusting forms. The cavities trapped faecal pellets and skeletal remains of the cryptofauna. Additional cavities were continually created by borers, producing a steady supply of sediment. Other sediment was carried into cavities by wave action and gravity, and also by feeding and respiratory currents set up by the cryptofauna. This sediment was gradually recrystallized and cemented within the cavity system. These processes took place less than 1 cm below the living surface of the algal cup reef (Ginsberg & Schroeder 1973). Such processes are not restricted to algal cup reefs, but are widespread in the reef system.

Cryptofauna may play an important role by utilizing the large amounts of mucus produced by corals and in recycling the detritus and organic material trapped in the mucus. This may explain why, where organic nutrients levels are elevated because of sewage effluent or high land run off, the biomass of cryptofauna increases in comparison to undisturbed sites (Brock & Brock 1977). By contrast Kohn & White (1977) could find no differences between a site subjected to heated effluent from a power station and sites remote from the effluent. However, no measurements were taken before the power station began discharging effluent. Further investigation is necessary to determine if measurement of biomass of cryptofauna have utility in coral reef management. They may well play a crucial role in the development and maintenance of reef systems, and so warrant further investigation.

ACKNOWLEDGEMENT

This work was funded by the Australian Research Grants Committee.

LITERATURE CITED

Borowitzka, M. A., 1981. Algae and grazing in coral reef ecosystems. *Endeavour* 5, 99-106.

Brander, K. M., A. A. McLeod and W. F. Humphreys, 1971. Comparison of species diversity and ecology of reef-living invertebrates on Aldabra Atoll and at Watamu, Keyna. *Symposia of the Zoological Society of London* 28, 397-431.

Brock, R. E. and J. H. Brock, 1977. A method for quantitatively assessing the infaunal community in coral rock. *Limnology and Oceanography* 22, 948-51.

Clausade, M., 1970. Importance et variations du peuplement mobile des cavitiés au sein des formations épirécifales, et modalités d'énchantillon nage en vue de son évaluation. *Recueil des Travaux de la Station Marine d'Endoume,* Supplement Hors Series, 10, 257-70.

Davies, P. J. and P. A. Hutchings, in press. Experimental bioerosion studies, Lizard Island, Great Barrier Reef. *Coral Reefs.*

Ebbs, N. K., 1966. The coral-inhabiting polychaetes of the northern Florida Reef tract. Part 1. Aphroditidae, Polynoidae, Amphinomidae, Eunicidae, and Lysaretidae. *Bulletin of Marine Science* 16, 485-555.

Ginsburg, R. N. and J. H. Schroeder, 1973. Growth and submarine fossilization of algal cup reefs, Bermuda. *Sedimentology* 20, 575-614.

Golubic, S., R. D. Perkins and K. J. Lucas, 1975. Boring microorganisms and microborings in carbonate substrates. In R. W. Frey (ed.), *The Study of Trace Fossils.* New York: Springer-Verlag, pp. 229-59.

Grassle, J. F., 1973. Variety in Coral Reef Communities. In O. A. Jones and R. Endean (eds.), *Biology and Geology of Coral Reefs.* Academic Press, vol. 2, Biology 1, pp. 247-70.

Hartman, O., 1954. Marine Annelida from the northern Marshall Islands, Bikini and nearby atolls, Marshall Islands. *United States Geological Survey Professional Paper* 260-Q, 617-44.

Hudson, J. H., 1977. Long-term bioerosion rates on a Florida reef: a new method. In D. L. Taylor (ed.), *Proceedings: Third International Coral Reef Symposium*. Miami: University of Miami, vol. 2, pp. 491-7.

Hutchings, P. A., 1974. A preliminary report on the density and distribution of invertebrates living in coral reefs. In, *Proceedings of the Second International Symposium on Coral Reefs*. Brisbane: Great Barrier Reef Committee, vol. 1, pp. 285-96.

Hutchings, P. A., 1978. Non-colonial cryptofauna. In D. R. Stoddart and R. E. Johannes (eds.), *Coral Reefs: Research Methods*. Paris: UNESCO, Monographs on Oceanographic Methodology, no. 5, pp. 251-63.

Hutchings, P. A., 1981. Polychaete recruitment onto dead coral substrates at Lizard Island, Great Barrier Reef, Australia. *Bulletin of Marine Science* 31, 410-23.

Hutchings, P. A. and P. B. Weate, 1977. Distribution and abundance of cryptofauna from Lizard Island, Great Barrier Reef. *Marine Research in Indonesia* 17, 99-112.

Hutchings, P. A. and P. B. Weate, 1978. Comments on the technique of acid dissolution of coral rock to extract endo-cryptolithic fauna. *Australian Zoologist* 19, 315-20.

Hutchings, P. A. and P. B. Weate, in press. Experimental recruitment of endo-cryptolithic communities at Lizard Island, Great Barrier Reef: preliminary results. *New Zealand Department of Scientific and Industrial Research Information Series* No. 37, 239-56.

Jackson, J. B. C., 1977. Competiton on marine hard substrate: The adaptive significance of solitary and colonial strategies. *American Naturalist* 111, 743-67.

Jackson, J. B. C., T. F. Goreau and W. D. Hartman, 1971. Recent brachiopod-coralline sponge communities and their paleoecological significance. *Science* 173, 623-5.

Kirsteuer, E., 1969. Qualitative and quantitative aspects of nemertean fauna in coral reefs of the Indian Ocean. In C. Mukundan and C. S. Gopinadha Pillai (eds.), *Proceedings of the Symposium on Corals and Coral Reefs*. The Marine Biological Association of India, Cochin, pp. 367-71.

Kirsteuer, E., 1978. Coral-associated Nemertina. In D. R. Stoddart and R. E. Johannes (eds.), *Coral Reefs: Research Methods*. Paris: UNESCO, Monographs on Oceanographic Methodology, no. 5, pp. 315-27.

Kobluk, D. R. and M. J. Risk, 1977. Rate and nature of infestation of a carbonate substratum by a boring alga. *Journal of Experimental and Marine Ecology* 27, 107-115.

Kohn, A. J. and M. C. Lloyd, 1973a. Polychaetes of truncated reef limestone substrates on Eastern Indian Ocean coral reefs: Diversity, abundance, and taxonomy. *Internationale Revue der gesamten Hydrobiologie* 58, 369-99.

Kohn, A. J. and M. C. Lloyd, 1973b. Marine Polychaete Annelids of Easter Island. *Internationale Revue der gesamten Hydrobiologie* 58, 691-712.

Kohn, A. J. and J. W. Nybakken, 1975. Ecology of *Conus* on Eastern Indian Ocean fringing reefs: diversity of species and resources utilization. *Marine Biology* 29, 211-34.

Kohn, A. J. and J. K. White, 1977. Polychaete annelids of an intertidal reef limestone platform at Tanguisson, Guam. *Micronesica* 13, 199-215.

MacGeachy, J. K., 1977. Factors controlling sponge boring in Barbados reef corals. In D. L. Taylor (ed.), *Proceedings: Third International Coral Reef Symposium*. Miami: University of Miami, vol. 2, pp. 477-85.

MacGeachy, J. K. and C. W. Stearn, 1976. Boring by macro-organisms in the coral *Montastrea annularis* on Barbados reefs. *Internationale Revue der gesamten Hydrobiologie* 61, 715-45.

McCloskey, L. R., 1970. The dynamics of the community associated with a marine scleractinian coral. *Internationale Revue der gesamten Hydrobiologie* 55, 13-81.

McLean, R. F., 1972. Nomenclature for rock-destroying organisms. *Nature* 240, 490.

Neumann, A. C., 1966. Observations on coastal erosion in Bermuda and measurement of the boring rate of the sponge *Cliona lampa*. *Limnology and Oceanography* 11, 92-108.

Patton, W. K., 1976. Animal associates of living reef corals. In O. A. Jones and R. Endean (eds.), *The Biology and Geology of Coral Reefs*. New York: Academic Press, vol. III, Biology 2, pp. 1-36.

Peyrot-Clausade, M., 1974. Ecological study of coral reef cryptobiotic communities. An analysis of the polychaete cryptofauna. In, *Proceedings of the Second International Symposium on Coral Reefs*. Brisbane: Great Barrier Reef Committee, vol. 2, pp. 269-83.

Peyrot-Clausade, M., 1979. Contribution a l'etude de la cryptofaune des platiers coralliens de la region de Tulear (Madagascar). *Annuaire de l'Institut oceanographie de Paris* 55, 71- 91.

Reichelt, R., 1979. Infaunal polychaetes of reef crest habitats at Heron Island, Great Barrier Reef. *Micronesica* 15, 297-307.

Reish, D. J., 1968. The Polychaetous annelids of the Marshall Islands. *Pacific Science* 22, 208-31.

Rice, M. E., 1969. Possible boring structures of sipunculids. *American Zoologist* 9, 803-12.

Rützler, K., 1978. Sponges in coral reefs. In D. R. Stoddart and R. E. Johannes (eds.), *Coral Reefs: Research Methods*. Paris: UNESCO, Monographs on Oceanographic Methodology, no 5, pp. 299-313.

Sale, P. F., P. S. McWilliam and D. T. Anderson, 1978. Faunal relationships among the near-reef zooplankton at three locations on Heron Reef, Great Barrier Reef, and seasonal changes in this fauna. *Marine Biology* 49, 133-45.

Scoffin, T. P., 1972. Fossilization of Bermuda patch reefs. *Science* 178, 1280-2.

Taylor, J. D., 1976. Habitats, abundance and diets of Muricacean gastropods at Aldabra Atoll. *Journal of Linnean Society of London (Zoology)* 59, 155-93.

Uebelacker, J. M., 1977. Cryptofaunal species/area relationship in the coral reef sponge *Gelliodes digitalis*. In D. L. Taylor (ed.), *Proceedings: Third International Coral Reef Symposium*. Miami: University of Miami, vol. 1, pp. 69-75.

Vivien, M. L. and M. Peyrot-Clausade, 1974. Comparative study of the feeding behaviour of three coral reef fishes (Holocentridae), with special reference to the Polychaeta of the reef cryptofauna as prey. In, *Proceedings of the Second International Symposium on Coral Reefs*. Brisbane: Great Barrier Reef Committee, vol. 2, pp. 179-92.

Warme, J. E. 1975. Borings as trace fossils, and the processes of marine bioerosion. In R. W. Frey (ed.), *A Synthesis of Principles, Problems and Procedures in Ichnology*. New York: Springer-Verlag, pp. 181-227.

Yonge, C. M., 1963. The biology of coral reefs. *Advances in Marine Biology* 1, 209-60.

13 Standards of Performance in Coral Reef Primary Production and Carbon Turnover

DONALD W. KINSEY
University of Georgia Marine Institute, Sapelo Island,
Georgia 31327, U.S.A.*

A REVIEW OF CARBON FLUX

Carbon flux through coral reef systems has been the subject of a number of studies, probably beginning with those of Sargent & Austin (1949, 1954). Most of these studies have reported on total community photosynthesis and respiration, and some of the more recent investigations have included observations of community calcification (see Smith, this volume).

This paper summarizes many of the findings of these metabolic studies, and emphasizes the considerable and somewhat unexpected uniformity that has been found in the carbon flux of a number of coral reef systems from different latitudes and with very differing biological makeup. A number of hypotheses of 'standard' performance are put forward. Additionally, Sournia (1977a) has produced an excellent general review concerning coral reef primary productivity. The review covers at length the kinds of information included in the subsequent tables (except for calcification), and discusses in detail the methodology used by various authors.

All studies considered here are based on the community metabolism of complete communities as reflected by changes in the chemistry of the overlying seawater. Specifically, the carbon dioxide fluxes have been determined, using both CO_2 and O_2 related techniques, and are expressed here in terms of: community gross photosynthesis (P) which is an estimate of all CO_2 fixed by all photosynthetic organisms in the reef community, community gross respiration (R) which is an estimate of all CO_2 released by all decomposition and respiration processes within the community including the respiration of the autotrophs themselves, and community gain in fixed inorganic carbonates (G) which is an estimate of the net precipitation of carbonates. From these three parameters a considerable amount can be inferred concerning fluxes of organic material, changes in standing-crop, and general community and reef growth phenomena. It is outside the scope of this paper to develop hypotheses of Holocene or future reef growth based on the 'standards' of inorganic carbon ($CaCO_3$) accumulation presented here. However, this concept is further developed by Smith (this volume).

*Present address Australian Institute of Marine Science,
P.M.B. No. 3, M.S.O., Townsville, Qld. 4810.

Notwithstanding the number of metabolic studies carried out, the data only loosely suggest consistent trends (table 1). This scatter is probably accentuated by several factors:

1. methodology varies considerably,

2. data handling and interpretation vary considerably, and

3. it is often difficult or nearly impossible to determine the relationships of communities and zonation because of inconsistencies in descriptive terminology. Table 1 summarizes metabolic data available from the published literature for reef flats or other shallow high activity areas. In addition to P, R and G, values for the diel excess production (E = P-R) and the ratio of gross photosynthesis to respiration (P/R) are tabulated. Units used are g C m^{-2} d^{-1} for CO_2 flux through the organic cycle, and kg $CaCO_3$ m^{-2} yr^{-1} for CO_2 flux through the inorganic cycles. These mixed units are used to conform with the most common units in the coral reef literature. Values in table 1 cover a reasonably wide range. However, considering the very loose degree of similarity in the communities and zones represented, and the frequently large differences in methods used, there is probably a surprising degree of consistency. Although the mean value for gross diel photosynthesis (P) and the mean value for gross diel respiration (R) are equal at 7.9 g C m^{-2} d^{-1}, suggesting an overall tendency towards a P/R of 1, it is apparent that there is considerable variation (or simply lack of agreement) in the autotrophic self-sufficiency (P/R) as determined for the individual systems (Mean P/R = 1.17).

The values in table 1 for the diel gain in carbonates (G) are very consistent. This consistency may, in part, reflect that the data originate from only two authors (Kinsey & Smith). This consistency was used by Smith & Kinsey (1976) to suggest the possibility of worldwide uniformity in the calcification rate of shallow reef flat areas (G = 4 kg $CaCO_3$ m^{-2} yr^{-1}).

There has been a tendency for the high values obtained for gross community metabolism of reef flats to be misquoted as representing the activity of 'coral reefs'. This is most unfortunate, as it has commonly allowed coral reefs to be considered as one of the world's most active natural ecosystems. It is very doubtful that this elevated status is justified. Table 2 gives the few metabolic rates available for complete reef systems (i.e. a reef system in which the only sources and sinks are the open sea or the atmosphere, and not other adjacent reef areas). It is apparent that overall activity is just as variable as that in table 1, but at a level more in keeping with that of most other natural ecosystems (Odum 1971). Presumably the large open atolls such as Enewetak would be much less active than the shallow systems reported because of the lower proportion of shallow patch reefs and reef flats. The one point that seems totally consistent in these total 'ecosystem' data is that E is very close to zero (P/R = 1). Thus, the somewhat variable P/R value of reef flats indicated in table 1 may nevertheless be a component of a balanced ecosystem. Kinsey (1979) has found that P/R is typically greater than 1 in high energy zones, establishing that these areas act as organic source zones. Further, low energy zones exhibit P/R typically less than 1 establishing that they act as organic sink zones. The straightforward interpretation of these findings is that excess production in reef perimeter zones is dislodged by turbulence and wave action, then carried downstream to be deposited on the low energy sand flats or in lagoons. There, such material is eventually respired (before or after ingestion by larger animals). The overall effect is that the system is in autotrophic balance (P/R = 1).

Reference	Location	P	-R	E	P/R	G
		(g C m^{-2}d^{-1})				(kg CaCO$_3$ m^{-2} yr^{-1})
Sargent & Austin (1949, 1954)	Rongelap atoll	4	3.5	+0.5	1.1	
Odum & Odum (1955)	Enewetak atoll	10	10	0	1.0	
Kohn & Helfrich (1957)	Kauai (Hawaii)	7.9	6.6	+1.3	1.2	
Odum et al. (1959)	Puerto Rico (various)	5-16	5-19		0.8-1.6	
Goreau et al. (1960)	Puerto Rico				0.7	
Emery (1962)*	Guam	4.5	7	-2.5	0.7	
Gordon & Kelly* (1962)	Oahu (Hawaii)	14	24	-10	0.6	
Jones (1963)	Florida			+0.8-2.0		
Ramachandran Nair Gopinadna Pillai (1972)	Indian Ocean (various)	4-9				
Qasim et al. (1972)	Laccadives	6.2	2.5	+3.7	2.5	
Kinsey (1972)	One Tree Is. (Australia)					4.6
Smith (1973)	Enewetak II	6	6	0	1	4
	Enewetak III	11.6	6.0	+5.6	1.9	4
Smith & Marsh (1973)	Enewetak II	6	6	0	1	
	Enewetak III	10.4	6.0	+4.4	1.7	
Kinsey & Domm (1974)	One Tree Is. (Australia)	7.5	6.8	+0.7	1.1	
Marsh (1974)	Guam	7.2	6.6	+0.6	1.1	
LIMER Team (1976)	Lizard Is. (Australia)					4
Sournia (1976)	Moorea (Tahiti)	7.2	8.4	-1.2	0.9	
Kinsey (1977)	One Tree Is.	7.2	7.4	-0.2	1.0	4.6
	Lizard Is.					3.7
Mean		7.9	7.9			4.2
S.D.		2.7	5.0			0.4
n		(16)	(15)			(6)

Note: Data reported as published. Conversion to g C made if necessary but other corrections not applied
* My interpretation of nuclear or non-specific results

Table 1   Published values for community metabolism in reef flat environments

Table 3 demonstrates that plankton metabolism in the water overlying reef communities has been found to be variable over two orders of magnitude. However, at its highest, it is still an order of magnitude lower than the activity of the total system. Thus, total system metabolism studies give data which are a reasonable approximation to the activities of the benthic community. The majority of the studies covered by table 3 do not include data for plankton respiration. Based on the one study (Kinsey 1978) where both P and R data are available, it seems that respiration may exceed photosynthesis in the water column over reef systems. The logical interpretation is that the elevated R is contributed by zooplankton and detritus (decomposing), both of local reef origin.

It should be stressed that most of the results in the three tables above are based on studies involving relatively few data taken in many cases during only one brief expedition. There is, therefore, little reason for confidence in many of the implied findings except for the fact that certain trends seem to be consistent.

Reference	Location	P	-R	E	P/R	G
		(g C m^{-2} d^{-1})				kg CaCO$_3$ m^{-2} yr^{-1}
Smith & Pesret (1974)	Fanning Is.			0.0	1	1
Smith & Jokiel (1975)	Canton Is.	6.0	5.9	0.06	1	0.5
Sournia & Richard (1976a)	Takapoto (Tuamotos)	4	4	0.0	1	
Kinsey (1977)	One Tree Is. (Australia)	2.3	2.3	−0.06	1	1.5
Kinsey & Davies (1979b)	Lizard Is. (Australia)	3.2	3.2	0.0	1	1.8

* It should be stressed that all the systems included here are relatively shallow with considerable lagoon patch reef development. It seems reasonable to anticipate that the large open atolls such as Enewetak will exhibit much lower overall activity

Note: The Bahama Banks data of Broecker & Takahashi (1966) have not been included as they represent an Atlantic system not clearly conforming to a reasonable definition of a coral reef

Table 2   Published values for community metabolism of 'complete' reef ecosystems*

	Reference	Location	P	-R	P/R
			(g C m^{-2} d^{-1})		
O$_2$	Based Estimations				
	Sargent & Austin (1949)	Rongelap Atoll	0.2		
	Jones (1963)	Florida	0.04−0.06		
	Motoda (1969)	Palau	0.08		
	Johannes et al. (1972)	Enewetak Atoll		0.003−0.03	
	Kinsey (1978)	One Tree Is. (Australia)	0−0.06	0.1−0.3	0−0.3
^{14}C	Estimations*				
	Gordon et al. (1971)	Fanning Is.	0.5		
	Sournia & Richard (1975a)	Moorea (Tahiti)	0.004−0.03		
	Sournia & Richard (1976b)	Vairao (Tahiti)	0.1−0.4		
	Sournia & Richard (1975b)	Tuamotos	0.1−0.3		
	Sournia & Richard (1976a)	Takapoto (Tuamotos)	0.02		
	Sournia (1977b)	Gulf of Elat	0.003−0.01		
	LIMER Team (1976)	Lizard Is. (Australia)	0.1		

*^{14}C estimates tend to lie somewhere between gross and net primary production. The tabulated values are therefore a low estimate for P

Table 3   Published values for planktonic metabolism in the water overlying reef systems

SEASONALITY

Some of the variability suggested by the data in table 1 can probably be explained by considering the fact that seasonality has not been taken into account in comparing the various studies. Kinsey (1977) stressed that seasonality was a major factor in community metabolism, though not in calcification. This variation might be attributed tentatively to the relatively extreme latitude of the study site reported (One Tree Island, 23°S).' However, subsequent studies (Kinsey 1979) have established that almost identical proportional seasonality exists at two other sites, Lizard Island (14°S) and Kaneohe Bay, Hawaii (21°N). Figure 1 indicates the seasonality pattern found for Lizard Island.

'STANDARD' PERFORMANCE

How comparable is the metabolic performance of typical unperturbed reefs — if there is an equivalent basis for comparison? Is there a standard of reef performance, departure from which can be taken as indicative of perturbation of some kind? Table 4 summarizes all those metabolic studies in the literature, including those reported by Kinsey (1979), where one can confidently claim that the reef flat reported conforms to the concept of a fully developed (at or near present-day low tide level), areally extensive (at least 100 m across), high activity, perimeter zone of the coral/algal kind. It does not include transects of 'homogenous' algal pavement or the very narrow (2–50 m) bands of heavy cover which may occur on the edges of many lagoonal patch reefs, unless they are included as part of larger high activity systems. Studies such as those of Sargent & Austin (1949), Odum & Odum (1955), Gordon & Kelly (1962), and Ramachanadran Nair & Gopinadha Pillai (1972) which are based on inadequate or doubtful data, or on data from extreme seasons, also have not been included in table 4. It is apparent that the biotically heterogeneous reef flats do exhibit a remarkably uniform 'standard' of performance, seemingly regardless of where they are (at least in the Pacific Ocean).

The proposed 'standard' reef flat is:
$$P = 7 \pm 1; \, P/R = 1 \pm 0.1; \, G = 4 \pm 1.$$
Because such reef flats are by far the most easily monitored area on any reef, the 'standard' provides an excellent basis for checking for the effects of stresses and perturbations. It seems inevitable that any reasonable interference will cause a shift in at least one of the parameters listed. This type of response has been found to a number of different perturbations in various coral reefs (Kinsey & Domm 1974; Smith *et al.* 1978; Kinsey 1979; Kinsey & Davies 1979a). In every case examined, the departure from 'standard' performance was marked and, in some cases, readily detectable before any visual change in the reef community composition.

Uniformity of all, or at least most, community metabolic functions seems likely to be a pronounced feature of comparisons made between coral reefs (Kinsey 1979). The principal obstruction to detecting this uniformity seems to be that of identifying an equivalent basis for comparison in the first place. This complication is frequently aggravated by the differences between authors in defining morphological features and community structure. There is also the further complication of differences in methods, experimental philosophy, and interpretation of results.

MODAL DISTRIBUTION OF COMMUNITY METABOLISM

If we accept that reef flat perimeter zones exhibit a predictable metabolic makeup of the community, then it is reasonble to seek similar uniformity in activity for other areas of the reef system.

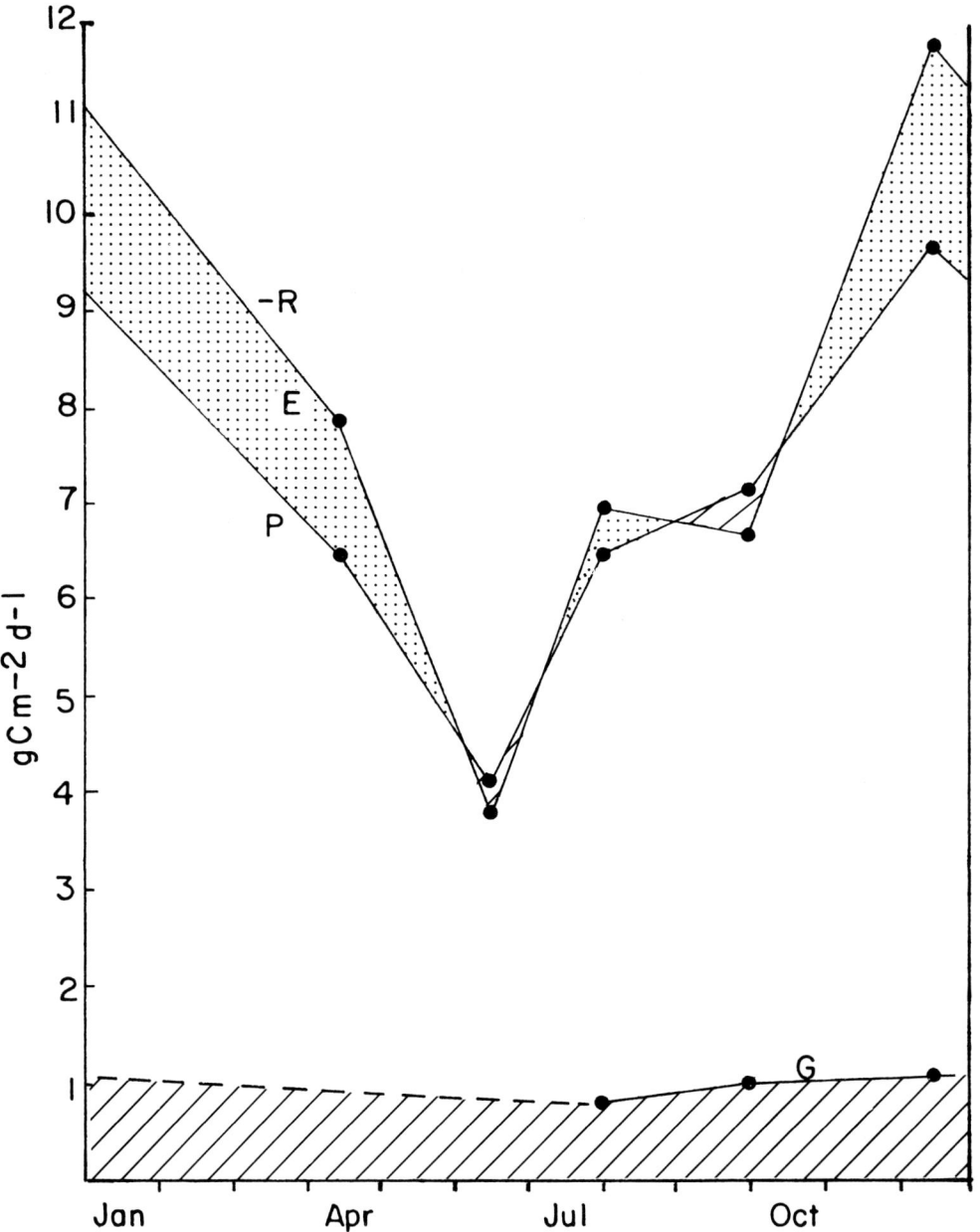

Figure 1   Daily metabolic rates as a function of the time of year for the seaward reef flat, Transect A, at Lizard Is. (data after Kinsey 1979). The cross-hatched areas represent net gain, and the stippled area net loss

Location	Reference	P (g C m^{-2}d^{-1})	P/R	G (kg CaCO$_3$ m^{-2} yr^{-1})
Enewetak II (seaward coral/algal)	Smith 1973	6.0	1	4
Kauai (seaward fringing reef)	Kohn & Helfrich* 1957	7.2	1.1	
Guam (seaward fringing reef)	Marsh 1974*	6.6	1.0	
Moorea (seaward fringing reef)	Sournia 1976	7.2	0.9	
One Tree Is. DK13 (seaward reef flat; coral/algal)	Kinsey 1977	7.2	1.0	4.6
Lizard Is. A2 (seaward reef flat; coral/algal)	Kinsey 1979	7.8	0.9	4.6
Lizard Is. D1 (lagoon reef flat; coral/algal)	Kinsey 1979	7.0	1.2	3.1
Mean		7.0	1.0	4.0
±S.D.		±0.6	±0.1	±0.7

*Values as listed vary slightly from those published because of recalculation using equivalent data handling to that in the present study

Table 4   The evidence for a 'standard' performance for extensive coral/algal reef flats

If any typical, well developed Indo-Pacific coral reef is examined on a scale of tens to hundreds of metres, it is apparent that the physiography is dominated by two levels. Simplistically, these can be described as reef flats just below the low tide level, and lagoons containing two to many metres of water (fig. 2). The lagoons, of course, may contain quite complex patch reef or line reef structures but these in turn conform to the same basic very shallow or deep distribution of levels. The slopes within this overall structure occupy a relatively small proportion of the total area (15% in the Central Kaneohe Bay Sector, Smith *et al.* 1978), and an appreciable contributor to this category is frequently the slopes formed by sediments prograding away from active zones on reef flats into lagoons. Thus active, biologically producing slopes are an even smaller part of the whole system, inside the outer-reef crest. However, the outer seaward slopes of the reef can be quite gradual and therefore may represent an important exception to the simple vertical bimodality stressed for the rest of the system.

If the shallow, physiographically dominant area, the reef flat is considered further and still on a scale of tens to hundreds of metres, it can readily be seen that there is usually a further bimodal distribution to be considered (fig. 2). This horizontal distribution is between the ubiquitous active perimeter zone, and the sand and rubble flat usually, but not always, leeward of the active zone. From the air, the boundary between these two zones is always very distinct.

If the active perimeter zones are considered on a smaller scale of metres or less, their uniform operational role in the reef system (stressed generally in table 1, and specifically in table 4) seems to be little supported by the very great variability evident in their biological makeup. The perimeter community can be anything from almost smooth algal pavement to a uniform cover with living corals. In these two extremes, the community is uniform and continuous and there is no apparent scope for any breakdown into further

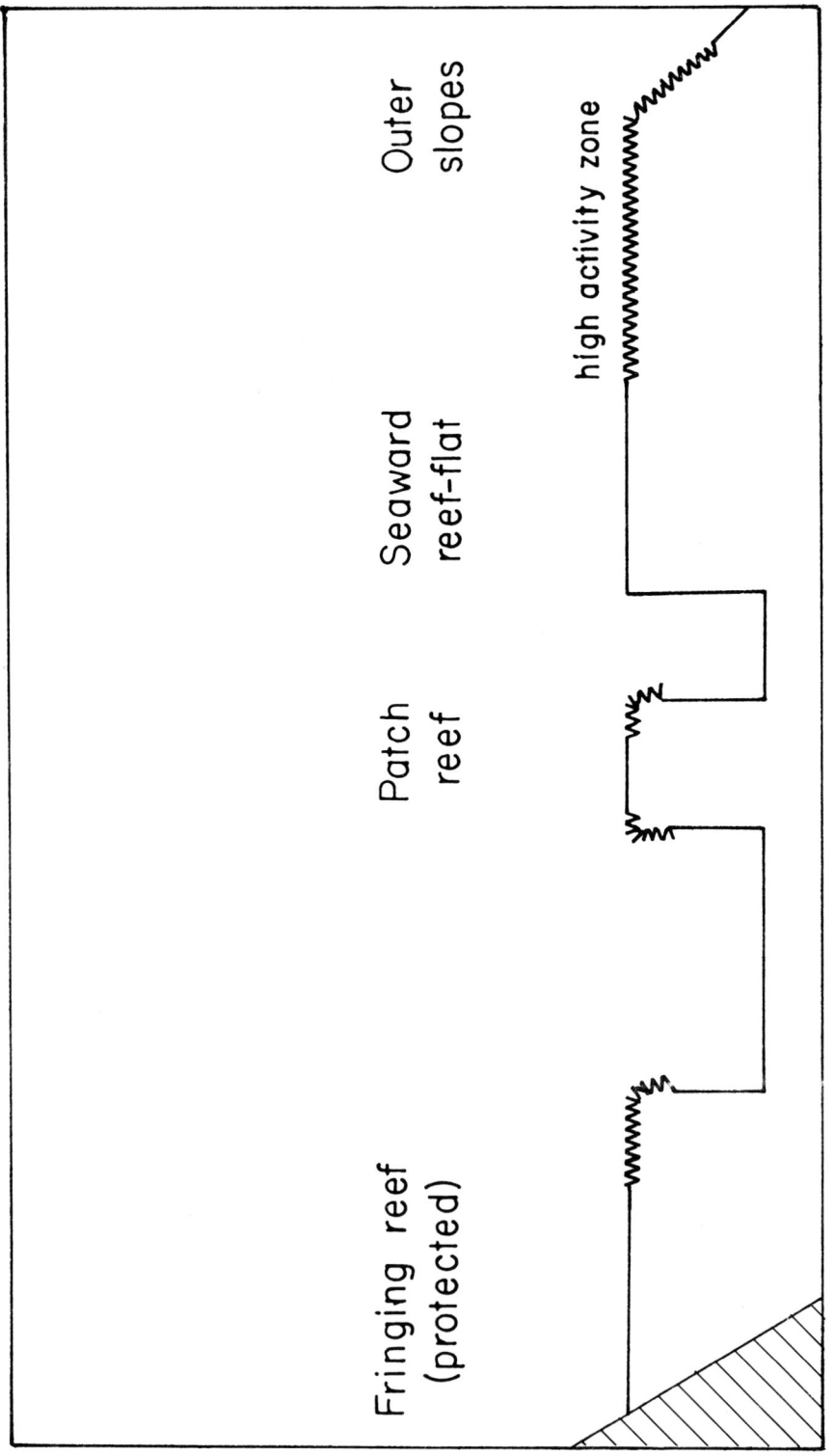

Figure 2   Diagrammatic representation of the vertical and horizontal bimodality of coral reef morphology and zonation. The outer slope is the principal exception to this concept. The zig-zag line represents areas of appreciable metabolic activity. The straight lines represent sediment bottom. The vertical faces do have high activity over a small area but it should be stressed that the diagram has greatly exaggerated vertical scale

bimodal distributions. However, in the more typcial case of mixed coralline algae/coral substratum on the reef flat, there is, in some situations, a lack of morphological and biological homogeneity which requires another level of bimodality to be considered (i.e. coral/algal formations interspersed with sand/rubble bottom).

An examination of published metabolic data, with all the preceding concepts in mind, raises some interesting possibilities. The bimodal distribution of coral reef calcification' activity has been recognized by Smith & Kinsey (1976) and Smith (1978). These papers accepted the typical reef perimeter calcification rate to be 4 kg $CaCO_3$ m^{-2} yr^{-1} and the more extensive lagoonal systems were considered to exhibit an approximate uniformity in the range 0.5–1.0 kg $CaCO_3$ m^{-2} yr^{-1}. Similarly it was established in tables 1 and 2 that primary production rates in the literature fall into two general groupings: high rates determined for shallow reef flats or fringing reefs (P = approximately 7–9 g C m^{-2} d^{-1}); and low rates determined for total or dominantly lagoonal systems. However, at the level of discrimination allowed by the information available in these published papers, the production data could hardly be classified as neatly bimodal.

All available metabolic data for specific reef zones, and for isolated reef formations, sand and rubble bottom, etc., have been considered (Kinsey 1979) in relation to the physiographic modality outlined above. This consideration has resulted in the proposition that many (and possibly all) unperturbed Pacific reef systems can be considered to have a metabolic performance which is bimodally distributed in a general sense (table 5), but which is really distributed into four and possibly five discrete modes (table 5) if the finer aspects of zonation outlined above are considered. Considering table 5, mode 1(a) is proposed to be that performance exhibited by individual coral/algal formations in the

Mode		P (g C m^{-2} d^{-1})	G (kg $CaCO_3$ m^{-2} yr^{-1})
1(a).	Coral/algal cover as discrete heads, or continuous where adequate water available	20	10
1(b).	Coral/algal cover with limited water available	10	6
1(c).	Algal pavemant	5	4
2.	Sand/rubble	1	0.5
OR GENERALIZED:			
1.	Reef flats and all extensive, present-day perimeter zones	5–10	3–5
2.	Sand/rubble	1	0.5

The outer reef slopes should perform as per the relevant category down to about 5 m. Below that depth, progressive attenuation occurs (Smith & Harrison 1977)

All lagoonal systems should be compounded, on an areal basis, from the above modes

An additional non-specific mode for algal/sea grass beds should really be considered

1(a), 1(b) and 1(c) modes are included in the generalized reef flat mode 1, but 1(a) and 1(b) are only ever a component of the whole, i.e., an extensive present-day reef flat will never have overall metabolic rates as high as P = 20 or P = 10, and G = 10 or G = 6
Concept after Kinsey (1979)

Table 5 Proposed modes for coral reef metabolism (each category refers to 100% cover)

typical present-day reef flat or patch reef perimeter. 1(a) is also the ultimate mode attainable by continuous coral/algal cover under adequate but not excessive water depth (<5 m). This is the situation probably occurring in many reefs during a rise in sea level, but relatively uncommon today except in some outer slope areas. 1(b) is the mode occurring in narrow bands on the outer edges of relatively low energy reef flats (e.g. patch reefs). 1(b) is assumed to be a vestigal, depth-limited present-day manifestation of 1(a). 1(c) is the normal mode occurring on the outer section (algal pavement) of present-day, standing sea level, high energy reef flats. Mode 2 is the metabolic performance to be expected of all sand/rubble areas whether they be the low activity bottom between the 1(a) mode reef flat formations, the inner inactive reef flat (sand flats), or the lagoon floor. In really extensive zones, present-day reefs can probably all be generalized to the overall modes 1 and 2. It should be stressed that the 'standard' reef flat proposed in the previous section is a specialized case of mode 1, as it specifically excludes algal pavements or very high activity edges occurring on patch reefs.

CONCLUSIONS

In general, it seems fairly certain that coral reef community metabolism and productivity will be found to conform to relatively narrow-range standard modes. Future research hopefully will clarify these hypotheses and establish the limiting conditions for conformity to the standard modes. More importantly, future research should attempt to find the factors controlling this conformity. The value of 'standards' to the investigation of ecosystem perturbations seems the most useful aspect of these findings.

LITERATURE CITED

Broecker, W. S. and T. Takahashi, 1966. Calcium carbonate precipitation on the Bahama Banks. *Journal of Geophysical Research* 71, 1575-602.

Emery, K. O., 1962. Marine geology of Guam. *United States Geological Survey Professional Paper* 403-B.

Gordon, D. C., R. O. Fournier and G. J. Krasnick, 1971. Note on the planktonic primary production in Fanning Island Lagoon. *Pacific Science* 25, 228-33.

Gordon, M. C. and H. M. Kelly, 1962. Primary productivity of a Hawaiian coral reef: A critique of flow respirometry in turbulent waters. *Ecology* 43, 473-80.

Goreau, T. F., V. T. Llauger, E. L. Mas and E. R. Seda, 1960. On the community structure, standing crop and oxygen balance of the lagoon at Cayo Turrumote. *Association of Island Marine Laboratories, 3rd Meeting* 8-9.

Johannes, R. E., J. Alberts, C. D'Elia, R. A. Kinzie, L. R. Pomeroy, W. Sottile, W. Wiebe, J. A. Marsh Jr., P. Helfrich, J. Maragos, J. Meyer, S. Smith, D. Crabtree, A. Roth, L. R. McCloskey, S. Betzer, N. Marshall, M. E. Q. Pilson, G. Telek, R. I. Clutter, W. D. Du Paul, K. L. Webb and J. M. Wells Jr., 1972. The metabolism of some coral reef communities: a team study of nutrient and energy flux at Eniwetok. *Bioscience* 22, 541-3.

Jones, J. A., 1963. Ecological studies of the southeastern Florida patch reefs. Part I. Diurnal and seasonal changes in the environment. *Bulletin of Marine Science* 13, 282-307.

Kinsey, D. W., 1972. Preliminary observations on community metabolism and primary productivity of the pseudo-atoll reef at One Tree Island, Great Barrier Reef. In C. Mukundan and C. S. Gopinadha Pillai (eds.), *Proceedings of the Symposium on Corals and Coral Reefs*. The Marine Biological Association of India, Cochin, pp. 13-32.

Kinsey, D. W., 1977. Seasonality and zonation in coral reef productivity and calcification. In D. L. Taylor (ed.), *Proceedings: Third International Coral Reef Symposium*. Miami: University of Miami, vol. 2, pp. 383-8.

Kinsey, D. W., 1978. Productivity and calcification estimates using slack-water periods and field enclosures. In D. R. Stoddart and R. E. Johannes (eds.), *Coral Reefs: Research Methods*. Paris: UNESCO, Monographs on Oceanographic Methodology, No. 5, pp. 439-68.

Kinsey, D. W., 1979. Carbon turnover and accumulation by coral reefs. Ph.D. dissertation, University of Hawaii, 248pp.

Kinsey, D. W. and P. J. Davies, 1979a. Effects of elevated nitrogen and phosphorous on coral reef growth. *Limnology and Oceanography* 24, 935-40.

Kinsey, D. W. and P. J. Davies, 1979b. Carton turnover, calcification and growth in coral reefs. In P. A. Trudinger and D. J. Swaine (eds.), *Biogeochemical Cycling of Mineral-Forming Elements*. Amsterdam: Elsevier Scientific Publishing Company, pp. 131-62.

Kinsey, D. W. and A. Domm, 1974. Effects of fertilization on a coral reef environment — primary production studies. In, *Proceedings of the Second International Symposium on Coral Reefs*. Brisbane: Great Barrier Reef Committee, vol. 1, pp. 49-66.

Kohn, A. J. and P. Helfrich, 1957. Primary organic productivity of a Hawaiian coral reef. *Limnology and Oceanography* 2, 241-51.

LIMER 1975 Expedition Team, 1976. Metabolic processes of coral reef communities at Lizard Island, Queensland. *Search* 7, 463-8.

Marsh, J. A., Jr., 1974. Preliminary observations on the productivity of a Guam reef flat community. In, *Proceedings of the Second International Symposium on Coral Reefs*. Brisbane: Great Barrier Reef Committee, vol. 1, pp. 139-45.

Motoda, S., 1969. An assessment of primary productivity of a coral reef lagoon in Palau, Western Caroline Islands based on the data obtained during 1935-37. *Records of Oceanographic Works in Japan* 10, 65-74.

Odum, E. P., 1971. Fundamentals of Ecology. Philadelphia: W. B. Saunders Company, 574pp.

Odum, H. T. and E. P. Odum, 1955. Trophic structure and productivity of a windward coral reef community on Eniwetok Atoll. *Ecological Monographs* 25, 291-320.

Odum, H. T., P. Burkholder and P. Rivero, 1959. Measurements of productivity of turtle grass flats, reefs and the Bahia Fosforescente of southern Puerto Rico. *Publications of the Institute of Marine Science* 6, 159-70.

Qasim, S. Z., P. M. A. Bhattathiri and C. V. G. Reddy, 1972. Primary production of an atoll in the Laccadives. *Internationale Revue der gesamten Hydrobiologie* 57, 207-25.

Ramachandran Nair, P. V. and C. S. Gopinadha Pillai, 1972. Primary productivity of some coral reefs in the Indian Seas. In C. Mukundan and C. S. Gopinadha Pillai (eds.), *Proceedings of the Symposium on Corals and Coral Reefs*. The Marine Biological Association of India, Cochin, pp. 33-42.

Sargent, M. C. and T. S. Austin, 1949. Organic productivity of an atoll. *Transactions of the American Geophysical Union* 30, 245-9.

Sargent, M. C. and T. S. Austin, 1954. Biologic economy of coral reefs, Bikini and nearby atolls, Marshall Islands. *United States Geological Survey Professional Paper* 260-E, 293-300.

Smith, S. V., 1973. Carbon dioxide dynamics: a record of organic carbon production, respiration, and calcification in the Eniwetok reef flat community. *Limnology and Oceanography* 18, 106-20.

Smith, S. V., 1978. Coral-reef area and the contributions of reefs to processes and resources of the world's oceans. *Nature* 273, 225-6.

Smith, S. V. and J. T. Harrison, 1977. Calcium carbonate production of the *Mare incognitum*, the upper windward reef slope, at Enewetok Atoll. *Science* 197, 556-9.

Smith, S. V. and P. L. Jokiel, 1975. Water composition and biogeochemical gradients in the Canton Atoll lagoon. 2. Budgets of phosphorus, nitrogen, carbon dioxide, and particulate materials. *Marine Science Communications* 1, 165-207.

Smith, S. V. and D. W. Kinsey, 1976. Calcium carbonate production, coral reef growth, and sea level change. *Science* 194, 937-9.

Smith, S. V. and J. A. Marsh, Jr., 1973. Organic carbon production on the wind-ward reef flat of Eniwetok Atoll. *Limnology and Oceanography* 18, 953-61.

Smith, S. V. and F. Pesret, 1974. Processes of carbon dioxide flux in the Fanning Island Lagoon. *Pacific Science* 28, 225-45.

Smith, S. V. *et al.* (10 others), 1978. Kaneohe Bay sewage relaxation experiment. Pre-diversion report. *Hawaii Institute of Marine Biology Report.* 166 pp.

Sournia, A., 1976. Oxygen metabolism of a fringing reef in French Polynesia. *Helgoländer wissenschaftliche Meeresuntersuchungen* 28, 401-10.

Sournia, A., 1977a. Analyse et bilan de la production primaire dans les récifs coralliens. *Annales de L'Institut Océanographique* 53, 47-74.

Sournia, A., 1977b. Notes on primary productivity of coastal waters in the Gulf of Elat (Red Sea). *Internationale Revue der gesamten Hydrobiologie* 62, 813-19.

Sournia, A. and M. Ricard, 1975a. Production primaire planctonique dans deux lagons de Polynésie française (île de Moorea et atoll de Takapoto). *Comptes Rendus Hebdomadaires Des Séances De l'Academie Des Sciences. Serie D* 280, 741-3.

Sournia, A. and M. Ricard, 1975b. Phytoplankton and primary productivity in Takapoto Atoll, Tuamotu Islands. *Micronescia* 11, 159-66.

Sournia, A. and M. Ricard, 1976a. Données sur l'hydrologie et la productivité du lagoon d'un atoll fermé (Takapoto, Iles Tumaotu). *Vie et Milieu* 26, 243-79.

Sournia, A. and M. Ricard, 1976b. Phytoplankton and its contribution to primary productivity in two coral reef areas of French Polynesia. *Journal of Experimental Marine Biology and Ecology* 21, 129-40.

14 The Primary Productivity of Plant Communities on Coral Reefs

A. W. D. Larkum

School of Biological Sciences, University of Sydney, N.S.W. 2006.

INTRODUCTION

Plants produce new organic matter by the process of photosynthesis. The total production of new organic matter is termed *gross production,* whereas that which remains after some has been respired by the plant to provide energy for its metabolism is termed *net production.* Net production over unit time is usually taken to be equivalent to the *primary productivity* (the rate of organic matter production measured over a period of days or months); see Westlake (1963) and Larkum (1981). Many workers use the terms gross primary productivity and net primary productivity but these are superfluous if primary productivity is defined as above.

Production of new organic material by agricultural crops is easily assessed within a reasonable period of time (days or months). In natural and especially in marine systems this is not possible because the newly produced organic matter quickly passes into the food chain. In such situations, primary productivity is often assessed by measuring rates of photosynthesis and respiration over short periods.

Net photosynthesis is defined as the rate of carbon dioxide fixation in the light without correction for respiratory release of carbon dioxide (it may therefore be negative at low light levels). *Gross photosynthesis* is defined as the rate of carbon dioxide fixation in the light *minus* the (negative) rate of carbon dioxide evolution in the dark (it is therefore always positive or zero). The rate of respiration, i.e., the rate of carbon dioxide evolution in the dark, is often assumed to be the same in both light and dark. This assumption is convenient for calculating primary productivity, but it may not always be true, especially where photorespiration is involved (see Jassby 1978; Larkum 1981).

Primary productivity may then be redefined as 'the gross amount of carbon dioxide fixed during the light period *minus* the respiratory carbon dioxide evolved over 24 h'. Figure 1 illustrates how the photosynthesis *v.* light intensity curve can be used together with the diel light intensity curve to estimate gross photosynthesis (cf. Marsh 1970). If oxygen exchange is measured, the values may be converted to carbon dioxide equivalent by using a conversion factor of 1.2 (Ryther 1956). This technique can only be used where a community is predominated by plants. When animals form a significant part of a community, as commonly occurs, the primary productivity is difficult to determine and *community production* is measured instead. This is defined as 'the amount of carbon dioxide fixed during the light period *minus* the amount of carbon dioxide respired by the community of plants and animals over 24 h' (see Kinsey, chapter 13 this volume).

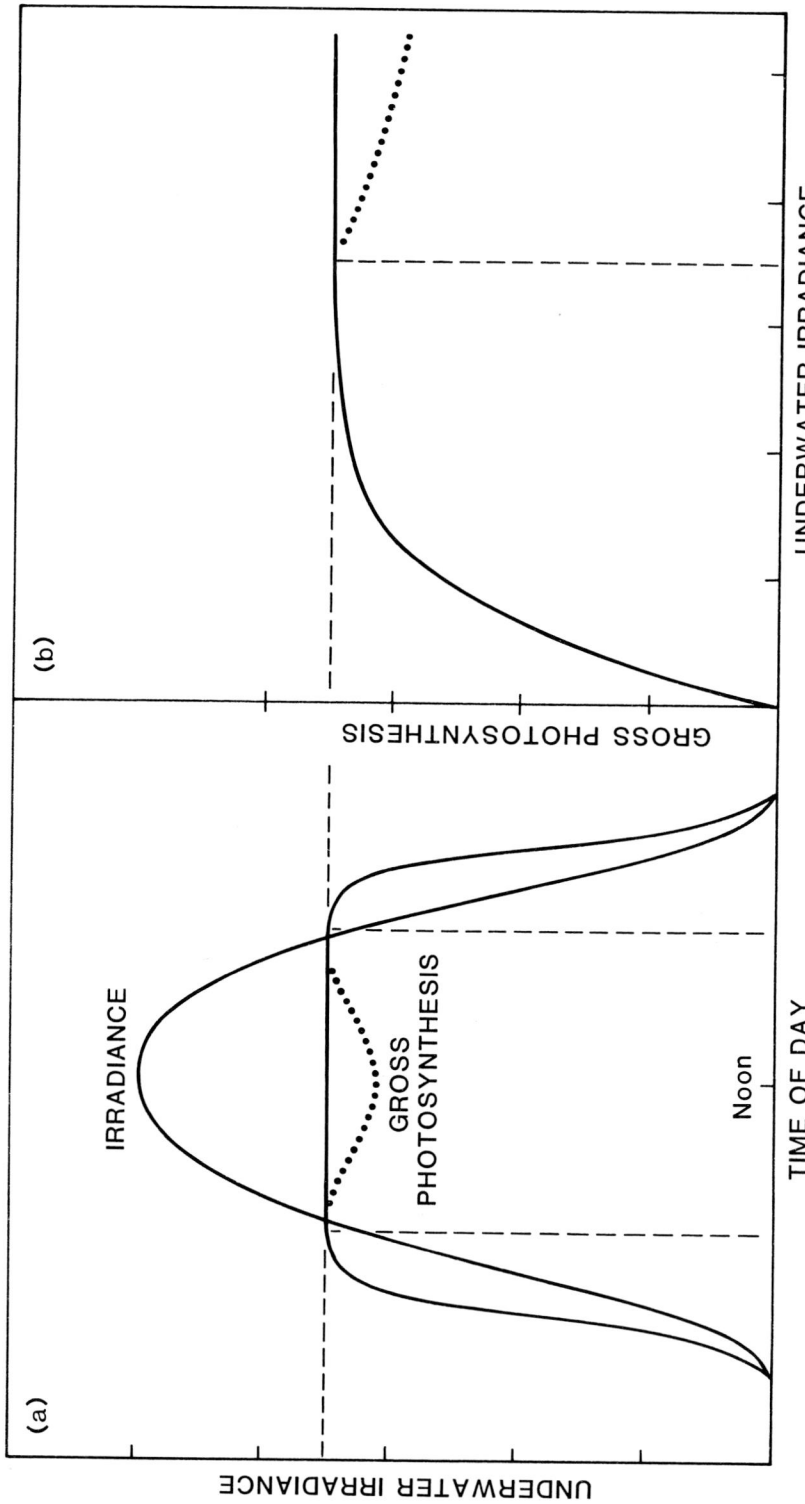

Figure 1   (a) Hypothetical curve for the daily pattern of underwater irradiance and gross photosynthesis.
b) Hypothetical curve for the relationship of gross photosynthesis to light intensity for plants in shallow water on a coral reef. The dashed lines indicate the point at which photosynthesis is light-saturated. The dotted lines apply when photoinhibition occurs, resulting in a midday depression in photosynthesis (see, for example, Sournia 1976b).

LIMITS TO PRIMARY PRODUCTIVITY

The highest sustainable yield from land plants is 90 tonnes dry matter ha^{-1} yr^{-1} (for Napier grass: Warren-Wilson 1967), which represents a mean daily primary productivity of 8 g C m^{-2} d^{-1} and a conversion efficiency of solar to chemical energy of 1.6% (Boardman & Larkum 1974). By reason of the low diffusion coefficient for gases in water, it seems unlikely that aquatic plants would have a primary productivity equal to the best land plants and this has been borne out in practice (see e.g. Tamiya 1957; Talling 1961; Fogg 1968). Over the past forty years, much work has been devoted to the efficiency of light energy conversion of plants. Algae have been a favoured experimental organism. This work has shown that green, brown and red algae are no more efficient in harvesting solar energy than higher plants (see 'Quantum Yield' in Rabinowitch 1951; Gaffron 1960; Radmer & Kok 1977).

Annual mean primary productivity of 8 g C m^{-2} d^{-1} is therefore the upper limit that can possibly be expected from a coral reef plant community. However, higher values can be achieved for shorter periods and, as with land plants, it is possible that under exceptional circumstances a conversion efficiency of 5% can occur, i.e., 40 g C m^{-2} d^{-1} (Talling 1961).

FACTORS LIMITING THE PRIMARY PRODUCTIVITY OF CORAL REEF COMMUNITIES

General Considerations

The major factor determining the primary productivity of a plant is its photosynthetic capacity (defined as the saturation level of the P v. I curve in figure 1). Also important are the ratio of photosynthetic rate to respiratory rate (P:R ratio) and the extent of grazing by animals (see Hatcher, chapter 10 this volume).

Raven *et al.* (1979) have suggested that the photosynthetic capacity of algae varies according to their ecological characteristics. According to Grime (1974, 1977), plants may be classified into three groups on ecological considerations: canopy-dominant, opportunist (ruderal) and stress tolerant; canopy-dominant algae overtop other algae to sequester the primary resource, light; opportunist algae grow on recently disturbed (cleared) surfaces; stress-tolerant algae grow in stressed situations, e.g., under low light. Raven *et al.* (1979) propose that those stress-tolerant algae which specialize as shade algae exhibit low photosynthetic capacity. Dominant and opportunistic types of algae exhibit medium and high photosynthetic capacity respectively. There are few dominant algae (in the sense of Grime) on coral reefs. Most of the benthic algal community are kept at an early successional stage by grazing. As a result, there are many opportunist algae with high photosynthetic capacity.

The maximum photosynthetic capacity of (opportunist) algae seems to be in the range of 200-300 μmoles CO_2 mg chlorophyll^{-1} h^{-1} (Downton *et al.* 1976; Burris 1977; Scott & Jitts 1977; Raven *et al.* 1979) and this is equally true for reef algae (see Downton *et al.* 1976; Scott & Jitts 1977). To obtain a rate of 8 g C m^{-2} d^{-1} with an assumed period of 10 h of light-saturated photosynthesis per day and a P/R ratio of 5, a film of algae having a chlorophyll density of about 100 mg choropylyy m^{-2} would suffice, provided that all the cells maintained maximum photosynthetic capacity. In actual fact, the chlorophyll densities of most communities are around 1 g chorophyll m^{-2} (Talling 1961) and this applies also to coral reefs (Odum *et al.* 1968). The tenfold difference is partly explained as a result of shading of lower layers of cells (Raven 1978), which exist under a different spectral regime, and partly as a result of other limiting conditions. It should also be borne in mind that the maximum photosynthetic capacity is rarely achieved in natural communities. The limits to photosynthetic capacity of marine plants have been discussed fully elsewhere (Morris 1974; Nielsen 1975; Larkum 1981). The most important factors are stirring, light intensity (submarine irradiance), temperature, nutrient supply, inorganic carbon supply and oxygen concentration.

There has been little work on the effect of these limiting factors on coral reef algae but it seems unlikely that light intensity, temperature and inorganic carbon supply are limiting in this environment. Nutrient supply, especially of nitrogen and phosphorus, is probably the major limitation to productivity (see below, also Crossland, Chapter 5 this volume).

Stirring will also be an important limitation in many regions of the reef — but the windward crest, front, and outer edge of a reef and areas of constant current activity will not be limited in this regard. Areas in the region of the crest-top and in the immediate back-reef may also be well stirred for much of the day since, depending on the height of the lagoon at low water, a fast stream of water may occur as water flows off the reef for a significant part of the tidal cycle.

Probably a very important limitation to the photosynthesis of algae in the lagoon, back-reef and crest is inhibition by oxygen (Downton *et al.* 1976; Turner & Brittain 1962). Photosynthesis can be inhibited by up to 60% at very high oxygen tensions. Reef waters probably never reach more than 300% air saturation (Kinsey & Kinsey 1967; Kinsey 1972; Larkum, unpub.) but localized concentrations around algae may be much higher resulting in a very significant inhibition of photosynthesis.

Nutrient Limitation

The levels of nitrogen in reef waters are $0.03-0.3$ μMolar ($0.2-2.0$ μg atom 1^{-1}). These are low and differ only slightly from the levels of surface tropical ocean waters. The levels of phosphorous (phosphate) in reef waters are $0.002-0.02$ μMolar. These are also low and also differ only slightly from the levels of surface ocean waters. The low levels of both these key elements suggest that they might limit productivity on coral reefs, as has been shown with other sytems. Some work has indeed shown that fertilization by nitrogen and phosphorus increased the oxygen production of a coral micro-atoll community (Kinsey 1974) and the photosynthesis of a coral reef alga, *Chlorodesmis fastigiata* (Larkum in Kinsey 1974; Larkum, unpub.). Despite the low levels of nutrients in reef waters, algal (and seagrass) communities on coral reefs have a high productivity (see above). It may well be that very efficient nutrient cycles exist in the coral reef communities similar to those in rain forests or seagrass communities. If this is so, fertilization with nitrogen and phosphorus may have very little effect on the overall primary productivity of a coral reef. The effect of fertilization might be to enhance the primary productivity of one component, say the phytoplankton, at the expense of another component, say the turf algae. This would alter the balance of component productivities, and possibly the structure of food webs of a reef, while leaving the overall productivity unaltered.

Grazing

Grazing may increase the productivity of communities as several previous studies have shown (Black 1965; Lewis 1977). It should therefore be possible, at least in theory, to define grazing pressure and to study the change of primary productivity along a gradient in grazing pressure. Along such a gradient, primary productivity should decrease in regions of highest and lowest grazing pressure, i.e., with over-grazed and under-grazed situations. The grazing pressure of herbivores on primary producers on coral reefs has not been investigated but Larkum, Borowitzka & Day (in prep.) have speculated that the system is nicely poised near the point of optimum grazing pressure. Grazing is discussed further by Hatcher (chapter 10 this volume).

PRIMARY PRODUCERS ON THE REEF

Large Benthic Reef Algae

Few large fleshy benthic algae are abundant on coral reefs. The reason for this is not entirely clear but an important factor is undoubtedly fish grazing (Randall 1961; Wanders 1977; Larkum, Borowitzka & Day, in prep.) and other herbivorous activity (Sammarco *et al.* 1974). There are few canopy dominants in the sense of Grime (1974, 1977). *Sargassum* spp. are the main representatives but little is known of their photosynthetic capacity or contribution to reef productivity. Opportunist algae are common both in the turf algae and, if they escape by growth (Dayton 1975), in the large benthic algae. Especially prominent amongst these are *Halimeda* spp. which are calcified and may contribute substantially to reef formation and especially to the sandy substrates of reefs (Finckh 1904; Emery *et al.* 1949). Hillis-Colinvaux (1974) estimated from field observations of cover, and laboratory experiments of oxygen exchange, a primary productivity of 2.3 g C m^{-2} d^{-1} for *H. opuntia* on rocks. *Laurencia* spp., *Hydroclathrus clathratus* and *Caulerpa* spp. may also be seasonally abundant but there are no reports on their contribution to the annual mean reef primary productivity. Sournia (1976b) has reported on the photosynthetic capacity of several fleshy algae. Laboratory experiments on the photosynthesis and respiration of algae under "ideal" conditions are useful for estimating primary productivity. However, natural conditions are not necessarily "ideal" and may lead to a very different *in situ* primary productivity (see Larkum 1981).

Crustose Coralline Algae

Crustose coralline algae are abundant in all parts of a reef (Adey & Vassar 1975; Littler & Doty 1975). They bind loose fragments of coral and are particularly important in consolidating the reef.

The contribution of these algae to reef productivity is not fully clear as yet. Marsh (1970) obtained a rate of 45 μmoles CO_2 mg chlorophyll^{-1} h^{-1} for growth of crustose coralline algae (mainly *Porolithon* spp.) and estimated an annual mean primary productivity of 0.66 g C m^{-2} d^{-1}. Littler & Doty (1975) estimated a primary productivity of 2.2–2.4 g C m^{-2} d^{-1} but recalculation of their data indicates a more realistic value of 0.94 g C m^{-2} d^{-1} (for *Poroliton onkodes*).

Considering the abundance of these algae on coral reefs it is very possible that they are very important primary producers in the coral reef environment. In making such a prediction one has to bear in mind the very large actual surface area of coral reef substrates. Dahl (1973) has shown that the actual surface area to projected surface area ratio may be as high as 5 in certain regions of coral reef.

Turf Algae

Two independent investigations, one in the Caribbean (Wanders 1976) and the other on the Great Barrier Reef (Larkum, Borowitzka & Day, in prep.), have indicated the very great importance of turf algae to overall reef productivity. The turf algae have a high photosynthetic rate (100-200 μmoles 0_2 mg chlorophyll^{-1} h^{-1}; Larkum, Borowitzka & Day, in prep.) and are intensively grazed by fish and invertebrates (Hatcher, chapter 10 this volume). They are abundant on all limestone substrates of the reef. If these reports of high productivity are correct, then turf algae must represent the largest component of primary production on coral reefs (cf. Lewis 1977).

Phytoplankton

The phytoplankton of coral reefs has been little studied. It has received much less attention than the zooplankton. There is no clear evidence, for example, whether the phytoplankton of reefs is distinct from the open ocean plankton, as is the case for the zooplankton (Sale *et al.* 1976, 1978). The few studies of photosynthetic and respiratory activity suggest very low rates (Sournia & Ricard 1976; Scott & Jitts 1976). The abundance of phytoplankton is often low (Jeffrey 1968; Sournia & Ricard 1976). This suggests that the overall primary productivity of reef phytoplankton is low. However, phytoplankton blooms are common in lagoons and other areas of coral reefs. It is probable, therefore, that much closer attention will have to be given to the stochastic nature of phytoplankton blooms, and to the turnover of phytoplankton in reef waters, before a proper assessment of their importance to reef productivity can be made (cf. Griffiths 1976).

Sand Algae

Where seagrasses are not abundant, as on Australian reefs, the lagoon sands provide substrates for the growth of many kinds of algae; the majority of these are unicellular or filamentous. The few investigations so far carried out indicate a rather low level of productivity of these algae (Plante-Cuny 1973; Sournia 1976). The lack of adequate light penetration into the substrate makes it unlikely that the thin film of algae at the surface would ever contribute significantly to the primary productivity of reefs. However, much more work is needed.

Zooxanthellae

The symbiotic dinoflagellate algae of hermatypic corals must play an enormous role in the overall production of a coral reef. However, due to the symbiotic nature of the coral association, this portion of reef primary productivity takes place largely in a closed compartment which only indirectly communicates with the rest of the reef, via the few fish that eat coral (Randall 1974), coral mucous and coral detritus. The many studies on coral metabolism indicate that the corals are slightly autotrophic or just in trophic balance; the photosynthesis of the zooxanthellae during the day accounting for the energetic needs of the host and alga (McCloskey *et al.* 1978). However, as McCloskey *et al.* (1978) point out, there is a great need for greater standardization and more research over a full range of environmental conditions and depths.

Seagrasses

Seagrasses are abundant in sheltered,lagoonal situations in the Caribbean (Randall 1965) and in the Indian Ocean (Qasim & Bhattathiri 1971; Randall 1965). On Australian coral reefs, seagrasses are a very minor component of the vegetation (Larkum 1976). The primary productivity of seagrasses is generally high: 2-4 g C m^{-2} d^{-1} (McRoy &

Plant community	Productivity (g C m^{-2}d^{-1})	Approx. cover on reefs (%)
Benthic algae	0.1 − 4	0.1 − 5
Turf algae	1 − 6	10 − 50
Zooxanthellae	0.6	10 − 50
Sand algae	0.1 − 0.5	10 − 50
Phytoplankton	0.1 − 0.5	10 − 50
Seagrasses	1 − 7	0 − 40

Table 1   Primary productivities of component plant communities on coral reefs and the distribution of these communities on reefs. Some of the data are discussed in the text. References are also given in the text

McMillan 1977). Where they occur on reefs, seagrasses have a high productivity (Qasim & Bhattathiri 1971; Patriquin 1974) and presumably contribute significantly to overall reef productivity.

Primary productivities and distributions of these different coral reef plant communities are summarized in table 1.

CONCLUSIONS

Theoretical considerations set an upper limit to the primary productivity of coral reefs at about 8 g C m^{-2} d^{-1}. The actual mean value may turn out to be considerably lower than this. Several areas of reefs (the outer edge, the crest and certain back-reefs) have the type of algal communities (large benthic algae and crustose coralline algae) and the conditions (high light intensity, vigorous stirring and possibly adequate nutrient supply) to sustain very high levels of productivity. However, other areas of reefs (lagoons and reticulated reefs) do not sustain this high productivity. The zoothanxellae of corals are probably very productive but, because of their symbiotic condition, their primary productivity is not easily measured. The high primary productivities measured for sections of reefs by gaseous exchange into flowing water (see Kinsey, chapter 13 this volume) must be seen as representative of the areas with highly productive plant communities. When the primary productivity of a coral reef as a whole is properly assessed it will probably turn out to be much lower than the latter values suggest. Some recent work seems to confirm this suggestion (Kinsey 1977; Rogers 1979). However, much more work is needed before any firm conclusions can be drawn.

LITERATURE CITED

Black, J. N., 1965. An analysis of the potential production of swards of subterranean clover *Trifolium subterraneum* L.) at Adelaide, South Australia. *Journal of Applied Ecology* 1, 3–18.

Boardman, N. K. and A. W. D. Larkum, 1974. Biological conversion of solar energy. In H. Messel and S. Butler (eds.), *Solar Energy*. Shakespeare Head Press, Sydney, pp. 125–44.

Burris, J. E., 1977. Photosynthesis, photorespiration and dark respiration in eight species of algae. *Marine Biology* 39, 371–79.

Dahl, A. L., 1973. Surface area in ecological analysis: quantification of benthic coral-reef algae. *Marine Biology* 23, 239–49.

Dayton, P. K., 1975. Experimental evaluation of ecological dominance in a rocky intertidal algal community. *Ecological Monographs* 45, 137–59.

Downton, W. J. S., D. G. Bishop, A. W. D. Larkum and C. B. Osmond, 1976. Oxygen inhibition of photosynthetic oxygen evolution in marine plants. *Australian Journal of Plant Physiology* 3, 73-9.

Emery, K. O., J. I. Tracey, Jr. and H. S. Ladd, 1949. Submarine geology and topography in the Northern Marshalls. *Transactions of the American Geophysical Union* 30, 55-8.

Finck, A. E., 1904. Biology of the reef-forming organisms at Funafuti Atoll. In, *The Atoll of Funafuti. Borings into a Coral Reef and the Results*. London, Royal Society Coral Reef Committee Report, pp. 125-50.

Fogg, G. E., 1968. Photosynthesis. London: English Universities Press, 116 pp.

Griffiths, D. J., 1976. The photosynthetic capacity of the phytoplankton in the waters of a coral reef. *Australian Journal of Plant Physiology* 3, 53-6.

Grime, J. P., 1974. Vegetation classification by reference to strategies. *Nature* 250, 26-31.

Grime, J. P., 1977. Evidence for the existence of three primary strategies in plants and its relevance to ecological and evolutionary theory. *American Naturalist* 111, 1169-94.

Hillis-Colinvaux, L., 1974. Productivity of the coral reef alga *Halimeda* (Order Siphonales). In, *Proceedings of the Second International Symposium on Coral Reefs*. Brisbane: Great Barrier Reef Committee, Vol. 1, pp. 35-42.

Jassby, A. D., 1978. Polarographic measurement of respiration following light-dark transitions. In J. A. Hellebust and J. S. Craigie (eds.), *Handbook of Phycological Methods: Physiological and Biochemical Methods*. Cambridge University Press, pp. 297-304.

Jeffrey, S. W., 1968. Photosynthetic pigments of the phytoplankton of some coral reef waters. *Limnology and Oceanography* 13, 350-5.

Kinsey, D. W., 1972. Preliminary observations on community metabolism and primary productivity of the pseudo-atoll reef at One Tree Island, Great Barrier Reef. In C. Mukundan and C. S. Gopinadha Pillai (eds.), *Proceedings of the Symposium on Corals and Coral Reefs*. The Marine Biological Association of India, Cochin, pp. 13-32.

Kinsey, D. W., 1974. Effects of fertilization on a coral reef environment — primary production studies. In, *Proceedings of the Second International Symposium on Coral Reefs*. Brisbane: Great Barrier Reef Committee, vol. 1, pp. 49-66.

Kinsey, D. W., 1977. Seasonality and zonation in coral reef productivity and calcification. In D. L. Taylor (Editor), *Proceedings: Third International Coral Reef Symposium*. Miami: University of Miami, vol. 2, pp. 383-8.

Kinsey, D. W. and B. E. Kinsey, 1967. Diurnal changes in oxygen content of the water over the coral reef platform at Heron Island. *Australian Journal of Marine and Freshwater Research* 18, 23-34.

Larkum, A. W. D., 1981. Marine primary productivity. In M. N. Clayton and R. J. King (eds.), *Marine Botany: an Australian Perspective* Melbourne: Longman Cheshire pp. 370-85.

Lewis, J. B., 1977. Processes of organic production on coral reefs. *Biological Reviews of the Cambridge Philosophical Society* 52, 305-47.

McCloskey, L. R., D. S. Wethey and J. W. Porter, 1978. Measurement and interpretation of photosynthesis and respiration in reef corals. In D. R. Stoddart and R. E. Johannes (eds.), *Coral Reefs: Research Methods*. Paris: UNESCO, Monographs on Oceanographic Methodology, Number 5, pp. 379-96.

McRoy, C. P. and C. McMillan, 1977. Ecology and physiology of seagrasses. In C. P. McRoy and C. Helfferich (eds.). *Seagrass Ecosystems: Scientific Perspective*. New York: Marcel Dekker. pp. 53-87.

Marsh, J. A., 1970. Primary productivity of reef-building calcareous red algae. *Ecology* 51, 255-63.

Morris, I., 1974. The limits to the productivity of the sea. *Science Progress* 61, 99-122.

Nielsen, E. S., 1975. Marine Photosynthesis with Special Emphasis on the Ecological Aspects. Elsevier Oceanography Series No. 13. Amsterdam: Elsevier Scientific Publications, 141 pp.

Odum, H. T., W. A. McConnell and W. Abbott, 1958. The chlorophyll "A" of communities. *Publications of the Institute of Marine Science University of Texas* 5, 65–97.

Patriquin, D. G., 1975. "Migration" of blowouts in seagrass beds at Barbados and Carriacou, West Indies, and its ecological and geological implications. *Aquatic Botany* 1, 163–89.

Plante-Cuny, M. R., 1973. Recherches sur la production primaire benthique en milieu marin tropical. I Variations de la production primaire et des teneurs en pigments photosynthétiques' sur quelques fonds sableux. Valeur des résultats obtenus par la méthode du ^{14}C. *Cahiers ORSTOM Serie Ocanographie* 11, 317–48.

Qasim, S. Z. and P. M. A. Bhattathiri, 1971. Primary production of a seagrass bed on Kavarati Atoll (Laccaclives). *Hydrobiologia* 38, 29–38.

Rabinowitch, E. I., 1951. Photosynthesis and Related Processes. New York: Interscience Publications, vol 2. Part 7.

Radmer, R. and B. Kok, 1977. Photosynthesis: limited yield, unlimited dreams. *Bioscience* 27, 599–605.

Randall, J. E., 1961. Overgrazing of algae by herbivorous marine fishes. *Ecology* 42, 812–14.

Randall, J. E., 1965. Grazing effect on sea grasses by herbivorous reef fishes in the West Indies. *Ecology* 46, 255–60.

Randall, J. E., 1974. The effect of fishes on coral reefs. In, *Proceedings of the Second International Symposium on Coral Reefs*. Great Barrier Reef Committee, Brisbane, Volume 1, pp. 159–66.

Raven, J. A., 1978. Photosynthesis in cells and tissues. In, *Proceedings of the Fourth International Congress on Photosynthesis*. London: Biochemical Society.

Raven, J. A., F. A. Smith and S. M. Glidewell, 1979. Photosynthetic capacities and biological strategies of giant-celled and small-celled macro-algae. *New Phytologist* 83, 299–309.

Rogers, C. S., 1979. The productivity of San Cristobal Reef, Puerto Rico. *Limnology and Oceanography* 24, 342–9.

Ryther, J. H., 1956. The measurement of primary production. *Limnology and Oceanography* 1, 72–84.

Sale, P. F., P. S. McWilliam and D. T. Anderson, 1976. Composition of the near-reef zooplankton at Heron Reef, Great Barrier Reef. *Marine Biology* 34, 59–66.

Sale, P. F., P. S. McWilliam and D. T. Anderson, 1978. Faunal relationships among near-reef zooplankton at three locations on Heron Reef, Great Barrier Reef, and seasonal changes in this fauna. *Marine Biology* 49, 133–45.

Sammarco, P. W., J. S. Levinton and J. C. Ogden, 1974. Grazing and control of coral reef community structure by *Diadema antillarum* Philippi (Echinodermata: Echiniodea): A preliminary study. *Journal of Marine Science* 32, 47–53.

Scott, P. D. and H. R. Jitts, 1977. Photosynthesis of phytoplankton and zooxanthellae on a coral reef. *Marine Biology* 41, 307–15.

Sournia, A., 1976a. Primary production of sands in the lagoon of an atoll and the role of foraminiferan symbionts. *Marine Biology* 37, 29–32.

Sournia, A., 1976b. Oxygen metabolism of a fringing reef in French Polynesia. *Helgoländer wissenschaftliche Meeresuntersuchungen* 28, 401–10.

Sournia, A. and M. Ricard, 1976. Phytoplankton and its contribution to primary productivity in two coral reef areas of French Polynesia. *Journal of Experimental Marine Biology and Ecology* 21, 129–40.

Talling, J. F., 1961. Photosynthesis under natural conditions. *Annual Review of Plant Physiology* 12, 133–54.

Tamiya, H., 1957. Mass culture of algae. *Annual Review of Plant Physiology* 8, 309–34.

Turner, J. S. and E. G. Brittain, 1962. Oxygen as a factor in photosynthesis. *Biological Reviews of the Cambridge Philosophical Society* 37, 130–7.

Wanders, J. B. W., 1976. The role of benthic algae in the shallow reef of Curacao (Netherlands Antilles). I. Primary production on the coral reef. *Aquatic Botany* 2, 235–70.

Warren-Wilson, J., 1967. The Collection and Processing of Field Data. Interscience, New York, 77pp.

Westlake, D. F., 1963. Comparisons of plant productivity. *Biological Reviews of the Cambridge Philosophical Society* 38, 385–425.

15 Marine Lithification in Coral Reefs

John F. Marshall
Bureau of Mineral Resources,
P.O. Box 378, Canberra City, A.C.T. 2601

INTRODUCTION

Lithification is the process by which unconsolidated sediment is converted into a coherent and solid rock. Several processes may be involved including compaction, compression and recrystallization, but the most important process occuring in lithification of coral reef sediments is cementation. Cementation on reefs involves deposition or precipitation of carbonate minerals in the spaces among the sedimentary particles, thereby binding them together. It has long been realized that lithification has an important role in preserving the structure of fossil reefs in the face of erosive processes. However, it has only recently been recognized that lithification occurs in modern reef sediments contemporary with their deposition.

Up to about ten years ago, most geologists working on both ancient and modern reef carbonates accepted that cementation in coral reefs was a freshwater phenomenon, occuring in vadose or fresh water phreatic environments (i.e. related to ground moisture or the presence of a water-table). This idea had persisted even though earlier workers on modern coral reefs had observed well cemented reef rock on Pacific atolls (Darwin 1851; Dana 1872; Wood-Jones 1910), and that Dana *(op. cit.)* considered these rocks to have been cemented in the marine environment. It is now obvious that the carbonate cements observed by Cullis (1904) in the upper part of the Funafuti borehole were also of marine origin (Bathurst 1971, p. 351).

Interest in modern coral reefs during the last decade has led to the recognition of substantial subtidal lithification contemporaneous with reef growth. Subtidal lithification has been reported from algal cup reefs (Ginsburg *et al.* 1971a,b; Schroeder 1972) and the fore-reef slope of Bermuda (Focke & Gebelin 1978), the fore-reef crest and slope of Jamaica (Land & Goreau 1970; Goreau & Land 1974; Land & Moore 1980), Belize (James *et al.* 1973; Ginsburg & James 1976; James *et al.* 1976; James & Ginsburg 1979), Barbados (Macintyre *et al.* 1968), Panama (Macintyre 1977), and the Red Sea (Friedman *et al.* 1974). Preliminary examination of core material from the southern Great Barrier Reef has revealed substantial marine lithification in certain environments (Marshall & Davies, 1981; Marshall, in press.).

NATURE OF OCCURRENCE OF MARINE LITHIFICATION

Marine lithification occurs dominantly on seaward margins. In the Belize reefs, James *et al.* (1976) found that cementation was restricted to local areas along the seaward margins of the reefs, but was entirely absent from the lagoonal reefs. The generalization is not absolute: in the southern Great Barrier Reef, lithification has also been reported in lagoonal patch reefs (Marshall, in press.).

The internal structure of lithified reef rock varies with the type of allochemical constituent (discrete organized carbonate particles such as skeletal fragments), the degree of boring and infill by internal sediment, and the degree of cementation. Corals and crustose coralline algae tend to dominate the framework, but encrusting foraminifera, vermetid gastropods, serpulids and bivalves are also important contributors. In some places, such as on the fore-reef slope of Belize (James & Ginsburg 1979), *Halimeda* may be an important framework component. Plate-like or branching corals are commonly coated with lithified crusts, and cavities between layers of coralline algae are filled with cemented internal sediment. The reef rock is often extensively riddled with borings by sponges, bivalves and other organisms. Although some borings are empty, most are partly or wholly filled with internal sediment and cement. The sediments range from unconsolidated to friable, to completely cemented. Reef rock commonly consists of a complex framework resulting from repeated alternations of boring, deposition of sediment within borings and cavities, and cementation. This overall process is often so pervasive that the original framework is almost completely obliterated (Macintyre 1977; James & Ginsburg, 1979).

CEMENTS

Mineralogy

High-magnesium calcite and aragonite are the only carbonate cements found within subtidal reef rock (Macintyre *et al.* 1968; Land & Goreau 1970; Ginsburg *et al.* 1971b; Friedman *et al.* 1974; Friedman 1975; James *et al.* 1976; Macintyre 1977; Focke & Gebelin 1978; James & Ginsburg 1979). High-Mg calcite is more abundant than aragonite, occurring within intraskeletal pores and as interskeletal cement, coating and binding grains together and filling the interstices of internal sediment. Aragonite is generally less common and is mainly present as an intraskeletal cement. However, Ginsburg & James (1976) and James & Ginsburg (1979) have described cavity-filling botryoidal aragonite and interskeletal aragonite from the fore-reef slope of Belize.

Textures and Fabrics

Aragonite cement (fig. 1a) occurs as acicular aggregates, commonly growing as syntaxial extensions of aragonitic substrates (e.g. coral), although in some cases a micritic envelope may be present (Friedman *et al.* 1974). The crystals are mostly straight-sided and elongate forms, which are either pointed or have abrupt, flat terminations (fig. 2a). Crystal length varies greatly, some are stubby (20-50 μm) or long (>200 μm). Crystal width is usually between 5 and 15 μm. The aragonite crystals project into voids either normal or slightly inclined to the substrate, and the crystal aggregate commonly fills the void space.

High-Mg calcite cement occurs in a variety of forms but it most commonly occurs as three types: (1) bladed spar, (2) micrite, (3) peloids. Macintyre (1977) introduced the term microsucrosic to describe a particular type of fine-grained high-Mg calcite cement, but this appears either to be rare or has elsewhere been reported as micrite.

Blades of high-Mg calcite spar grow as isopachous rinds, 10-500 μm thick, either around sand grains or lining cavities and intraskeletal pores (fig. 1a). Commonly two

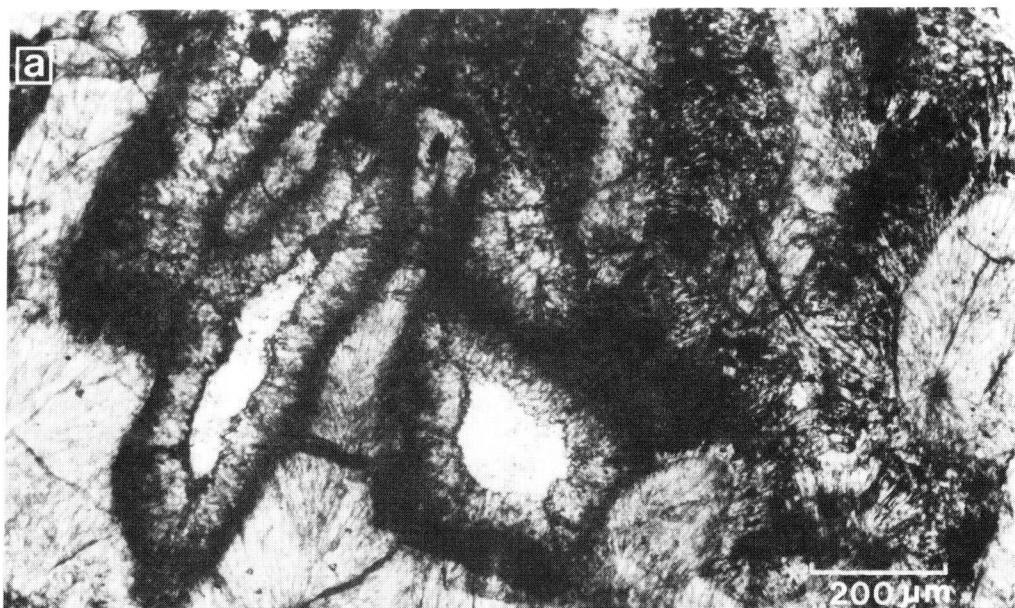

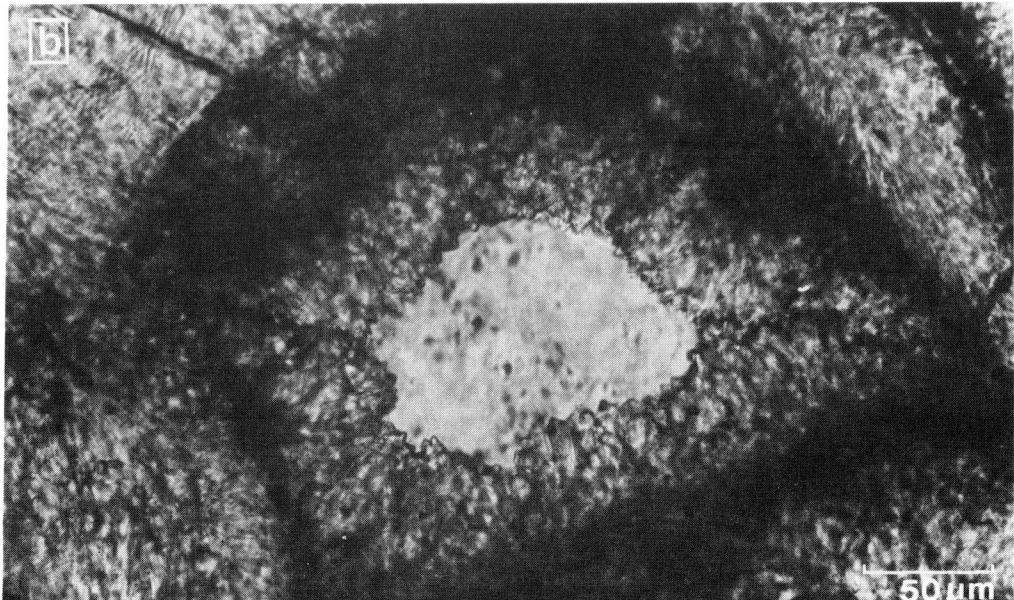

Figure 1   Types of submarine cements:

(a) Intraskeletal development of aragonite and bladed high-Mg calcite cements within a coral. The aragonite cement (right) has grown syntaxially from the coral wall and completely fills the voids. The high-Mg calcite spar (left) forms isopachous rims lining the coral void. Note the irregular micritic layer (dark) between the coral wall and the bladed spar.

(b) Growth of high-Mg calcite bladed spar on coral walls with thin, diffuse micrite layer (dark) between them. Note the typical dog-tooth terminations of bladed spar projecting into the void.

distinct layers are present; an inner micrite zone and an outer zone of bladed spar (Macintyre *et al.* 1968; James *et al.* 1976; James & Ginsburg 1979; Marshall & Davies 1981). The inner micrite zone is usually less than 20 μm thick and commonly grades diffusely and unevenly outward into the bladed spar (fig. 1a). The crystals of bladed spar are from 10 to 100 μm long and from 2 to 20 μm wide, and have their long axes perpendicular to the grain surface or cavity wall. The larger crystals occur as elongate, flared blades with rhombic terminations.

Micrite or micro-crystalline carbonate cement is usually the most abundant type of high-Mg calcite cement present in coral reefs. It generally lines or fills the interstices between internal sediment particles (fig. 1c). Micritic cement is most common in silt or fine sand-size sediments or where there are mixtures of silt and sand. It rarely occurs as a void-filling cement around or between sand-size grains. Sand-size internal sediment is usually cemented by high-Mg calcite spar (James *et al.* 1976; James & Ginsburg 1979). Under the microscope, micrite appears either clear and translucent or dark and semi-opaque, depending on the amount of fine-grained internal sediment and degree of crystallization (fig. 1c). This type of cement is best examined with a scanning electron microscope which commonly shows crystals that form three-sided pyramids with curved crystal faces (fig. 2b). Crystal size ranges from 0.5 to 8.0 μm, but is usually in the range 2 to 4 μm.

Peloids are spherical or sub-spherical bodies that consist of a mass of micrite crystals (fig. 1d). The peloids have diameters between 10-100 μm, most commonly 20-60 μm. The size of the individual micrite crystals is extremely small, usually 0.5 to 2.0 μm. Often, the peloids have a rim of coarser, more crystalline, high-Mg calcite cement that projects into and may fill the void spaces between the peloids (figs. 1d, 2b). There has been some discussion on the origin of these more crystalline rims (Alexandersson 1978; Macintyre 1978), but it is generally accepted that they represent a secondary stage of slower precipitation of coarser high-Mg calcite cement. In cases where coarser rims do not occur, the inter-peloidal voids may be filled with a crystalline mosaic that is similar in texture to the peloids (Alexandersson 1978).

GEOCHEMISTRY

High-Mg calcite cements generally contain 14-19 mole percent $MgCO_3$, although Friedman *et al.* (1974) report values up to 22 mole percent $MgCO_3$. Micrite appears to have slightly lower values than the peloids or bladed spar, but this is probably a result of numerous inclusions of fine-grained skeletal particles. Most peloids are relatively uncontaminated and Macintyre (1977) reported a value of 18.8 ± 0.4 mol percent $MgCO_3$ for sediment-free peloids. The strontium content of the aragonite cement is usually within the range 8000-9500 ppm; this is typical for aragonite precipitated from sea water (Kinsman 1979).

Stable isotope values of the cements have ratios that fall within the range reported for skeletons of marine organisms. Land (1971) considered that this suggests enrichment in ^{13}C and ^{18}O over what might be expected to precipitate at equilibrium from sea water.

PARAGENESIS

The paragenesis, or sequence of cementation, is most commonly aragonite, followed by high-Mg calcite. This is typical within skeletal chambers (Ginsburg *et al.* 1971; James *et*

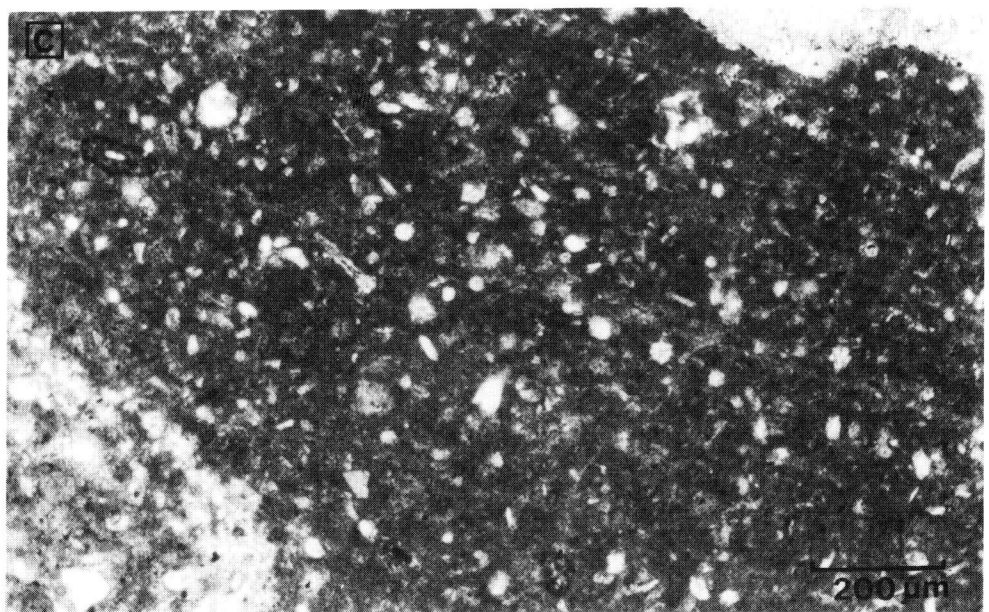

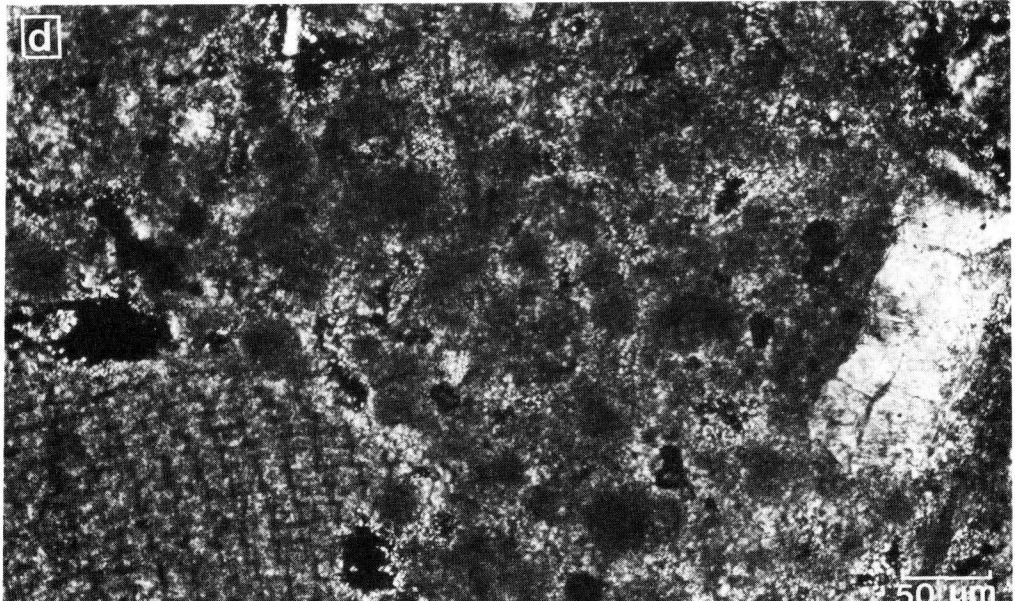

(c) Micrite cement within lithified internal sediment containing numerous silt-size fragments.
(d) Peloids (dark, subspherical bodies in central part of photograph) with translucent crystalline rims of high-Mg calcite cement around and between them.

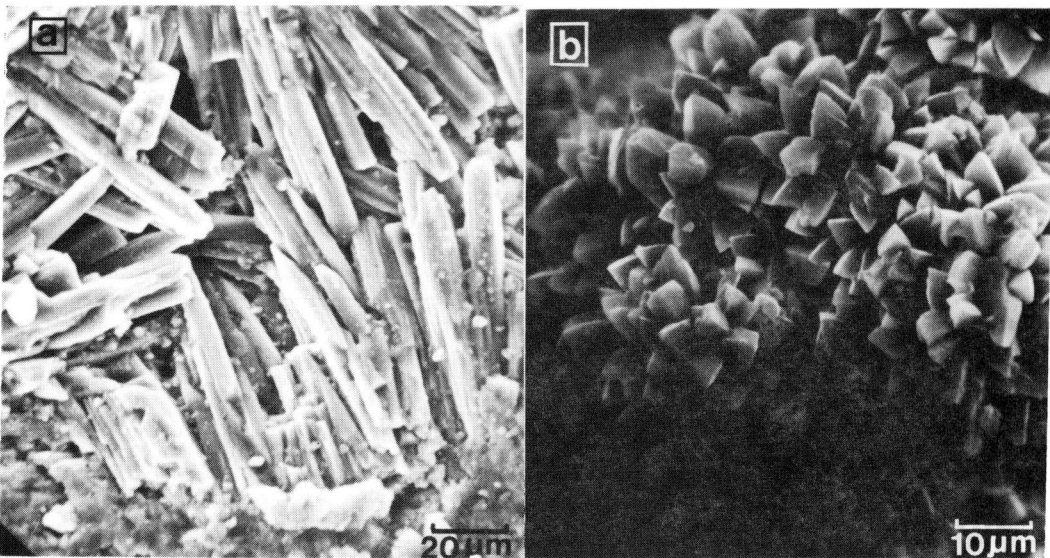

Figure 2   (a) Fibrous crystals of aragonite cement projecting into and almost filling an intraskeletal void.

       (b) Crystals of high-Mg calcite cement which, in this case, form the outer rim of a peloid

al. 1976; Macintyre 1977). However, James & Ginsburg (1979) have observed that high-Mg calcite micrite can be the first cement and that high-Mg calcite and aragonite do form together.

Of the different types of high-Mg calcite cement, micrite (including peloids) generally forms before bladed spar. James *et al.* (1976) consider that there is a complex relationship between grain size (and thus pore size) and the type of high-Mg calcite cement. They consider that the silt-size internal sediments are cemented soon after deposition by micrite, whereas mixtures of silt- and sand-size sediments are cemented slightly later by micrite or by a combination of micrite and bladed spar. Only when the fine-grained sediments are well lithified do sand-size sediments become cemented by bladed spar.

Radiocarbon dating of framework and lithified reef rock show comparatively young ages, most being less than 3000 yr BP (Ginsburg *et al.* 1971b; Goreau & Land 1974; James *et al.* 1976; Focke & Gebelein 1978). However, dating of core from within the reef at Galeta (Macintyre 1977) and the deep fore-reef of Belize (James & Ginsburg 1979) shows that cementation was also occurring contemporaneously with reef growth during the last rise of sea level. Submarine lithification tends to be a near-surface phenomenon. Macintyre (1977) found decreased lithification below the surface of individual corals in the Galeta cores, and suggested that this was caused by porosity destruction by near-surface cementation. Goreau & Land (1974) found that material blasted from the reef face had begun to lithify when examined a year later. Therefore, most cementation post-dates the recent rise of sea level and the age-depth relationship suggests that cementation is contemporaneous with reef growth.

MODELS OF LITHIFICATION

Two models have been proposed to explain marine lithification in coral reefs. These are:

1. a physico-chemical model, in which super-saturation of sea water with respect to calcium carbonate is sufficient to cause precipitation and
2. a biological model in which organisms play a crucial role in effecting precipitation.

The physico-chemical model suggests that the inorganic precipitation of calcium carbonate requires no substantial modification to the composition of sea water. A suitable pumping mechanism is required to push sea water through the pore system to precipitate the calcium carbonate cement. Such a pumping mechanism is provided by the continuous wave surge that impinges on seaward margins, the environment where most cementation takes place. Ginsburg *et al.* (1971b) consider that changes in total pressure resulting from such strong wave surge could lead to a reduction in the carbon dioxide content of the water and cause precipitation. Focke & Gebelein (1978) maintain that although biological processes may have an influence, lithification is controlled mainly by a combination of super-saturation, an efficient pumping mechanism (they suggest oceanic currents), and low accumulation rates.

A principal argument for biological precipitation stems from Alexandersson's (1974) work from the North Sea where marine cementation occurs in waters that are undersaturated with respect of calcium carbonate. Alexandersson considered that cementation was effected by the metabolic activity of coralline algae. Friedman *et al.* (1974) and Friedman (1975) suggest that photosynthesis may raise pH within thin or monomolecular layers around carbonate materials and cause precipitation of calcium carbonate. The carbon and oxygen isotope enrichment measured by Land & Goreau (1970) in the cements of Jamaican reef rock has also been put forward to suggest biological control of cementation.

More needs to be known about these processes before the mechanism of marine lithification can be understood. Both laboratory and field experiments are required to test both models, and there is a need to examine marine cements from many different localities. Although the physico-chemical model is attractive it does not readily explain the dominance of high-Mg calcite as a marine cement. From a physico-chemical viewpoint it has been demonstrated that this is the least likely mineralogy (e.g. Bathurst 1971, pp. 243–50).

On the other other hand, the localization of marine cements to the high energy areas of the reef would seem to preclude an exclusive biological origin, which one would assume would be operative over the entire reef. It is apparent that a reconciliation of the two models is necessary before a suitable mechanism is found that would explain the origin of these cements.

ROLE OF MARINE LITHIFICATION IN THE MAINTENANCE OF CORAL REEFS

With respect to the binding of the reef framework, marine lithification (to some extent subordinate to the role of encrusting coralline algae) is important because, as has been stressed earlier, most lithification takes place on the seaward margins, an environment that is constantly subjected to the erosive forces of wave surge and water movement. It is somewhat of a paradox that these erosive forces, together with the action of endolithic organisms, produce most of the sediment that fills the spaces between the framework which when cemented forms the wave resistant margins of the reef. The process of marine lithification also leads to the formation of new substrates for subsequent colonization by reef organisms. It is in these respects that marine lithification plays an important role in the maintenance of coral reefs.

ACKNOWLEDGEMENT

Published with the permission of the Director, Bureau of Mineral Resources.

GLOSSARY

Allochem   A descriptive term for materials that have formed by chemical or biochemical precipitation, and which are organized into descrete carbonate aggregates. In coral reefs, the term usually refers to skeletal particles.

Aragonite   A metastable, orthorhombic form of $CaCO_3$.

High-Mg Calcite   A metastable, trigonal form of $CaCO_3$ that contains of the order of 11 to 19 mol percent $MgCO_3$.

Internal Sediment   Clastic or chemical sediments derived from the surface of, or within, a reef framework and deposited within cavities formed in the host rock.

Lithification   The conversion of an unconsolidated sediment into a coherent and solid rock, involving processes such as cementation and compaction.

Micrite   A descriptive term for the semiopaque, crystalline, interstitial component of a limestone, consisting of chemically precipitated calcite whose crystals have diameters of less than 4 μm.

Paragenesis   The sequential order of mineral formation.

Peloid   In general this refers to an allochem formed of crypto-crystalline or micro-crystalline material irrespective of size or origin, whereas in reef rock they have been loosely defined as spherical or subspherical aggregates of micritic high-Mg calcite that have precipitated within the reef substrate.

Spar (sparite)   A descriptive term for calcite crystals that are larger than micrite.

Syntaxial   An overgrowth on a crystal whereby the original crystal and the overgrowth form a single larger crystal with the same crystallographic axes.

LITERATURE CITED

Alexandersson, T., 1974. Carbonate cementation in coralline algal nodules in the Skagerrak, North Sea: biochemical precipitation in undersaturated waters. *Journal of Sedimentary Petrology* 44, 7–26.

Alexandersson, E. T., 1978. Distribution of submarine cements in a modern Caribbean fringing reef, Galeta Point, Panama. *Journal of Sedimentary Petrology* 48, 665–8.

Bathurst, R. G. C., 1971. Carbonate Sediments and their Diagenesis. Amsterdam: Elsevier Scientific Publishing Company, 620 pp.

Cullis, C. G., 1904. The mineralogical changes observed in the cores of the Funafuti borings. In T. G. Bonney (ed.), *The Atoll of Funafuti*. The Royal Society of London, pp. 392–420.

Dana, J. D., 1872. Corals and coral islands. Dodd & Mead, New York.

Darwin, C. R., 1842. The Structure and Distribution of Coral Reefs. London: Smith, Elder and Co., 214pp.

Focke, J. W. and C. D. Gebelein, 1978. Marine lithification of reef rock and rhodolites at a fore-reef slope locality (-50m) off Bermuda. *Geologie en Mijnbouw* 57, 163–71.

Friedman, G. M., 1975. The making and unmaking of limestones or the downs and ups of porosity. *Journal of Sedimentary Petrology* 45, 379–98.

Friedman, G. M., A. J. Amiel and N. Schneidermann, 1974. Submarine cementation in reefs: Example from the Red Sea. *Journal of Sedimentary Petrology* 44, 816–25.

Ginsburg, R. N., E. A. Shinn and J. H. Schroeder, 1967. Submarine cementation and internal sedimentation within Bermuda reefs. *Geological Society of America, Special Paper* 115, 78–9.

Ginsburg, R. N., D. S. Marszalek and N. Schneidermann, 1971a. Ultra-structure of carbonate cements in a Holocene algal reef or Bermuda. *Journal of Sedimentary Petrology* 41, 472–80.

Ginsburg, R. N., J. H. Schroeder and E. A. Shinn, 1971b. Recent synsedimentary cementation in subtidal Bermuda reefs. In O. P. Bricker (ed.), *Carbonate Cements*. Baltimore: John Hopkins University Press, pp. 54-8.

Ginsburg, R. N. and N. P. James, 1976. Submarine botryoidal aragonite in Holocene reef limestones, Belize. *Geology* 4, 431-6.

Goreau, T. F. and L. S. Land, 1974. Fore-reef morphology and depositional processes, North Jamaica. *Society of Economic Paleontologists and Mineralogists, Special Publication* 18, 77-89.

James, N. P., R. N. Ginsburg, D. S. Marszalek and P. W. Choquette, 1973. Subsea cementation of shallow British Honduras reefs (abstract). *American Association of Petroleum Geologists Bulletin* 57, 786.

James, N. P., R. N. Ginsburg, D. S. Marszalek and P. W. Choquette, 1976. Facies and fabric specificity of early subsea cements in shallow Belize (British Honduras) reefs. *Journal of Sedimentary Petrology* 46, 523-44.

James, N. P. and R. N. Ginsburg, 1979. The seaward margin of Belize barrier and atoll reefs. *International Association of Sedimentologists, Special Publication* 3, 191pp.

Kinsman, D. J. J., 1969. Interpretation of Sr^{2+} concentrations in carbonate minerals and rocks. *Journal of Sedimentary Petrology* 39, 486-508.

Land, L. S., 1971. Submarine lithification of Jamaican reefs. In O. P. Bricker (ed.), *Carbonate Cements*. Baltimore: John Hopkins University Press, pp. 59-62.

Land, L. S. and T. F. Goreau, 1970. Submarine lithification of Jamaican reefs. *Journal of Sedimentary Petrology* 40, 457-62.

Land, L. S. and C. H. Moore, 1980. Lithification, micritization and syndepositional diagenesis of biolithites on the Jamaican island slope. *Journal of Sedimentary Petrology* 50, 357-70.

Macintyre, I. G., 1977. Distribution of submarine cements in a modern Caribbean fringing reef, Galeta Point, Panama. *Journal of Sedimentary Petrology* 47, 503-16.

Macintyre, I. G., 1978. Distribution of submarine cements in a modern Caribbean fringing reef, Galeta Point, Panama: reply. *Journal of Sedimentary Petrology* 48, 669-70.

Macintyre, I. G., E. W. Mountjoy and B. F. D'Anglejan, 1968. An occurrence of submarine cementation of carbonate sediments off the west coast of Barbados, W.I. *Journal of Sedimentary Petrology* 38, 660-4.

Marshall, J. F. and P. J. Davies, 1981. Submarine lithification on windward reef slopes: Capricorn-Bunker group, southern Great Barrier Reef. *Journal of Sedimentary Petrology*, 51, 953-60.

Marshall, J. F., in press. Submarine cementation in a high-energy platform reef: One Tree Reef, southern Great Barrier Reef. *Journal of Sedimentary Petrology*.

Schroeder, J. H., 1972. Fabrics and sequences of submarine carbonate cements in Holocene Bermuda cup reefs. Geologische Rundschau 61, 708-30.

Wood-Jones, F., 1910. Coral and Atolls. Lovell Reeve and Company London 23, 392 pp.

16 Coral Reef Calcification

S. V. Smith

Hawaii Institute of Marine Biology, P.O. Box 1346, Kaneohe, Hawaii 96744.

CORAL REEF CALCIUM CARBONATE AS A METABOLIC PRODUCT

The coral reef metabolic process with the largest end product is biological calcium carbonate production, or calcification[1]. It is possible that some $CaCO_3$ precipitation in coral reef environments may occur from strictly inorganic processes (e.g. Cloud 1962; Broecker & Takahashi 1966), although it seems more likely that such muds originate from skeletal breakdown (Neumann & Land 1975). The mud of Pacific atoll lagoons seems unquestionably to be skeletal breakdown products (e.g. Weber & Woodhead 1972; Fütterer 1974). Other carbonate muds are fine-grained biogenic precipitates (Scholle & Kling 1972). It thus seems clear that lime muds are dominantly if not entirely biogenic.

There is some $CaCO_3$ precipitation that does not represent the direct accretion of skeletal materials. For example, a variety of non-skeletal grains characterize some carbonate sediments (Purdy 1963), and also calcareous cements are found within the interstices of reef materials. Bricker (1971) edited a book which presaged the appearance of numerous detailed studies of carbonate cements (see also Milliman 1974). It seems likely that even these materials may be largely biological in their origins.

Most $CaCO_3$ in coral reefs is clearly material precipitated by the biota that inhabit reefs. We may thus consider coral reef calcification to be largely a metabolic process.

CALCIFYING COMPONENTS OF CORAL REEFS

It is difficult to assign quantitative importance to each of the biota that produces $CaCO_3$ on coral reefs. Various attempts at estimating abundance of sedimentary products from the taxa have not been entirely satisfactory because these assessments do not weigh individual sediment samples by the accumulation rate represented by these samples. Moreover, fine-grained components, which dominate the total sediment volume of many reefs, are not identified in many investigations. Finally, there is sufficient variability among reefs that generalizable patterns of abundance can only be rather crudely assigned. We can, nevertheless, suggest that corals, coralline red algae and calcareous green algae are probably the three dominant components of coral reefs. Foraminifera, molluscs and echinoderms are generally next in abundance and may even be dominant in individual samples. Other biogenic calcareous materials (bryozoans, worms, etc) are usually present but rarely abundant.

[1] The term 'calcium carbonate' includes those cations (especially Mg and Sr) that are co-precipitated with Ca^{2+}. These co-precipitated cations can represent a substantial fraction of the total cations in the 'calcium carbonate'.

Chemical and mineralogical composition is somewhat easier to generalize and tends to confirm the generalities which have been offered about the dominance of biogenic materials. Well over 50% of most reef sediment is aragonite. Most of the remaining material is high-magnesium calcite. For comparison with the biological generalizations, both corals and calcareous green algae precipitate aragonite; coralline red algae precipitate high-magnesium calcite. Low-magnesium calcite is third in abundance, with dolomite being rare in unaltered sedimentary materials.

The topic of calcifying components of coral reefs is abuntantly reviewed by Milliman (1974).

ESTIMATORS OF CORAL REEF CALCIFICATION

The remainder of this review is largely concerned with the quantity, rather than quality, of calcium carbonate deposited by coral reefs. Moreover, calcium carbonate production is treated in this review at the level of the reef system or communities therein. The reader interested in physiological or organismic aspects of calcification is referred to the reviews in this volume by Chalker (corals) and Borowitzka (algae).

Because of the diverse interest in the $CaCO_3$ of coral reefs, it is not surprising that a variety of estimators of $CaCO_3$ production rate have been developed. These basically come down to three general classes of estimates, with variations on each theme.

Geological Estimator

This estimate consists of some measure of the thickness of accumulated $CaCO_3$ divided by the estimated time increment over which that material accumulated. In the terminology of Chave *et al.* (1972), this estimate is 'net $CaCO_3$ production'. The estimate represents the amount of material actually produced within the area of interest, minus any losses to export or dissolution, plus additions from import. This might also be termed 'reef growth' or 'net accretion'. Reef growth is most often noted in terms of vertical accretion. This estimate is most relevant to purely geological studies of reef stratigraphy and acknowledges the intimate relationship between reef growth and changing sea level.

Results from this approach, including some of the apparently conflicting results between it and the 'chemical estimates' (discussed below) are covered in the review by Davies (this volume). With recently developed portable underwater drilling rigs, this method is presently yielding a great deal of valuable information.

Biological Estimator

This estimate consists of the product of standing-crop (expressed in terms of percent cover or mass of $CaCO_3$ per unit area — 'calcimass') and some estimate of biological growth or turnover rate. This estimate is analogous to the 'harvest' method of estimating organic carbon production (Odum 1971). The method can give a good approximation of $CaCO_3$ production if a few calcifying organisms dominate and their growth rate can be reasonably approximated. However, effective use of the method can be very tedious and not particularly rewarding if the significant calcifying biota within an environment are diverse or otherwise difficult to enumerate and/or if the growth of these biota is not reasonably approximated by available data.

The method is 'gross production' in the terms of Chave *et al.* (1972) if standing-crop is really used to normalize the data or 'potential production' if (as has been done by some investigators) the taxa in question are either implicitly or explicitly assumed to cover the entire reef area. Most often the rates from this method are expressed in terms of mass accretion rate. Results tend to vary widely; at least part of the wide range of biological estimate of $CaCO_3$ production has arisen because authors have failed to make this distinction between gross production and potential production.

The method is the oldest of the various production rate estimators, and the early literature is reviewed by Chave *et al.* (1972). The most careful and exhaustive effort to use the biological estimator is probably the study of a reef area in Barbados (Stearn & Scoffin 1977; Stearn *et al.* 1977). Among other things, that effort reminds us of the inherent tedium of the method. There is not sufficient recent literature on this subject to justify updating the review by Chave *et al.* (1972) beyond reference to the Barbados studies.

Chemical Estimator

This estimate takes advantage of the chemical stoichiometric equivalence which must exist between the accumulation of $CaCO_3$ precipitated from the water and the removal of dissolved materials from the water. One might anticipate that measuring the calcium depletion from seawater would be the most appropriate chemical estimator of calcification (with appropriate attention to co-precipitated cations). Because calcium is abundant in seawater, the depletion of this element in all but the most extreme circumstances is insufficient to yield to analysis. As discussed in various papers (e.g. Broecker & Takahashi 1966; Smith 1973, 1978a; Kinsey 1978a, 1979; etc), the depression of total alkalinity can also be used in principle and is measurable in practice under many circumstances.[2]

The alkalinity depression estimate yields a value which represents net $CaCO_3$ precipitation minus dissolution. The values so obtained differ, conceptually at least, from the biological 'gross production' cited above, because gross production does not address itself to $CaCO_3$ loss through dissolution. It is likely (Smith 1973) that the results of the alkalinity depression technique in most reef systems are close to gross production of $CaCO_3$ (but see Kinsey 1979, for an example to the contrary in a heavily stressed ecosystem).

The values also differ from the geological 'net production', because the geological approach accounts for mechanical gains or losses. Because alkalinity is a property of the dissolved materials in seawater, influxes and effluxes of particulate materials are not measured. This effect is almost invariably large within individual reef environments, but to the extent that reefs retain most of their sediments due to trapping within the ecosystem, the effect probably 'averages out' across the entire reef. Such sediment trapping undoubtedly varies widely from reef to reef (e.g. contrast Land 1979, with Smith *et al.* 1971). I suggest that the preponderance of reefs, including especially both barrier reefs and atolls, have silled basins which trap the bulk of the reef sedimentary products.

The term 'net calcification' is a useful expression for this net precipitation minus dissolution. Of the three general estimators of reef calcium carbonate production, net

---

[2] There have been recent suggestions among some coral reef scientists that the alkalinity depression technique does not work or could only be used by two persons (S. V. Smith and D. W. Kinsey). The review provides me with a forum to respond to those suggestions. As with any method, the alkalinity depression technique poses problems, uncertainties, etc. (see, for example, Gaines & Pilson 1972; Smith & Key 1975; Brewer & Goldman 1976; Kinsey 1978b); however it appears safe to state that the recent problems being encountered by various investigators using the alkalinity depression technique have largely arisen from lack of rigid adherence to analytical detail. During September and October 1979 I demonstrated to several persons that, with proper adherence to that detail, the method is both fast and readily usable by analysts not previously trained in the proper use of the technique. New users of this approach are urged to follow Smith & Kinsey (1978) in detail before 'improving' the technique.

calcification comes the closest to fitting the central theme of this review (coral reef calcification as a community metabolic process). It therefore provides the focus for the remainder of the analysis presented here.

CORAL REEF CALCIFICATION RATES

Table 1 summarizes published or otherwise available estimates of net calcification for coral reefs. Some generalities emerge. As summarized by Smith & Kinsey (1976), two common rates (an average of 4 kg $CaCO_3$ m^{-2} yr^{-1} and an average 0.8 kg m^{-2} yr^{-1}) appear most frequently in the table. Recently, Kinsey (1979) and Smith (1981) have recognized a higher rate (approximately 10 kg m^{-2} yr^{-1}). It is noteworthy that these rates do not show quite the extreme exhibited by the biological estimates. Moreover, the results suggest a general pattern of where the various rates might be encountered (Smith & Kinsey 1976; Smith & Harrison 1977, Kinsey 1979; Smith, in press).

Smith (1978b) suggested that 90–95% of the area of coral reefs calcifies at the 0.8 kg rate, and 5–10% calcifies at the 4 kg rate. These proportionalities can be modified to allow for the 10 kg rate, to suggest that perhaps 1–2% of the reef area calcifies at this extreme rate; 4–8% at the intermediate rate; and 90–95% at the slow rate. Rearrangement of these numbers suggests that coral reefs produce an average of about 1.0 to 1.2 kg $CaCO_3$ m^{-2} yr^{-1}.

MASS AND VERTICAL ACCRETION COMPARED

As an ecological parameter, calcification expressed as a rate of mass accretion seems intuitively reasonable. It is, for example, easily related to organic carbon production, which is usually reported in terms of biomass or carbon accretions. However, a coral reef is a three-dimensional structure which grows from some water depth towards sea level, and the metabolic process of calcification is responsible for this upward growth. Because there is an intimate relation between reef growth and sea level, the measure of vertical growth is also useful. It is useful to restate the chemical (or biological) estimator in a manner analogous with the geological estimator.

The aforementioned mass accretion rates can be restated as vertical accretion rates if some assumptions or limits are given:

1. the material being deposited has a density close to that of aragonite (2.9 g cm^{-3}),

2. local gains or losses of sedimentary materials precipitated 'average out' across the entire reef so that it is meaningful to assume an 'average condition' of total sediment retention and

3. the average porosity of the precipitated materials is about 50%, with the possibility of excursions between 0% and 90% porosity. With these boundary conditions, figure 1 is a summary comparison between mass accretion and vertical accretion. At 50% porosity, average reef growth (1 kg m^{-2} yr^{-1}) would be slightly less than 1 mm yr^{-1}; intermediate growth (4 kg m^{-2} yr^{-1}) which characterizes most of today's reef margins is about 3 mm yr^{-1}; and fast reef growth (10 kg m^{-2} yr^{-1}) is 7 mm yr^{-1}.

Adey (1978) and others have reported net vertical accretion rates on some Caribbean reefs in excess of 10 mm yr^{-1}. At the extreme of 90% porosity (which at least temporarily occurs in some very open-work reefs), the intermediate calcification rate could yield such vertical accretion. It seems likely that this extreme porosity would yield inadequate strength (without secondary depositional and/or diagenetic infilling) to sustain such vertical accretion under most circumstances. The fast mass accretion rates can yield the

Location	Environment	Rate (kg CaCO$_3$ m^{-2} yr^{-1})	Reference
Abrolhos Islands	Coral bank	12	Smith (1981)
Johnston Atoll	Back-reef, heavy coral	9.6	Kinsey (1979)
Central Kaneohe Bay	Coral zone	8.8	Kinsey (1979)
Johnston Atoll	Lagoon, heavy coral cover	6.4	Kinsey (1979)
One Tree Island	Reef flat, coral zone	4.6	Kinsey (1979)
Johnston Atoll	Coral/algal pavement	4.4	Kinsey (1979)
One Tree Island	Algal pavement	4.0	Kinsey (1979)
Enewetak Atoll	Reef flat, coral-algal community	4	Smith (1973)
Enewetak Atoll	Reef flat, algal turf	4	Smith (1973)
Lizard Island	Lagoon-entrance pinnacle	3.6	Kinsey (1979)
Lizard Island	Lagoon reef flat	3.1*	Kinsey (1979)
Lizard Island	Seaward reef flat	2.7*	Kinsey (1979)
Kaneohe Bay	Coral algal ocean reef	2.6*	Webb (1977)
Enewetak Atoll	Windward forereef	1–2	Smith & Harrison (1977)
Johnston Atoll	Lagoon, reticulated reefs	1.5*	Kinsey (1979)
One Tree Island	Lagoon reticulum	1.5*	Kinsey (1979)
Central Kaneohe Bay	Sand flat	1.2	Kinsey (1979)
Fanning Atoll	Lagoon	1	Smith & Pesret (1974)
Canton Atoll	Lagoon	0.5	Smith & Jokiel (1978)
Bahama Banks	Carbonate shoals	0.5	Broecker & Takahashi (1966)
One Tree Island	Open Lagoon	0.5	Kinsey (1979)
One Tree Island	Reef flat, sand/rubble	0.4	Kinsey (1979)
Lizard Island	Sand/algal flats	0.3	Kinsey (1979)

* May represent a mixture of areas with intermediate and slow calcification rates

Table 1   Coral reef calcification rate, as determined by the alkalinity depression method

documented high vertical accretion rates at 75% porosity. Some reef morphologies (e.g. thickets of the staghorn coral, *Acropora palmata*) may be able to build relatively strong yet open structures of this kind, but the chemical estimates would suggest that such fast mass accretion rates are presently rare features on coral reefs.

Because most published data derived by the geological estimate come from the Caribbean Sea and most of the chemical data come from the Pacific Ocean, it is not clear whether there is truly a conflict between the two sets of results or the higher rates from the geological estimates can be explained by the algebraic arguments I have just posed. The major place of overlapping data to date is the Great Barrier Reef, where in fact the two methods yield quite comparable results (Davies & Kinsey 1977; Davies & Marshall 1979; Davies, this volume).

A LOOK TO FUTURE RESEARCH ON CORAL REEF CALCIFICATION

Until the last decade, estimates of coral reef CaCO$_3$ production were sparse; moreover there was very little basis, either methodological or conceptual, for intercomparison among estimates. We simply had too little information about CaCO$_3$ production on reefs to offer generalizable expectations.

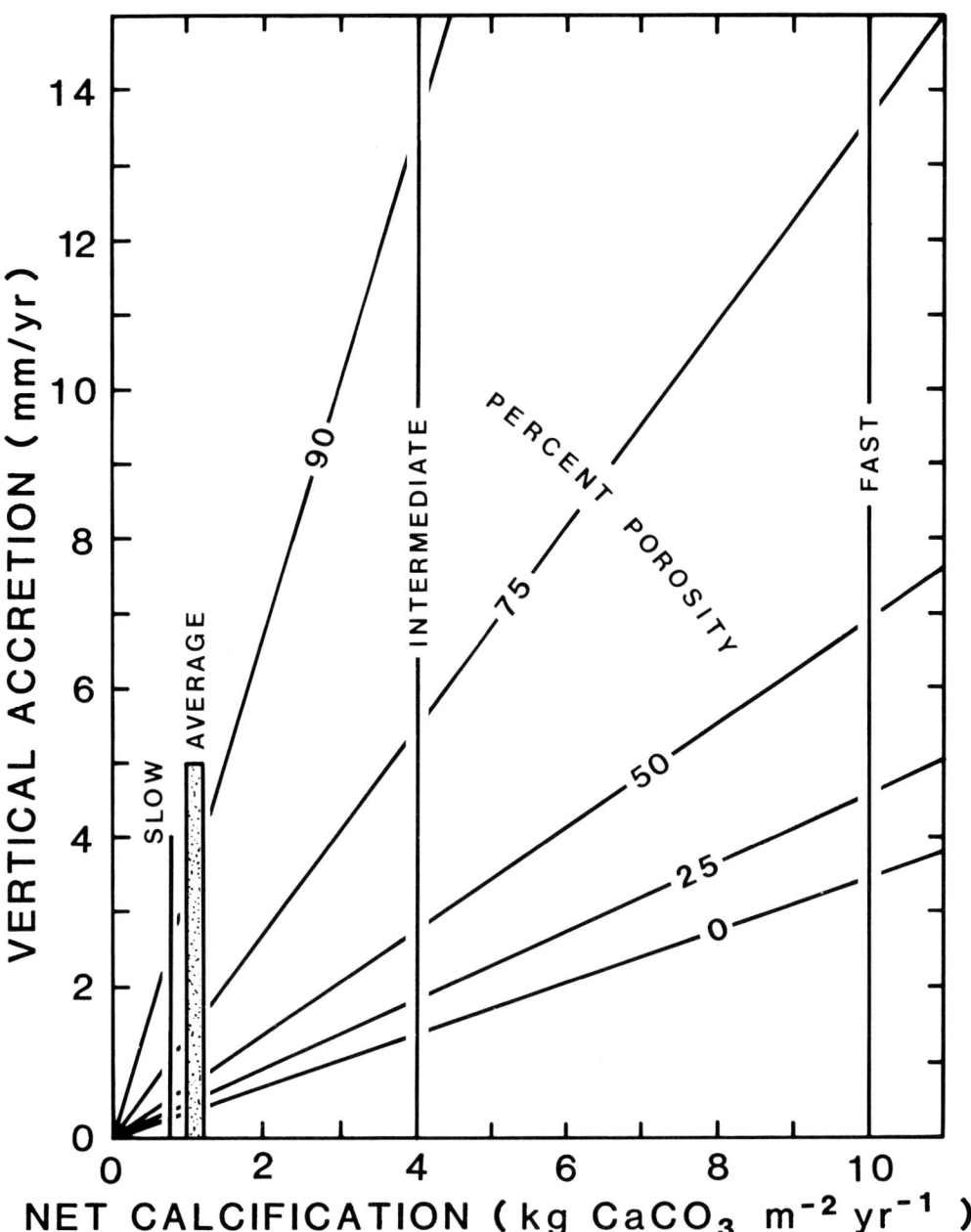

Figure 1   Nomograph which relates the various modes of net calcification rate to vertical accretion as a function of porosity (as noted), aragonite mineral density of 2·9 g cm⁻³, and no sediment gains to or losses from the calcification site

The last decade has seen the emergence of numerous estimates of $CaCO_3$ production, primarily by the geological and chemical methods discussed above. From these two sets of estimates are two sets of results which seem at once internally consistent and somewhat at odds with one another.

The 'conflicting' results are actually the basis for optimism, because they suggest that our data base of coral reef growth and calcification rates is becoming sufficient that we can venture generalizable statements of expectations. We have the tools to describe and quantify short-term calcification and long-term growth sufficiently for a more detailed conceptual understanding of reef genesis and maintenance to be possible during the next decade.

LITERATURE CITED

Adey, W. H., 1978. Coral reef morphogenesis: A multi dimensional model. *Science* 202, 831–7.

Brewer, P. G. and J. C. Goldman, 1976. Alkalinity changes generated by phytoplankton growth. *Limnology and Oceanography* 21, 108–17.

Bricker, O. P. (ed.), 1971 *Carbonate Cements*. Baltimore: Johns Hopkins Press, 376pp.

Broecker, W. S. and T. Takahashi, 1966. Calcium carbonate precipitation on the Bahama Banks. *Journal of Geophysical Research* 71, 1575–602.

Chave, K. E., S. V. Smith and K. J. Roy, 1972. Carbonate production by coral reefs. *Marine Geology* 12, 123–40.

Cloud, P. E., Jr., 1962. Behaviour of calcium carbonate in sea water. *Geochimicia et Cosmochimica Acta* 26, 867–84.

Davies, P. J. and D. W. Kinsey, 1977. Aspects of Holocene reef growth — One Tree Island, Great Barrier Reef. *Marine Geology* 24, M1–M11.

Davies, P. J. and J. F. Marshall, 1979. Aspects of Holocene reef growth — substrate age and accretion rate. *Search* 10, 276–9.

Fütterer, D. K., 1974. Significance of the boring sponge *Cliona* for the origin of fine grained material of carbonate sediments. *Journal of Sedimentary Petrology* 44, 79–84.

Gaines, A. G., Jr. and M. E. Q. Pilson, 1972. Anoxic water in the Pettaquamscutt River. *Limnology and Oceanography* 17, 42–9.

Kinsey, D. W., 1978a. Productivity and calcification estimates using slack-water periods and field enclosures. In D. R. Stoddart and R. E. Johannes (eds.), *Coral Reefs: Research Methods*. Paris: UNESCO, Monographs on Oceanographic Methodology, No 5, pp. 439–68.

Kinsey, D. W., 1978b. Alkalinity changes and coral reef calcification. *Limnology and Oceanography* 23, 989–91.

Kinsey, D. W., 1979. Carbon turnover and accumulation by coral reefs. Ph.D. dissertation, University of Hawaii, 248pp.

Land, L. S. 1979. The fate of reef-derived sediment on the north Jamaican Island slope. *Marine Geology* 29, 55–71.

Milliman, J. D., 1974. Marine Carbonates. Springer-Verlag, Berlin, 375pp.

Neumann, A. C. and L. S. Land, 1975. Lime mud deposition and calcareous algae in the Bight of Abaco, Bahamas: a budget. *Journal of Sedimentary Petrology* 45, 763–85.

Odum, E. P., 1971. Fundamentals of Ecology. W. B. Saunders Company, Philadelphia, 574pp.

Purdy, E. G., 1963. Recent calcium carbonate facies of the Great Bahama Bank. I. Petrography and reaction groups. 2. Sedimentary facies. *Journal of Geology* 71, 344–55; 472–97.

Scholle, P. A. and S. A. Kling, 1972. Southern British Honduras: lagoonal coccolith ooze. *Journal of Sedimentary Petrology* 42, 195–204.

Smith, S. V., 1973. Carbon dioxide dynamics: a record of organic carbon production, respiration, and calcification in the Eniwetok reef flat community. *Limnology and Oceanography* 18, 106–20.

Smith, S. V., 1978a. Alkalinity depletion to estimate the calcification of coral reefs in flowing waters. In D. R. Stoddart and R. E. Johannes (eds.), *Coral reefs: Research Methods*. Paris: UNESCO, Monographs on Oceanographic Methodology, No 5, pp. 397-404.

Smith, S. V., 1978b. Coral-reef area and the contributions of reefs to processes and resources of the world's oceans. *Nature* 273, 225-6.

Smith, S. V., 1981. The Houtman Abrolhos Islands: carbon metabolism of coral reefs at high latitude. *Limnology and Oceanography* 26, 612-21.

Smith, S. V. and J. T. Harrison, 1977. Calcium carbonate production of the *mare incognitum*, the upper windward reef slope, at Enewetak Atoll. *Science* 197, 556-9.

Smith, S. V. and P. L. Jokiel, 1978. Water composition and biogeochemical gradients in the Canton Atoll lagoon. *Atoll Research Bulletin* 221, 15-53.

Smith, S. V. and G. S. Key, 1975. Carbon dioxide and metabolism in marine environments. *Limnology and Oceanography* 20, 493-5.

Smith, S. V. and D. W. Kinsey, 1976. Calcium carbonate production, coral reef growth, and sea level change. *Science* 194, 937-9.

Smith, S. V. and D. W. Kinsey, 1978. Calcification and organic carbon metabolism as indicated by carbon dioxide. In D. R. Stoddart and R. E. Johannes (eds.), *Coral Reefs: Research Methods*. Paris: UNESCO, Monographs on Oceanographic Methodology, No 5, pp. 469-84.

Smith, S. V. and F. Pesret, 1974. Processes of carbon dioxide flux in the Fanning Island Lagoon. *Pacific Science* 28, 225-46.

Smith, S. V., K. J. Roy, H. G. Schiesser, G. L. Shepherd and K. E. Chave, 1971. Flux of suspended calcium carbonate ($CaCO_3$), Fanning Island Lagoon. *Pacific Science* 25, 206-21.

Stearn, C. W. and T. P. Scoffin, 1977. Carbonate budget of a fringing reef, Barbados. In D. L. Taylor (ed.), *Proceedings: Third International Coral Reef Symposium*. Miami: University of Miami, vol. 2, pp. 471-6.

Stearn, C. W., T. P. Scoffin and W. Martindale, 1977. Calcium carbonate budget of a fringing reef on the west coast of Barbados. I — Zonation and productivity. *Bulletin of Marine Science* 27, 479-510.

Webb, M. D., 1979. Pathways of CO_2, O_2, CO, and CH_4 in water flowing over a coral reef, Kaneohe Bay, Oahu, Hawaii. M.Sc. thesis, University of Hawaii, 107pp.

Weber, J. N. and P. M. J. Woodhead, 1972. Carbonate lagoon and beach sediments of Tarawa Atoll, Gilbert Islands. *Atoll Research Bulletin 157*, 1-29.

17 Relationships Between Reef Fishes and Coral Reefs

Michael Sutton
Department of Zoology, School of Biological Sciences,
University of Sydney, N.S.W. 2006.*

INTRODUCTION

Within the past five years, major reviews of the general ecology of coral reef fishes have been published by Ehrlich (1975), Goldman & Talbot (1976) and Sale (1981). Rather than summarize these works, this paper will examine specifically the interrelationships between fishes associated with coral reefs and their physiographic and biotic environments. Three topics will be discussed:
1. the ways in which reef fishes influence the growth, maintenance and change of reefs,
2. the effect of reef structure and other reef-dwelling organisms on the fish fauna and
3. the influence of factors extrinsic to the reef environment on reef fishes. The underlying theme of this paper is a synergistic approach to the ecology of coral reef fishes, emphasizing that coral reefs and their fish faunas must be considered together, in terms of their influence on each other.

EFFECT OF FISHES ON CORAL REEFS

Coral reef fishes appear to affect their environment principally through their feeding activities, which are quite diverse. Table 1 summarizes the results of four major studies of the food habits of reef fishes. This table represents a re-analysis of the published data, hence the categories and figures may not correspond exactly to those presented by the respective authors. Biomass figures for each feeding category were not available in most cases; see Bardach (1959), Bakus (1966; 1972) and Goldman & Talbot (1976).

The majority of fishes in each study were classed as carnivores. They comprised 60-68% of all species examined, probably reflecting the main thrust of teleostean evolution over the past 150 million years (Gosline 1971; Hobson 1974). Predominant among the carnivores were those specialized in various ways for feeding on benthic invertebrates (primarily crustaceans), although occasionally taking fish as well. These benthic carnivores, particularly the plectognaths (triggerfishes, puffers and others), have an effect in reducing the shells of the various invertebrate groups to coarse sediment

---

* Present address: National Park Service, Virgin Islands National Park, Cruz Bay, St. John, U.S. Virgin Islands, 00830, U.S.A.

Reference	Hiatt & Strasburg (1960)	Randall (1967)	Hobson (1974)	Goldman & Talbot (unpublished data)
Location	Marshall Islands	Virgin Islands; Puerto Rico	Hawaiian Islands	One Tree Island, Great Barrier Reef
Number of species examined	225	212	107	449
Herbivores	33(15)	27(13)	5(5)	134(30)
Planktivores	10(4)	27(13)	18(17)	45(10)
Benthic carnivores	125(56)	90(42)	56(52)	233(52)
Mid-water carnivores	27(12)	47(22)	12(11)	37(8)
Coral feeders	9(4)	0	6(6)	*
Omnivores	16(7)	21(10)	9(8)	—
Detritus feeders	4(2)	0	1(1)	—
Scavengers	1(0)	0	0	—

* All species in this study were placed in one of the first four categories; coral feeders and omnivores were apparently listed as herbivores.

Table 1. Feeding categories of coral reef fishes: numbers of species and percentage of total (in brackets) in each category.

(Randall 1974). Predaceous reef fishes may thus influence the evolution of external morphologies of their invertebrate prey (Vermeij 1978; Palmer 1979).

Herbivores make up a lesser proportion of the total number of reef fish species than do the carnivores but may have far more important, and certainly more direct, effects on the reef. The impact of feeding by grazers, such as the scarids (parrotfishes) and acanthurids (surgeonfishes), has been reviewed by Hatcher (this volume). Such large herbivores bring about changes not only through grazing on filamentous and fleshy algae, but by converting large amounts of coral limestone into sediments and dispersing them around the reef. According to Randall (1974), these grazers probably also consume many newly settled invertebrates and small coral colonies as a result of their constant scraping. Conversely, they may enhance the suitability of reef surfaces for the settlement and growth of larvae by reducing the algal biomass (Vine 1974; Birkeland 1977).

Browsing herbivores seem to have lesser impact, although there are notable exceptions. Kaufman (1977) suggested that *Eupomacentrus planifrons,* a common western Atlantic damselfish, affects reef-building corals by establishing feeding territories in which it kills the coral and cultivates a 'lawn' of filamentous algae (described by Brawley & Adey 1977). This process, Kaufman suggests, changes the outcome of competitive interactions among the corals and may alter the distributions and abundance of certain species of corals. In addition, the freshly killed coral is exposed to the effects of boring organisms, such as clionid sponges, which invade and break down the skeletal structure. Potts (1977) invoked a different mechanism to explain the suppression of growth and survivorship of coral colonies within the territories of the damselfish *Dischistodus perspicillatus* at Heron Island on the Great Barrier Reef. The damselfish excluded large herbivores from its territory, thus reducing the overall grazing pressure within the defended boundaries. This resulted in the death of coral due to competition from an increased standing-crop of algae. There was no evidence in this case that the damselfish directly affected the corals. Potts concluded: 'the abundance of *D. perspicillatus* is sufficient to explain the paucity of small *Acropora palifera* colonies in an otherwise favorable habitat'. In another example, Williams (1979) found that *Eupomacentrus planifrons* directly affected the distribution of herbivorous sea urchins through interference behavior in back-reef communities at Discovery Bay, Jamaica. Herbivorous fishes, such as these damselfishes, may have more important effects on reef structure and growth than previously suspected; this possibility requires further investigation.

A small number of fishes feed directly on coral, a potentially rich source of energy (Benson & Muscatine 1974). In a worldwide survey, Randall (1974) cited 57 species in ten families (Ephippidae, Chaetodontidae, Pomacentridae, Labridae, Scaridae, Bleniidae, Acanthuridae, Balistidae, Diodontidae and Tetraodontidae) as fishes that ingested parts of living scleractinian corals. There was considerable variation in both the importance of coral in the diets of the fishes and the method by which this food source was taken. Some species browsed individual polyps (e.g. butterflyfishes), and others bit off whole pieces of coral (the tetraodontids). For some, such as the parrotfishes, ingestion of live coral was only incidental during grazing for benthic algae, their principal food source. With the possible exception of the butterflyfishes (Chaetodontidae), none of the species that feed exclusively on living coral is common, and thus, these species may not have great impact on reefs. However, Neudecker (1977; 1979), conducted experiments that suggested that predation by fishes was an important factor limiting the distribution of *Pocillopora damicornis* to shallow water in Guam.

Most coral-feeding chaetodontids, although some are common, probably have little effect on corals because they browse only individual polyps, which are easily regenerated by a colony. Nevertheless, Reese (1977) urged the study of corals and coral-feeding fishes in a co-evolutionary framework, postulating: 'irritation caused by the feeding of fishes, especially when parts of the corallite are damaged, will affect skeletal structure and possibly the growth pattern of the entire colony. The behaviour and physiology of the polyps may also be affected'. The importance of coral-feeding fishes and the possibility of co-evolution in this system remains to be investigated.

The reef fishes with other food habits may also have an influence on the reef community. Planktivorous fishes may have an effect on the abundance and behaviour of near-reef zooplankton, including the larvae of corals (Emery 1968). These fishes may be responsible for the import of nutrients and energy from the open ocean into the reef system (Odum & Odum 1955; Stevenson 1972). There is some evidence that planktivores may feed to a certain extent on coral mucus and plant detritus suspended in the water column, thus retaining nutrients and energy within the reef system (Johannes 1967; Coles & Strathmann 1973). Other fishes are true omnivores, taking a wide variety of food items. A few species, such as sharks, act as scavengers; their influence has not been investigated.

In summary, the primary impact reef fishes have on their environment appears to be manifested in their adaptations for feeding. Randall (1974) stated: 'except for aesthetic aspects and serving as prey for other resident animals, fishes are not beneficial to reefs. They use them for shelter, reproduction, and as a source of food. It is in the latter role that their impact is greatest'. Carnivores prey on both vertebrates and invertebrates, and can significantly affect the behaviour, distribution, and abundance of these animals. Herbivores have similar impacts on marine plants, and in addition, serve as agents of bioerosion. It is likely that many observed adaptations of reef organisms have evolved in response to selection pressures exerted by coral reef fishes.

EFFECTS OF CORAL REEFS ON REEF FISHES

Within-Habitat Effects

Most fishes resident on coral reefs depend on the reef for shelter as well as food. Consequently, the nature of the substratum may influence the number of individuals that can live in an area, the number of species found there, and the species composition. Many authors have commented on the apparent relationship between the complexity of reef topography and the distribution and abundance of reef fishes (e.g. Bardach 1959; Hiatt & Strasburg 1960; Randall 1963; Talbot 1965; Vivien 1973; Jones & Chase 1975; Goldman & Talbot 1976; Gladfelter & Gladfelter 1978). The few quantitative studies (Risk 1972; Talbot & Goldman 1972; Luckhurst & Luckhurst 1978; Molles 1978; Talbot *et al.* 1978) all report statistically significant positive correlations between within-habitat species richness or diversity of reef fishes and various aspects of substratum heterogeneity, but acknowledge that the relationship is highly complex and not easily analyzed.

Few significant correlations have yet been found between the total number of fishes and the substratum characteristics of a reef. Risk (1972) measured the topographic complexity of reefs in the US Virgin Islands and found that it was significantly correlated with the species diversity of reef fishes present. There was no correlation, however, between any substratum index and the total number of fishes over the substratum.

Working at One Tree Island on the Great Barrier Reef, Talbot *et al.* (1978) constructed artificial reefs of cement blocks with varying hole sizes and allowed them to be colonized by reef fishes in order to test for effects of structural complexity on species diversity. They found only a weak relationship between the number of species and habitat intricacy (hole size). However, the number of species on these aritificial reefs was significantly lower than the number on nearby natural reefs of similar size. The authors attributed this variation to the 'much simpler habitat structure and possibly lower carrying capacity of the artificial reefs'. They did not correlate reef complexity with the number of fish occupants, probably because of the great variation in numbers over time. There were significant differences in species composition among the artificial reefs, which were attributed to habitat selection at that time of settlement by colonizing fishes rather than interspecific competition between resident adults.

In a study of natural and model patch reefs in the Gulf of California, Molles (1978) found a significant positive correlation between reef height (an aspect of heterogeneity) and fish species richness and diversity. However, he found no correlation between the size-diversity of substratum interspaces (analogous to hole size) and the number of fish species on either artificial or natural reefs.

Luckhurst & Luckhurst (1978), working on reefs in the Lesser Antilles of the Carribbean, developed an index of topographic complexity they called 'substrate rugosity'. They also sampled three parameters of the local reef fish community: species richness, diversity and the total number of fishes. Highly significant positive correlations were found at both study sites between fish species richness and 'substrate rugosity'. The correlations of rugosity, vertizial relief, and the percentage cover of the substratum with fish diversity and total numbers yielded inconclusive or non-significant results. It is likely that such correlations and indices of complexity are simplistic, thus masking 'a complex of factors determining the responses of species to the substrate with which they are associated' (Luckhurst & Luckhurst 1978). It is becoming clear that the fine structure of coral reefs has at least some influence, in both proximate and evolutionary terms, on the species richness, composition, diversity, and perhaps abundance of reef fishes.

Between-Habitat Effects

On a larger scale, Goldman & Talbot (1976) compared the species composition of reef fishes in eight habitats at One Tree Reef, Great Barrier Reef. Of the 395 species that were collected across the reef, only 7% were found in all eight habitats, and about half (49%) were restricted to a single habitat. The authors stated only: 'a certain degree of, habitat selection was thus discernible'. More recent data reveal a lesser degree of local endemism (B. C. Russell, pers. comm.). There was considerable variation in both the total number of species collected in, and restricted to, each habitat. In a similar study at Fanning Island, Chave & Eckert (1974) defined seven reef habitats, from the lagoon shoreline to the outer reef slope. They found a total of 214 species of fishes, only 14 (6%) of which were seen more than once in at least six of the seven habitats.

Neither pair of authors suggested tht the habitat associations implicit in their data were solely the result of the different topography of each habitat; they realized that many other factors must affect the distribution of fishes across the reef. However, physical factors which may affect habitat selection among fishes, such as current speed and wave action, are themselves dependent somewhat on reef structure. Thus, reef physiography can be visualized as an important constituent of an entire spectrum of factors which must influence the distribution and abundance of coral reef fishes.

Other Effects

Many other factors in addition to habitat structure affect the fish faunas of coral reefs. For example, the behaviour, distribution, and abundance of many fishes is tied to those of their preferred food (Goldman & Talbot 1976). The diversity of other organisms on the reef, particularly the benthic invertebrates, may have contributed to the development of a wide variety of feeding specializations and hence to the evolution of diversity among reef fishes (Hobson 1974). Some reef fishes have specific behavioral requirements which can drastically affect their distributions. For example, anemonefishes of the genus *Amphiprion* are limited to dwelling in association with the anthozoans for which they are named (Allen 1972). Several gobies occur only where there are certain alpheid shrimps which dig the burrows in which both species live (Cummins 1979). The damselfish *Dascyllus aruanus* is found on the Great Barrier Reef only in coral colonies suitable for its use as shelter (Sale 1972).

EXTRINSIC EFFECTS

It is becoming evident that some of the factors having perhaps the greatest effects on the distribution and abundance of coral reef fishes have little to do with reefs (q.v. Sale 1977). Because most reef fishes have a pelagic larval phase, they are subject to unpredictable events in the ocean plankton that almost certainly have important effects on patterns of settlement of larvae on reefs (Leis & Miller 1976). In turn, these chance patterns of juvenile recruitment may eventually have profound influences on the distribution and abundance of adult populations (Sale & Dybdahl 1978). Kept in check by predation, catastrophe, and seasonally supplied with an unpredictable number of recruits, many reef fish populations may never reach an equilibrium number. Indeed, it has been suggested that this is the principal mechanism for the maintenance of high diversities in communities of coral reef fishes (Sale 1977; Talbot *et al.* 1978).

There is an inherent circularity in the reasoning of the previous pages; reef fishes affect their habitats but are themselves limited by many factors within and outside the same environment. Perhaps this fact is the best illustration of the main point of this paper: that we must consider an entire range of intrinsic and extrinsic factors as having effects on the distribution, abundance and evolution of coral reef fishes, just as these animals should never be omitted in any analysis of the same properties of other reef organisms and processes.

ACKNOWLEDGEMENTS

The ideas in this paper were developed in part during discussions with J. D. Bell, B. G. Hatcher, B. C. Russell, P. F. Sale, R. E. Thresher, D. McB. Williams and M. P. Wunderlich, all of whom also kindly read and criticized the manuscript. I am grateful to B. C. Russell and H. P. A. Sweatman for lending me the data from explosive samples of the fish fauna at One Tree Reef. The author is supported by a University of Sydney Postgraduate Research Studentship, Great Barrier Reef Committee Postgraduate Research Grant and by the Great Barrier Reef Marine Park Authority.

LITERATURE CITED

Allen, G. R., 1972. The Anemonefishes: Their Classification and Biology. Neptune City, New Jersey: TFH Publications.

Bakus, G. J., 1966. Some relationships of fishes to benthic organisms on coral reefs. *Nature* 210, 280-4.

Bakus, G. J., 1972. Effects of the feeding habits of coral reef fishes on the benthic biota. In C. Mukundan and C. S. Gopinadha Pillai (eds.), *Proceedings of the Symposium on Corals and Coral Reefs*. The Marine Biological Association of India, Cochin, pp. 445-8.

Bardach, J. E., 1959. The summer standing crop of fish on a shallow Bermuda reef. *Limnology and Oceanography* 4, 77-85.

Benson, A. A. and L. Muscatine, 1974. Wax in coral mucus: Energy transfer from corals to reef fishes. *Limnology and Oceanography* 19, 810-4.

Birkeland, C., 1977. The importance of rate of biomass accumulation in early successional stages of benthic communities to the survival of coral recruits. In D. L. Taylor (ed.), *Proceedings: Third International Coral Reef Symposium*. Miami: University of Miami, vol. 1, pp. 15-21.

Brawley, S. H. and W. H. Adey, 1977. Territorial behavior of three spot damselfish *(Eupomacentrus planifrons)* increases reef biomass and productivity. *Environmental Biology of Fishes* 2, 45-51.

Chave, E. H. and D. B. Eckert, 1974. Ecological aspects of the distribution of fishes at Fanning Island. *Pacific Science* 28, 297-317.

Coles, S. L. and R. Strathmann, 1973. Observations on coral mucus "flocs" and their potential trophic significance. *Limnology and Oceanography* 18, 673-8.

Cummins, R. A., 1979. Ecology of gobiid fishes associated with alpheid shrimps at One Tree Reef. Ph.D. dissertation, University of Sydney.

Ehrlich, P. R. 1975. The population biology of coral reef fishes. *Annual Review of Ecology and Systematics* 6, 211-47.

Emery, A. R., 1968. Preliminary observations on coral reef plankton. *Limnology and Oceanography* 13, 293-303.

Gladfelter, W. B. and E. H. Gladfelter, 1978. Fish community structure as a function of habitat structure on West Indian patch reefs. *Revista de Biologia Tropical* 26 (Suppl. 1), 65-84.

Goldman, B. and F. H. Talbot, 1976. Aspects of the ecology of coral reef fishes. In O. A. Jones and R. Endean (ed.), *Biology and Geology of Coral Reefs*. New York: Academic Press, vol. 3, Biology 2, pp. 125-54.

Gosline, W. A., 1971. *Functional Morphology and Classification of Teleostean Fishes*. Honolulu: University of Hawaii Press, 197pp.

Hiatt, R. W. and D. W. Strasburg, 1960. Ecological relationships of the fish fauna on coral reefs of the Marshall Islands. *Ecological Monographs* 30, 65–127.

Hobson, E. S., 1974. Feeding relationships of teleostean fishes on coral reefs in Kona, Hawaii. *Fishery Bulletin* 72, 915–1031.

Johannes, R. E., 1967. Ecology of organic aggregates in the vicinity of a coral reef. *Limnology and Oceanography* 12, 189–95.

Jones, R. S. and J. A. Chase, 1975. Community structure and distribution of fishes in an enclosed high island lagoon in Guam. *Micronesica* 11, 127–48.

Kaufman, L., 1977. The three spot damselfish: effects on benthic biota of Caribbean coral reefs. In D. L. Taylor (ed.), *Proceedings: Third International Coral Reef Symposium*. Miami: University of Miami, vol. 1, pp. 559–64.

Leis, J. M. and J. M. Miller, 1976. Offshore distribution patterns of Hawaiian fish larvae. *Marine Biology* 36, 359–67.

Luckhurst, B. E. and K. Luckhurst, 1978. Analysis of the influence of substrate variables on coral reef fish communities. *Marine Biology* 49, 317–23.

Molles, M. C., Jr., 1978. Fish species diversity on model and natural reef patches: experimental insular biogeography. *Ecological Monographs* 48, 289–305.

Neudecker, S., 1977. Transplant experiments to test the effect of fish grazing on coral distribution. In D. L. Taylor (ed.), *Proceedings: Third International Coral Reef Symposium*. Miami: University of Miami, vol. 1, pp. 317–23.

Neudecker, S., 1979. Effects of grazing and browsing fishes on the zonation of corals in Guam. *Ecology* 60, 666–72.

Odum, H. T. and E. P. Odum, 1955. Trophic structure and productivity of a windward coral reef community on Eniwetok Atoll. *Ecological Monographs* 25, 291–320.

Palmer, A. R., 1979. Fish predation and the evolution of gastropod shell sculpture: Experimental and geographic evidence. *Evolution* 33, 697–713.

Potts, D. C., 1977. Suppression of coral populations by filamentous algae within damselfish territories. *Journal of Experimental Marine biology and Ecology* 28, 207–16.

Randall, J. E. 1963. An analysis of the fish populations of artificial and natural reefs in the Virgin Islands. *Caribbean Journal of Science* 3, 31–47.

Randall, J. E., 1967. Food habits of reef fishes of the West Indies. *Studies in Tropical Oceanography* 5, 665–847.

Randall, J. E., 1974. The effect of fishes on coral reefs. In, *Proceedings of the Second International Symposium on Coral Reefs*. Brisbane: Great Barrier Reef Committee, vol. 1, pp. 159–66.

Reese, E. S., 1977. Coevolution of corals and coral feeding fishes of the family Chaetodontidae. In D. L. Taylor (ed.), *Proceedings: Third International Coral Reef Symposium*. Miami: University of Miami, vol. 1, pp. 267–74.

Risk, M. J., 1972. Fish diversity on a coral reef in the Virgin Islands. *Atoll Research Bulletin* 153, 1–6.

Sale, P. F., 1972. Influence of corals in the dispersion of the pomacentrid fish *Dascyllus aruanus*. *Ecology* 53, 741–4.

Sale, P. F., 1977. Maintenance of high diversity in coral reef fish communities. *American Naturalist* 111, 337–59.

Sale, P. F., 1980. Ecology of fish on coral reefs. *Oceanography and Marine Biology Annual Review* 18, 367–421.

Sale, P. F. and R. Dybdahl, 1978. Determinants of community structure in isolated coral heads at lagoonal and reef slope sites. *Oecologia* 34, 57–74.

Stevenson, R. A., Jr., 1972. Regulation of feeding behavior of the bicolor damselfish *(Eupomacentrus partitus* Poey) by environmental factors. In H. E. Winn and B. L. Olla (eds.), *Behavior of Marine Animals, Volume 2: Vertebrates*. New York: Plenum Press, pp. 278–302.

Talbot, F. H. 1965. A description of the coral structure of Tutia Reef (Tanganyika Territory, East Africa) and its fish fauna. *Proceedings of the Zoological Society of London* 145, 431–70.

Talbot, F. H. and B. Goldman, 1972. A preliminary report on the diversity and feeding relationships of the reef fishes of One Tree Island, Great Barrier Reef. In C. Mukundan and C. S. Gopinadha Pillai (eds), *Proceedings of the Symposium on Corals and Coral Reefs*. Cochin: The Marine Biological Association of India, pp 425–42.

Talbot, F. H., B. C. Russell and G. R. V. Anderson, 1978. Coral reef fish communities: unstable, high-diversity systems? *Ecological Monographs* 48, 425–40.

Vermeij, G. J., 1978. Biogeography and Adaptation: Patterns of Marine Life. Cambridge, Massachusetts: Harvard Univeristy Press, 332pp.

Vine, P. J., 1974. Effects of algal grazing and aggressive behavior of the fishes *Pomacentrus lividus* and *Acanthurus sohal* on coral-reef ecology. *Marine Biology* 24, 131–6.

Vivien, M. L., 1973. Contribution a la connaissance de l'ethologie alimentaire de l'ichtyofaune du platier interne des recifs corraliens de Tulear (Madagascar). *Tethys,* Supplement 5, 221–308.

Williams, A. H., 1979. Interference behavior and ecology of threespot damselfish *(Eupomacentrus planifrons)*. *Oecologia* 38, 223–30.

18 Measurements of Rates of Erosion of Reefs and Reef Limestones

Stephen T. Trudgill*

Department of Geography, University of Sheffield, S10 2TN, U.K.

INTRODUCTION

Measurements of rates of erosion on coral reefs are important since the form of a reef at any moment in time is a function of the balance between constructive and destructive forces. Knowledge of erosion rates thus assists in the interpretation of the past evolution of reef form, the present morphology and the prediction of future developments.

Three distinct situations exist on reefs depending upon the nature of the substrate under discussion:

1. where coral and algal growth is active;
2. where dead carbonate skeletons exist as loose particles and,
3. where the skeletons are cemented into rocks by chemical precipitation.

In the first situation, reef morphology is a function of the balance between active growth of coral, algae and shells and simultaneous erosion of parts of the substrate. In the second situation, the loose fragments, or clasts, of biogenic carbonates, which have been derived from the living material by mechanical fracture, are moved around in response to wave energy. This material includes boulders, shingle and sands. It is easily moved by high energy events such as storms and cyclones, and thus reef morphology is often related to low frequency, high magnitude erosional events. In the third situation, lithified material is not actively accreting in any way. The substrate may be lithified reef material formed *in situ*, or lithified clastic material. Scoffin & McLean (1978) recognise *in situ* reef rock, beach rock, rampart rock (from shingle ridges) and boulder rock. Here the morphology of the landform is related solely to the way in which the forces of erosion act upon the inherent structures of the substrate.

Material of any origin may be eroded from one area and transported to another area where it may be deposited. Thus, a full understanding of reef morphology requires knowledge of not only growth and erosion but also of transport and deposition processes.

The study of erosion rates of terrestrially exposed reef limestone is important since much recent work on the evolution of the morphology of reefs stresses the role of inheritance of reef form from pre-existing subaerial forms Purdy (1974). Much of the argument rests on the speed of limestone erosion during periods of Pleistocene emersion.

Having outlined the principal situations applying to reef erosion, it will be necessary to outline the principal processes of erosion before discussing how erosion rates may be measured.

*Visiting fellow, Australian National University, Canberra, July–September 1979.

EROSION PROCESSES

Erosion processes can be thought of under three main headings: mechanical, biological and chemical.

Mechanical processes involve:
1.   the fracture and movement of carbonate material by wave action;
2.   abrasive action of clasts upon stable substrate and also the attrition of each clast with adjacent clasts and
3.   salt weathering, i.e., the growth of salt crystals in interstices in the rock during water evaporation and the physical movement of carbonate grains by crystal growth.

Biological processes involve grazing by fish and molluscs on epilithic and endolithic algae, which also removes considerable carbonate substrate. In addition, many organisms bore into carbonate substrate (e.g. Bromley 1978; Schneider 1976), directly removing material and also making substrates weaker and more prone to mechanical fracture under wave impact.

Chemical processes involve the dissolution of carbonate by rain water and sea water. It is doubtful if the latter process operates without some biological involvement since aqueous carbonate equilibria in sea water are closely related to biological activity.

Of these three sets of processes, recent work has suggested that biological processes may be quantitatively the most important as they operate continuously over long periods of time. However, mechanical processes may operate periodically with devastating effect because of storms and cyclones. Chemical erosion is possible in inshore environments, especially nocturnally when CO_2 production by respiration exceeds CO_2 uptake by photosynthesis; however, the process is usually subordinate to biological processes.

In terrestrially exposed limestones, two principal processes operate: dissolution in rain, pool or soil water, and boring by endolithic algae.

MEASUREMENTS OF EROSION RATES

Erosion rates may be assessed by the study of either the overall retreat of a surface or by the study of the operation of individual agencies. Surface retreat has been measured with reference to datum points over the long term, around ten years (Hodgkin 1964; Evans 1970), or over a shorter period, around one or two years, by use of a micrometer (Trudgill 1976a,b; Kirk 1977). The bioerosion rates of sponges, echinoids and grazing molluscs have been assessed by several authors, listed below, who studied the production of carbonate detritus from lithic substrate. Cavity formation has been studied by other authors who used test coral blocks. Rates of boring of molluscs have been studied by comparing the depths of borings with estimates of age of the organisms. In addition, the relative rates of abrasion, bioerosion and dissolution have been studied by observations under control conditions where different combinations of processes exist. Finally, the effects of catastrophic events have been studied by comparing reef morphology on old maps with that existing after an event (Stoddart 1978).

Erosion rates have mostly been measured on reef limestones which are not actively growing. Very few measurements are available for growing reefs, or for areas dominated by clastic materials.

Two related methods for the measurement of surface retreat are reported in the literature. Firstly, Hodgkin (1964) used a series of steel pegs driven into eolianite limestone in Western Australia and observed a mean rate of retreat of 0.5–1.0 mm yr^{-1} over nine years. A micro-erosion meter was used by Trudgill (1976a) to measure the surface retreat of intertidally exposed limestones in Aldabra Atoll in the Indian Ocean. A micro-erosion meter consists of a micrometer dial gauge set in a tripod framework which rests upon three reference studs set into the rock. A measure of the height of the rock surface relative to the studs is taken and repeated at intervals in order to assess the rate of rock surface lowering relative to the studs (High & Hanna 1970). On Aldabra, the substrate was lithified reef rock of Aldabra Limestone (Brathwaite *et al.* 1973). A summary of the erosion data is presented in table 1. It can be seen that erosion rates increased with exposure to south-east trade winds from around 0.5-1.0 to 2.0-4.0 mm yr[-1].

Exposure*	Coastal Form	upper intertidal	Erosion Rate (mm.yr[-1]) mid intertidal	lower intertidal
Low	undercut notch	0.6	1.0	0.5
Moderate	vertical cliff	0.8	1.5	0.7
Moderate	steep cliff	2.2	2.0	0.9
High	steep ramp	3.3	2.8	1.5
High	low ramp	4.0	3.0	2.0

*Highly exposed sites face the dominant S.E. wind

Table 1   Marine erosion rates for coral reef limestone, Aldabra Atoll, micro-erosion meter results, Trudgill (1976a)

In addition to these data on surface lowering, evidence was provided for the relative roles of bioerosion, dissolution and abrasion. Sets of weight-loss limestone tablets were used (Trudgill 1975) where only dissolution occurred (continual submersion) and where solution plus abrasion occurred. The data were compared with data for surface lowering where bioerosion and dissolution were operative. The tablets were weighed accurately before field emplacement and re-weighed subsequently and weight loss was taken as a measure of erosion. The conclusions are presented in table 2.

	Abrasion	Grazing	Dissolution and Other Processes*
Sand present	33%	36%	31%
Sand absent	—	64%	36%

*These include salt weathering, wetting and drying and wave action

Table 2   Relative importance of marine erosion agencies, undercut notch on coast of low exposure, Aldabra Atoll, Indian Ocean, Trudgill (1976a)

These data apply to fossil Pleistocene carbonate rocks. Little is known about situations where carbonate material is actively being cemented by an encrustation of corals and corralline algae. The most comprehensive studies are those of Stearn & Scoffin (1977) on a fringing reef in Barbados. They suggested that corals and algae fix about 160 metric tons of calcium carbonate annually over an area of about 10 800 m². Studies of boring by

Locality	Substrate	Erosion Type/Agent	Rate	Author
I Growing Reefs				
Bermuda	Reef	Fish	2-3 tons ha^{-1} yr^{-1}	Bardach 1961
Bermuda	Reef	Bioerosion (Fish, clionids)	1.3 mm yr^{-1}	Bromley 1978
Mariana Islands	Atoll	Fish	1.1–1.6 tons m^{-2} yr^{-1}	Cloud 1959
Orpheus Island, G.B.R.	Reef	*Tridacna crocea*	100 cm^3 m^{-2} yr^{-1}	Hamner & Jones 1976
Florida	Reef	*Cliona* boring	746–4303 mm^3 reworked	Hein & Risk 1975
Florida	Reef	Bioerosion, esp. *Cliona*	1 m coral head in 150 years	Hudson 1977
Barbados	Fringing Reef	*Diadema*	97 tonnes sediment ha^{-1} yr^{-1} 40% reworked	Hunter 1977
Barbados	Fringing Reef	*Cliona* boring	80–377 g m^{-2} yr^{-1}	Stearn & Scoffin 1977
II Carbonate Rocks				
Puerto Rico	Reef limestone	Intertidal notch retreat	1.0 mm yr^{-1}	Kaye 1959
Red Sea	Coral Reef Limestone	Surface lowering	Surface lowering *cf. Tetraclita squamosa* 1 mm yr^{-1} if 10–15 yr old	MacFadyen 1930
Barbados	Beachrock	*Echinometra* boring	4.9 cm yr^{-1}; 9.96 cc yr^{-1} 24.0 g yr^{-1}	McLean 1967
Barbados	Beachrock	*Acmaea* grazing	1.5 mm yr^{-1}; 0.99 cc yr^{-1} 2.4 g yr^{-1}	McLean 1967
Barbados	Beachrock	*Littorina ziczac*	0.4 cm^3 yr^{-1}	McLean 1967
Barbados	Beachrock	*L. meleaguis*	0.15 cm^3 yr^{-1}	McLean 1967
Barbados	Beachrock	*Nodilittorina tuberculata*	0.6 cm^3 yr^{-1}	McLean 1967
Barbados	Beachrock	*Nerita tesselata*	0.4 cm^3 yr^{-1}	McLean 1967
Barbados	Beachrock	*Nerita versicolor*	0.8 cm^3 yr^{-1}	McLean 1967
Barbados	Beachrock	*Cittarium pica*	1.3 cm^3 yr^{-1}	McLean 1967
Barbados	Beachrock	*Acmaea*	2.0 cm^3 yr^{-1}	McLean 1967
Barbados	Beachrock	*Fisurella*	5.0 cm^3 yr^{-1}	McLean 1967
Barbados	Beachrock	*Anacthopleura*	13.0 cm^3 yr^{-1}	McLean 1967
Barbados	Beachrock	*Chiton*	8.0 cm^3 yr^{-1}	McLean 1967
Barbados	Beachrock	*Echinometra lucunter*	14.0 cm^3 yr^{-1}	McLean 1967
Barbados	Beachrock	Surface grazers	1–2 mm yr^{-1}	McLean 1967
Heron Island, G.B.R.	Beachrock	*Acanthozostera*	18.0 cm^3 yr^{-1}; 0.5 mm yr^{-1}	McLean 1967
Virgin Islands	Reef Limestone	Sponge boring	Up to 7 kg m^{-2} yr^{-1}	Moore & Shedd 1977
Bermuda	Eolianite	*Cliona* boring	1.0–1.4 cm yr^{-1}	Neumann 1966
G.B.R.	Beachrock	*Lithophaga*	1.5 cm yr^{-1}	Otter 1937; in McLean 1974
Bikini Atoll	Beachrock	Surface lowering	0.3 mm yr^{-1}	Revelle & Emery 1957
S.W. Australia	Reef Limestone	Surface lowering	270–670 cm^3 yr^{-1} 100 cm^{-2}	Revelle & Fairbridge 1957
Bermuda	Iceland spar calcite	Sponge boring	7 kg m^{-2} yr^{-1}	Rützler 1975
Heron Island, G.B.R.	Beachrock	Surface lowering	0.5 mm yr^{-1}	Stephenson 1961
Aldabra	Reef limestone	Intertidal surface retreat	0.5–4.0 mm yr^{-1}(micro-erosion meter)	Trudgill 1976a
Aldabra	Reef limestone	*Lithophaga* boring	0.9 cm yr^{-1}; 0.87 cc yr^{-1}	Trudgill 1976a
Aldabra	Reef limestone	*Lithotrya* boring	0.8 cm yr^{-1}; 0.78 cc yr^{-1}	Trudgill 1976a
Aldabra	Reef limestone	Subaerial surface lowering	0.26 mm yr^{-1}	Trudgill 1976a
Oman	Reef limestone	*Lithophaga*	0.0025 m yr^{-1}	Vita-Finzi & Cornelius 1973

Table 3 Erosion rates measured on reefs and reef limestones (listed alphabetically by author)

sponges, barnacles and bivalves suggest that they annually removed about 1.5 metric tons from live corals and 23.5 metric tons from algal encrusted dead coral surfaces. Grazing by parrot fish removed 1 metric ton and the sea urchin *Diadema antillarum* removed 163 metric tons. It should be emphasized that only 37% of the area under discussion was covered by live coral and 22% of the area was dead coral, and the remaining 40% was coralline algae. The data given above should be qualified by the observations that in addition to simply removing substrate *Diadema* also re-works sand material and erodes living coral and coralline algae and, in addition, other processes may operate which have not been investigated.

One problem which emerges from work on the erosion rates on reefs is that some data are presented in terms of mm per year and others in terms of volume or weight per year. These figures are not comparable unless the area over which removal takes place and the density of the material removed are known. Some of the available works are summarized in table 3.

DISCUSSION

It can be seen that research on erosion rates is patchy and that work on one particular aspect has been undertaken in one place but that on another aspect is in a different locality. Measurements of rates of net erosion and the apportionment of the overall rate to individual agencies are not available for any one area of actively growing coral reef. Such works as do exist suggest two things: first, the erosion rates are relatively rapid and second, that biological processes are often the most important. Further work on sediment production, net erosion (where re-working of sediment is distinguished from the production of new sediment), on rates of biocavity production (Carriker 1969) and on surface retreat are necessary. This work should usually be carried out at one site both on a short-term basis (most easily with experimentally observed coral blocks) and also on a long-term basis, so the effects of episodic catastrophic events can be observed.

ACKNOWLEDGEMENTS

I would like to thank Dr Roger McLean for helpful comments on the manuscript and assistance with literature searches.

LITERATURE CITED

Bardach, J. E., 1961. Transport of calcareous fragments by reef fishes. *Science* 133, 98–9.

Braithwaite, C. J. R., J. D. Taylor and W. J. Kennedy, 1973. The evolution of an atoll: the depositional and erosional history of Aldabra. *Philosophical Transactions of the Royal Society of London, Series B* 266, 307–40.

Bromley, R. G., 1978. Bioerosion of Bermuda reefs. *Palaeogeography, Palaeoclimatology, Palaeoecology* 23, 169–97.

Carriker, M. R. (ed.), 1969. The penetration of $CaCO_3$ substrate by lower plants and invertebrates. *American Zoologist* 9, 629–30.

Cloud, P. E., Jr., 1959. Geology of Saipan, Mariana Islands. Part 4. Submarine topography and shoal water ecology. *United States Geological Survey Professional Paper* 280-K, 361–445.

Evans, J. W., 1970. A method of measurement of the rate of intertidal erosion. *Bulletin of Marine Science* 20, 305–14.

Hamner, W. M. and M. S. Jones, 1976. Distribution, burrowing and growth rates of the clam *Tridacna crocea* on interior reef flats. Formation of structures resembling micro atolls. *Oecologia* 24, 207–27.

Hein, F. J. and M. J. Risk, 1975. Bioerosion of coral heads: inner patch reefs, Florida reef tract. *Bulletin of Marine Science* 25, 133–8.

High, C. and F. K. Hanna, 1970. A method for the direct measurement of erosion of rock surfaces. *British Geomorphological Research Group, Technical Bulletin* 5.

Hodgkin, E. P., 1964. Rate of erosion of intertidal limestone. *Zeitschrift für Geomorphologie* 8, 385 –92.

Hudson, J. H., 1977. Long-term bioerosion rates on a Florida reef: a new method. In D. L. Taylor (ed.), *Proceedings: Third International Coral Reef Symposium.* Miami: University of Miami, vol. 2, pp. 491–7.

Hunter, I. G., 1977. Sediment production by *Diadema antillarum* on a Barbados fringing reef. In D. L. Taylor (ed.), *Proceedings: Third International Coral Reef Symposium.* Miami: University of Miami, vol. 2, pp. 105–9.

Kaye, C. A., 1959. Shoreline features and Quaternary shoreline changes in Puerto Rico. *United States Geological Survey Professional Paper* 317–23.

Kirk, R. M., 1977. Rates and forms of erosion on intertidal platforms at Kaikoura Peninsula, South Island, New Zealand. *New Zealand Journal of Geology and Geophysics* 20, 571–613.

MacFadyen, W. A., 1930. The undercutting of coral reef limestone on the east of some islands of the Red Sea. *Geographical Journal* 75, 27–37.

McLean, R. F., 1967. Measurement of beach rock erosion by some tropical marine gastropods. *Bulletins of Marine Science* 17, 551–61.

McLean, R. F. 1974. Geologic significance of bioerosion of beach rock. In, *Proceedings of the Second International Symposium on Coral Reefs.* Brisbane: Great Barrier Reef Committee, vol. 2, pp. 401–8.

Moore, C. H., Jr. and W. W. Shedd, 1977. Effective rates of sponge bioerosion as a function of carbonate production. In D. L. Taylor (ed.), *Proceedings: Third International Coral Reef Symposium.* Miami: University of Miami, vol 2, pp. 499–505.

Neumann, A. C., 1966. Observations on coastal erosion in Bermuda and measurements of the boring rate of the sponge, *Cliona lampa. Limnology and Oceanography* 11, 92–108.

Otter, G. W., 1937. Rock-destroying organisms in relation to coral reefs. *Scientific Report of the Great Barrier Reef Expedition, 1928–29, British Museum (Natural History)* 1, 323–52.

Purdy, E. G., 1974. Reef configurations: cause and effect. *Society of Economic Paleontologists and Mineralogists. Special Publication* 18, 9–76.

Revelle, R. and K. O. Emery, 1957. Chemical erosion of beach rock and exposed reef rock: Bikini and nearby atolls, Marshall Islands. *United States Geological Survey Professional Paper* 260-T, 699–709.

Revelle, R. and R. Fairbridge, 1957. Carbonates and carbon dioxide. In J. Hedgepeth (ed.), *Treatise on Marine Ecology and Palaeoecology.* Geological Society of America, 67, pp. 239 –96.

Schneider, J., 1976. Biological and inorganic factors in the destruction of limestone coasts. *Contributions to Sedimentology* 6.

Scoffin, T. P. and R. F. McLean, 1978. Exposed limestones of the Northern Province of the Great Barrier Reef. *Philosophical Transactions of the Royal Society of London, Series A* 291, 119 –38.

Stearn, C. W. and T. P. Scoffin, 1977. Carbonate budget of a fringing reef, Barbados. In D. L. Taylor (ed.), *Proceedings: Third International Coral Reef Symposium.* Miami: University of Miami, vol. 2, pp. 471–6.

Stephenson, W. and R. B. Searle, 1960. Experimental studies on the ecology in intertidal environments at Heron Island. I. Exclusion of fish from beach rock. *Australian Journal of Marine and Freshwater Research* 11, 241–67.

Stoddart, D. R., 1978. Forty-five years of change at Low Wooded Isles, northern Great Barrier Reef. *Philosophical Transactions of the Royal Society of London, Series A* 291, 47–58.

Trudgill, S. T., 1975. Measurement of erosional weight loss of rock tablets. *British Geomorphological Research Group Technical Bulletin* 17, 13-19.

Trudgill, S. T., 1976a. The marine erosion of limestone on Aldabra Atoll, Indian Ocean. *Zeitschrift für Geomorphologie,* Supplementband, 26, 164-200.

Trudgill, S. T., 1976b. The sub-aerial and sub-soil erosion of limestones on Aldabra Atoll, Indian Ocean. *Zeitschrift für Geomorphologie,* Supplementband, 26, 201-10.

Vita-Finzi, C. and P. F. S. Cornelius, 1973. Cliff sapping by molluscs in Oman. *Journal of Sedimentary Petrology* 43, 31-2.

19 Role of Sponges in Coral Reef Structural Processes

Clive R. Wilkinson
Australian Institute of Marine Science,
P.M.B. No. 3, M.S.O., Townsville, Qld. 4810.

INTRODUCTION

This review summarizes the role that sponges play in coral reef geological processes, including reef construction, bioerosion and subsequent reformation of coral substrate into rock and sediment. Sponges may be divided for our purposes into those which produce fused, calcareous skeletons and those without a massive skeleton, i.e., containing either siliceous and or calcareous spicules, or no mineral component at all. Sponges with fused calcareous skeletons are not prominent today but, in earlier geological eras, they were particularly significant in forming large carbonate reefs. Although spiculated sponges have always been prevalent, they have contributed little to reef accretion. However, their role in bioerosion and subsequent modification of reefal morphology is particularly important. These aspects will be accentuated in this review; for information on the role of sponges in the nutrient balance of reefs consult Reiswig (1971a) and Vacelet (1979a); for more general information, the text of Bergquist (1978) is recommended.

REEF CONSTRUCTION

Fossil Reefs

Sponges were among the first metazoans to occur in the geological record. They were prominent in the early Cambrian, and sponges that resemble the largest class of present day sponges, the Demospongiae, have been reported from the Cambrian (Finks 1970; Ziegler & Rietschel 1970). These sponges contained discrete siliceous spicules and did not contribute significantly to reef formation.

The first reefs that were constructed primarily by sponges occurred in the later Ordovician period. These were formed by large populations of sponges with massive, fused, calcareous skeletons — the Stromatoporoidea (Finks 1970), which have been shown to be closely related to present day sclerosponges (Hartman & Goreau 1970; Hartman 1979). Reefs were formed in the earlier Cambrian by the Archaeocyatha which are regarded as being calcareous algae, although they do resemble later calcareous sponges (Zhuravleva 1970; Zhuravleva & Miagkova 1979). Throughout the whole Palaeozoic, sponges and corals alternated as major contributors to reef formation (table 1). Massive reefs were formed by calcareous sponges towards the end of the Palaeozoic

Period	Sponges	Other Reef-constructing Organisms
Permian	Sphinctozoa	Algae
	Pharetronida	Bryozoa
		Hydrocorallines
Carboniferous		Algae
		Corals
Devonian	Stromatoporoidea	Algae
		Corals
Silurian	Stromatoporoidea	Bryozoa
		Algae
		Corals
Ordovician	Stromatoporoidea	Bryozoa
	Demospongia	Algae
	(Receptaculitida)	Corals
Cambrian	(Archaeocyatha)	Algae

Table 1   Role of sponges in reef construction during the Palaeozoic compared with other major reef-constructing organisms. Italicized periods indicate sponge dominance in reef construction. Those organisms with questioned affinities to the Porifera are in parentheses (Ex. Finks 1970)

(Permian) and in the early Mesozoic (Triassic). During these periods the dominant reef-forming sponges were the Pharetronida and Sphinctozoa. 'Living fossil' forms of both of these groups have recently been found in cryptic habitats (Vacelet 1967, 1970, 1977, 1979b; Hartman 1979), but their present sizes and populations are insignificant compared with those in earlier periods. Both groups have solid calcareous skeletons: the Pharetronida have siliceous spicules embedded in the calcium carbonate and the Sphinctozoa form a series of sub-spherical perforated chambers around the tissue (Vacelet 1977, 1979b).

During the later Cretaceous period, sponges were particularly abundant in benthic ecosystems (Finks & Hill 1967) and it has been estimated that there were three times as many demosponge genera as there are today (Bergquist 1978). However, these sponges did not contribute much to reef construction as their skeletons contained only a small proportion of isolated siliceous spicules.

Modern Reefs

Sponges play a minor role in modern reef construction compared with corals and coralline algae. The only sponges that directly contribute to reef growth are the calcareous sclerosponges. On Caribbean reefs, these sponges are only prevalent where the amount of coral cover is low, such as in recesses and caverns in shallow water and on steep reef slopes in deeper water (Hartman & Goreau 1970). On some slopes the sclerosponges, particularly *Ceratoporella,* are the primary framework builders (Lang *et al.* 1975). They also consolidate those areas where the rate of coral and algal calcification is low and where the effects of bioeroding organisms are most marked (Hartman & Goreau 1970).

Tropical demosponges often physically support corals, preventing collapse after their basal structures have been bioeroded (Goreau & Hartman 1963). Sponges are also important in reef framework consolidation as they hold corals and rubble together during sediment infilling and lithification (Wulff & Buss 1979). In competing with corals, some sponges cause distinct morphological changes and sometimes death of corals (Goreau & Hartman 1966; Bryan 1973; Wilkinson 1978).

Siliceous sponges are more numerous in coral reefs than calcareous sponges but their mineral skeletal content is comparatively low. In tropical seas, there are no large deposits of siliceous spicules like those found under the Antarctic ice shelf (Dayton *et al.* 1974) and siliceous spicules constitute a minor component of reef sediments (Rützler & Macintyre 1978). Although siliceous spicules do eventually erode and dissolve, a large proportion of the total silica in many reef environments is tied up in sediment spicules and those in living sponges. Therefore, this may limit the growth of some siliceous sponge populations (Rützler & Macintyre 1978).

BIOEROSION

The term 'bioerosion' was coined by Neumann (1966) to describe the activities of a large number of marine organisms which erode calcium carbonate substrate. The word 'calcibiocavitology' was proposed for the study of these animals, but has not gained recognition (Carriker & Smith 1969).

Bioeroding organisms include borers such as fungi, algae, sponges, polychaetes, molluscs and barnacles plus reef grazers and browsers such as molluscs, echinoderms and fish. Carriker and Smith (1969) attempted to define 'burrowers' as organisms that excavate a substrate for a living space whereas 'borers' penetrated the substrate to obtain nutrient. This terminology has not been generally accepted and the more commonly used term, 'borer' is employed here to describe bioeroding sponges.

The boring activities of some sponge genera have been long recognized (see Rützler & Rieger 1973). However, it is only recently that their effects on coral reefs have been recognized. Boring sponges erode large quantities of coral substrate resulting in its collapse and the production of large quantities of rubble and sediment which are then redistributed throughout the reef. Characteristic of these sediments are sponge 'chips' — small, discrete chips of substrate which are removed from coral rock during sponge bioerosion. These chips constitute a significant proportion of reef sediments and lithified coral rock.

Mechanisms of Bioerosion

A limited number of siliceous sponges are known to bore into coral substrate. The most important genera are *Cliona, Anthosigmella* and *Spheciospongia* of the order Hadromerida and *Siphonodictyon* of the order Haplosclerida. Although these sponges use a similar mechanism to erode calcium carbonate, it was only recently (Warburton 1958) that Nassonow's (1883) original observation that sponges cause erosion via the production of discrete chips was confirmed. A number of mechanisms have been proposed to explain sponge bioerosion, including chemical dissolution, mechanical abrasion and muscle-like excavation as well as combinations of these methods (see reviews in Goreau & Hartman 1963; Rützler & Rieger 1973).

The following description of sponge boring is summarized from Rützler & Rieger (1973) and Pomponi (1979). The etching cells of both newly settled larvae and mature sponges flatten one end against the substrate with the long axis perpendicular to the surface. Etching commences at the periphery of the cell where a filopodial sheet of membrane-bound cytoplasm, approximately 0.2 μm thick, extends into the substrate. The filopodial processes coalesce within the substrate and cut out a hemispherical chip. This mechanism is reminiscent of the action of an 'orange peel' grab sampler for collecting soft sediments. According to Pomponi (1979) the substrate is dissolved by the combined action of a lysosomal enzyme system which dissolves organic matter, and carbonic anhydrase regulated acid secretion at the periphery of the filopodial sheet. When the process is complete, both the chip of substrate and the spent etching cell are transported through the sponge and expelled.

Sponge boring results in solubilizing about 2 to 3% of the solid substrate, the rest is dispersed as chips. These chips are readily recognized in sediment because they are circular or oval shaped discs, ranging in size from 30-80 μm in diameter and 15-57 μm in depth. One surface is hemispherical and regular whereas the surface initially adjacent to the body of the sponge is faceted due to scalloping by previous etching cells (Rützler & Rieger 1973: Fütterer 1974).

Boring sponges excavate large chambers, 0.5-1.5 mm in diameter from which branch smaller galleries (Ward & Risk 1977; Bromley 1978). The total depth of penetration varies between 0.5 mm and 80 mm but usually does not exceed 20 mm. The depth and rate of penetration also depend on the porosity of the substrate, with porous corals usually being more readily bored (Neumann 1966; MacGeachy 1977).

Effects of Bioerosion

According to Goreau & Hartman (1963), boring sponges have a marked effect on the morphology and ecology of coral reefs. Boring sponges are ubiquitous on coral reefs and almost all live or dead corals show present or previous evidence of sponge bioerosion. However, boring sponges are difficult to detect as only a small proportion of the tissue is exposed — that tissue which contains the inhalant and exhalant papillae. Furthermore, many boring sponges occur in cryptic recesses and among coral rubble. Only one estimation of boring sponge populations has been made on a coral reef. Rützler (1975) estimated mean Caribbean populations as 16 g dry weight m^{-2} with localized areas being as much as 14 times higher (235 g dry weight m^{-2}).

The effects of sponge boring are most obvious on deep slopes where the density of living coral is already low and where there are large areas of bare substrate available for sponge larval settlement. Sponge bioerosion accentuates the steepness of these slopes and results in the accumulation of large amounts of talus at the bottom. This talus is itself subsequently bored (Goreau & Hartman 1963). In shallow areas, the rate of sponge bioerosion is probably similar to that in deep water, however, the rapid rates of calcification by corals and coralline algae counteract these effects of bioerosion (Goreau & Hartman 1963).

Much of the damage caused to corals during storms has been attributed to weakening of basal structures by bioerosion (Bertram 1936; Goreau & Hartman 1963; Hein & Risk 1975). Boring sponges gain access to these basal structures because they are usually not covered by living tissue and extensive bioerosion results in the eventual collapse of the coral. In branching corals, the structural weakening is marked and can be measured by applying engineering principles (Tunnicliffe 1979). It has been suggested that sponge boring is important in asexual reproduction of some branching corals, e.g., *Acropora cervicornis,* as the broken branches lodge in crevices and regenerate new colonies (Tunnicliffe 1979).

Chips formed by sponge boring activities constitute a significant proportion of fine silt sediments (silt: 4-62 μm sized particles) on and around coral reefs. In the Persian Gulf, 2-3% of all sediments consist of sponge chips and as much as 30% of the fine sediment in the lagoon of Fanning Island of the Pacific Line Islands are sponge chips (Fütterer 1974). Halley *et al.* (1977) estimated that 20-40% of silt sediments on patch reefs of Belize

Author	Rate (kg m^{-2} yr^{-1})	Depth (mm yr^{-1})	Method
a Sponges			
Neumann (1966)	20–25	14	small block transplant; 100 days
Rützler (1975)	0.25–3	0.1–1	small block transplant; 12 months
Hein & Risk (1975)	0.17	—	X-ray of core samples
Bak (1976)	2.6–3.3 #	—	underwater weighing of transplants
Moore & Shedd (1977)	0.19–3.29	—	optical-thin sections of cores
	0.20–0.51	—	sediment collection in traps
Stearn & Scoffin (1977)	0.20–0.51	—	X-ray of slabbed corals
Scoffin *et al.* (1980)	1.35	—	X-ray of slabbed corals
b Other Organisms			
Hudson (1977)	3.0–13.4*	15–67	total bioerosion
Hunter (1977)	9.7	—	*Diadema antillarum*
Ogden (1977)	0.49	—	Parrot fish
	4.6	—	*Diadema antillarum*
	3.9	—	*Echinometra*
Stearn & Scoffin (1977)	3.7–9	—	*Diadema*
	0.036	—	Parrot fish
Russo (1980)	0.5–30	—	*Echinometra*
	0.8–.32	—	*Echinometra, Echinostrephus*
Hein & Risk (1975)	0.18	—	spionid worms
	0.04	—	mytilid molluscs

\# Measured for 100 days—converted to 365 days
* Estimated using Specific Gravity of CaCO$_3$ = 2 g cm^{-3}

Table 2   Rates of bioerosion by sponges and other organisms

consisted of sponge chips. These fine sediments accumulate within the interstices of the reef and constitute a significant proportion of the material which is subsequently lithified to form consolidated reef rock (Marshall, this volume).

Rates of Bioerosion

Boring sponges are probably the most important bioeroding organisms on coral reefs in terms of effects and rates of destruction of substrate (Goreau & Hartman 1963; MacGeachy 1977; Warme 1977). A wide variety of techniques has been used to estimate sponge bioerosion, including X-ray analysis of natural substrate (Moore & Shedd 1977; Stearn & Scoffin 1977) X-rays of core samples (Hein & Risk 1975), underwater weighing of natural substrate (Bak 1976), collection and weighing of sediment (Moore & Shedd 1977) and periodic examination of transplanted artificial substrate blocks (Neumann 1966; Rützler 1975). The wide range of bioerosion rates in table 2 is probably indicative of both the diversity of techniques employed and genuine inter-habitat and inter-reefal differences. Rützler (1975) has reported that there are considerable variations in sponge populations between reefs and between zones on the same reef.

All studies and rate estimations of bioerosion have been performed in the Caribbean with the exception of Otter (1937) who examined bioeroding organisms at Low Isles on the Great Barrier Reef. Some of the bioerosion estimates suggest that reefs are eroding faster than they are growing (table 2). These bioerosion rates may be excessive because rates for individual organisms obtained from different locations have been added or because short duration experiments were used. In many instances, reported bioerosion rates may exceed the mean calcification rate for Pacific reefs of 1 kg $CaCO_3$ m^{-2} yr^{-1} (Smith 1978) and are often higher than the upper model rate of 4 kg m^{-2} yr^{-1} (Smith & Kinsey 1976). Higher calcification rates (e.g. 9 kg m^{-2} yr^{-1}, Smith, this volume; 15 kg m^{-2} yr^{-1}, Stearn *et al.* 1977) have been reported but these originate from small, very active areas (Smith, this volume). Likewise, many of the bioerosion rates for individual organisms originate from zones of high activity (e.g. *Diadema*, Stearn & Scoffin 1977). Bioerosion rates determined from short duration experiments (Neumann 1966) were found to be erroneous by Rützler (1975) who showed that sponge bioerosion is rapid for the first three to six months, while the sponge is becoming established within the substrate, but subsequent bioerosion proceeds much more slowly.

It is not advisable to combine bioerosion rates for different organisms and from different reefs and compare these with calcification rates from different regions. Two integrated studies have been performed on a fringing reef off Barbados. In the first, Stearn & Scoffin (1977) estimated that annual bioerosion was 189 tonnes and accretion was 163 tonnes — a net annual erosion rate of 16%. Whereas in the second study (Scoffin *et al.* 1980), annual calcification was measured at 206 tonnes and bioerosion was 123 tonnes — a net annual accretion rate of 40%. Although these estimates have large inherent variabilities, they indicate that bioerosion is a particularly significant process.

Few conclusions can be drawn from published bioerosion and calcification rates with respect to reef growth. More detailed and accurate estimates are necessary over wider geographical regions before it is possible to predict accurately the net rate and direction of reef growth and reaction of reefs to sea-level change.

FACTORS AFFECTING BIOEROSION

Available Substrate

Only a small proportion of the calcium carbonate veneer of a coral reef is actively bored by sponges at any time. The rest of the substrate is either occupied by other organisms or is unsuitable for sponge growth.

Little is known of substrate preference or behaviour of sponge larvae. Generally, they have a short planktonic phase (Bergquist *et al.* 1970; Ayling 1980) and larvae have specific behavioural responses to optimize their chances of locating a suitable habitat, e.g., cryptic sponge larvae avoid light. Boring sponge larvae appear to prefer a surface that is free of other encrusting organisms (Hartman 1977; MacGeachy & Stearn 1976) and boring sponges are rarely found among or immediately underneath living coral polyps (Goreau & Hartman 1963; Bak 1976). Some species, however, grow alongside or over living corals (Rützler & Rieger 1973), whereas others appear to grow within specific corals (Hartman 1977; Tunnicliffe 1979).

A suitable substrate for boring sponge larvae is one which is not occupied or is occupied by an organism susceptible to sponge attachment when the larvae are ready to settle. A suitable substrate for sponge development is one which is deep enough to enable the sponge to penetrate; is physically stable; is remote from excessive sediment; and is within a region where there is sufficient water movement. Suitable substrate is continually created by physical and biological disturbances, although suitable substrate is rarely abundant and lack of substrate is a major limiting factor for coral reef biota (Connell 1973; Glynn 1973). Physical disturbances (ranging from storm damage to occasional dislodgement of coral rock) are stochastic in nature — unpredictable and intermittent.

Biological disturbances, however, are more regular and frequent. They are caused by grazing and browsing animals such as fish, echinoderms and molluscs and by the death of corals due to stress, disease (Antonius 1977) or because of the over-topping actions of competing organisms. The echinoids *Diadema* and *Echinometra* (table 2; Ogden 1977; Russo 1980) and the asteroid *Acanthaster planci* (Endean 1973) are particularly important grazers, as are large populations of algal-grazing and coral-eating fish (Randall 1967; Ogden 1977) and grazing molluscs (McLean 1967).

The shape of the coral itself is also a factor influencing sponge bioerosion. The basal parts of living, branching and plate-like coral colonies not covered by coral polyps are more susceptible to sponge boring than more massive or solid forms (Goreau & Hartman 1963; MacGeachy 1977; Tunnicliffe 1979). Examples of these effects occur on the steep slopes of Jamaica where the predominant plate-like corals are extensively bored and often collapse (Goreau & Hartman 1963). Dead branching or plate-like corals are reduced to sediment more rapidly than solid forms because there is greater surface area for larval attachment and subsequent boring is more effective because sponges rarely penetrate deeper than 2 cm (Rützler 1975).

A positive correlation has been demonstrated between high rates of coral growth and rapid sponge bioerosion (Rützler 1975). The high sponge activity was attributed to both the ready availability of nutrient in shallow, turbulent regions and also to the availability of more substrate. The rate of substrate production (calcification) is greater in shallow, illuminated, turbulent regions (Smith & Kinsey 1976).

Predation and Competition

Predation and competition are important during all stages of sponge development. Little is known about predation on sponge larvae except that the larvae enter the plankton and would be subject to predation by plankton feeders, including corals. There are, presumably, specific predators of newly settled and juvenile sponges such as copepods and fish. Browsers and grazers such as echinoderms, molluscs and fish are important, though inadvertent, predators of juvenile sponges. However, little is known of the extent of this predation. Guida (1976) reported many specific predators of mature clionids in oyster beds. Such predators include molluscs, crustaceans and echinoderms, as well as numerous casual predators. A number of coral reef fish are known to include sponges in their diet (Randall & Hartman 1968).

Newly settled sponges have to compete with a wide range of organisms for space on and within the substrate. The sponges burrow into the substrate and erect papillae above competing organisms thereby ensuring access to the ambient water and reducing the effects of competition (Jackson 1977). Competition also exists within the substrate. Bromley & Tendal (1973) showed that different strains of the same sponge species display avoidance reactions in the same substrate. Similar reactions may also occur between sponges and other boring organisms, such as algae (McCloskey 1970).

The advantages of boring are that the sponges escape the major effects of predation while obtaining support for the tissue. However, Ward & Risk (1977) suggested that sponges may use organic matter embedded in the coral skeleton or possibly remove dissolved organic matter in the pore water (see Wilkinson & Garrone 1980).

Sponge Growth and Nutrition

Sponges are essentially filter feeders, taking both particulate and dissolved nutrient from the water. Although coral reef waters are not nutrient-rich, there is adequate particulate, colloidal and dissolved organic matter in such waters for sponge nutrition, provided the flow of water to the sponges is not restricted (Johannes 1967; Reiswig 1971a). Sponge growth is reduced in caves and in recesses where water movement is limited and Wilkinson & Vacelet (1979) have demonstrated reduced sponge growth in diminished ambient currents.

Boring sponges erect inhalant and exhalant papillae above the substrate (Rützler 1975). Competing organisms restrict water flow to boring sponges by either overgrowing the papillae or by diverting currents away from the substrate surface (MacGeachy 1977). Therefore, boring sponges are at a disadvantage when compared with erect sponges which have adequate access to ambient water.

Sponges have two mechanisms for inducing water flow through their tissue. Flagellated cells expend energy to drive water through the canal system, and water may flow passively through sponges in regions where there are ambient currents. This occurs when the exhalant oscules are raised above the inhalant ostia, thereby creating a current inducing negative pressure at the ostia (Vogel 1977). Boring sponges are more prevalent where there is adequate water circulation than in caves and deeper water (Goreau & Hartman 1963; Rützler 1975). The growth and boring rates of sponges are affected both by the available nutrient and by inherent species-specific differences (Pang 1973).

It has been suggested that light stimulates the growth of *Cliona lampa* (Rützler 1975) and *Anthosigmella varians* (Vincente 1978), although no explanatory mechanisms were suggested. Wilkinson & Vacelet (1979) demonstrated light-enhanced growth of certain non-boring sponges containing photosynthetic symbionts, but others without such symbionts were inhibited by light. Ultra-violet light damage was reported to be the reason for this inhibition of sponge growth (Jokiel 1980). The cyanobacterial (blue-green algal) symbionts in coral reef sponges have been shown to fix carbon dioxide, and possibly nitrogen, and to translocate at least one product, glycerol, to the host (Wilkinson 1979a,b). Photosynthetic nutrient production and transfer has not been demonstrated in boring sponges although some *Cliona* Species (Pang 1973; Rützler 1974) and *Anthosigmella varians* (J. C. Lang, pers. comm.) are reported to have zooxanthella-like symbionts. These sponges grow in well illuminated habitats (Pang 1973; Vincente 1978) and it is possible that they derive additional nutrient from symbionts.

Sediment load

Sponge growth is inhibited by high intermittent sediment loads or by continuous low levels of sediment because considerable amounts of energy are required to clear occluded canals and tissues (Reiswig 1971b). Sediment causes a reduction in pumping rates (Gerrodette & Flechsig 1979) and excessive sediment causes sponges to contract and cease pumping (feeding) for extended periods (Reiswig 1971b). Therefore, sponges are not prevalent at the base of steep slopes, in deeper drainage channels, in heavily silted lagoons or on the floor of caves.

CONCLUSION

Sponges do not contribute greatly to current reef growth but they are important in the restructuring and subsequent consolidation of the reef structure. Corals and coralline algae are responsible for the addition of large quantities of calcium carbonate to the reef environment (Smith 1978). Much of this added substrate is unconsolidated and fragile and is subsequently reduced to rubble by physical forces. This reduction is accentuated if the substrate has been eroded previously by boring organisms. These boring organisms, particularly sponges, further erode the substrate with the production of sediment and silt. Physical forces and biological agents, such as fish (Glynn *et al.* 1972), redistribute this sediment over the reef or beyond it. The sediments are eventually converted into more stable substrate through lithification (Marshall, this volume).

The taxonomy of boring sponges and the mechanisms by which they bore calcium carbonate are well documented. The effects and rates of these boring activities require further study. It is evident that a large proportion of the calcium carbonate on a coral reef has been, or will be, bored by sponges, resulting in extensive morphological changes to the reef as well as the production of fine sediments. Thus, the net result of sponge boring is not only destruction but a restructuring and, indirectly, a consolidation of the calcium carbonate substrate. Much of the erect, lattice-like coral skeletal carbonate is reduced to sediment and infilled coral rock.

LITERATURE CITED

Antonius, A., 1977. Coral mortality in reefs: a problem for science and management. In D. L. Taylor (ed.), *Proceedings: Third International Coral Reef Symposium*. Miami: University of Miami, vol. 2, pp. 617–23.

Ayling, A. L., 1980. Patterns of sexuality, asexual reproduction and recruitment in some subtidal marine demospongiae. *Biological Bulletin* 158, 271–82.

Bak, R. P. M., 1976. The growth of coral colonies and the importance of crustose coralline algae and burrowing sponges in relation with carbonate accumulation. *Netherlands Journal of Sea Research* 10, 285–337.

Bergquist, P. R., 1978. Sponges. Hutchinson and Co., London, 268pp.

Bergquist, P. R., M. E. Sinclair and J. J. Hogg, 1970. Adaptation to intertidal existence: Reproductive cycles and larval behaviour in Demospongiae. In W. G. Fry (ed.), *Symposia of the Zoological Society of London. The Biology of the Porifera*. no. 25, pp. 247–71.

Bertram, G. C. L., 1936. Some aspects of the breakdown of coral at Ghardaga, Red Sea. *Proceedings of the Zoological Society of London* 136, 1011–26.

Bromley, R. G., 1978. Bioerosion of Berumda reefs. *Palaeogeography, Palaeoclimatology, Palaeoecology* 23, 169–97.

Bromley, R. G. and O. S. Tendal, 1973. Example of substrate competition and phototropism between two clionid sponges. *Journal of Zoology* 169, 151–5.

Bryan, P. G., 1973. Growth rate, toxicity, and distribution of the encrusting sponge *Terpios* sp. (Hadromerida: Suberitidae) in Guam, Mariana Islands. *Micronesica* 9, 237–42.

Carriker, M. R. and E. H. Smith, 1969. Comparative calcibiocavitology: Summary and conclusions. *American Zoologist* 9, 1011–20.

Connell, J. H., 1973. Population ecology of reef building corals. In O.A. Jones and R. Endean (eds.), *Biology and Geology of Coral Reefs*. New York: Academic Press, vol. 2, Biology 1, pp. 205–45.

Dayton, P. K., G. A. Robilliard, R. T. Paine and L. B. Dayton, 1974. Biological accommodation in the benthic community at McMurdo Sound, Antarctica. *Ecological Monographs* 44, 105–28.

Endean, R., 1973. Population explosions of *Acanthaster planci* and associated destruction of hermatypic corals in the Indo-West Pacific region. In O. A. Jones and R. Endean (eds.), *Biology and Geology of Coral Reefs*. New York: Academic Press, vol. 2, Biology 1, pp. 390–438.

Finks, R. M., 1970. The evolution and ecologic history of sponges during Palaeozoic times. In W. G. Fry (ed.), *Symposia of the Zoological Society of London. The Biology of the Porifera*. no. 25, pp. 3–22.

Finks, R. M. and D. Hill, 1967. Porifera and Archaeocyatha. In *The Fossil Record. A Symposium with Documentation*. London: Geological Society of London, pp. 333–45.

Fütterer, D. K., 1974. Significance of the boring sponge *Cliona* for the origin of fine grained material of carbonate sediments. *Journal of Sedimentary Petrology* 44, 79-84.

Gerrodette, T. and A. O. Flechsig, 1979. Sediment-induced reduction in the pumping rate of the tropical sponge *Verongia lacunosa*. *Marine Biology* 55, 103-10.

Glynn, P. W., 1973. Aspects of the ecology of coral reefs in the western Atlantic region. In O. A. Jones and R. Endean (eds.), *Biology and Geology of Coral Reefs*. New York: Academic Press, vol. 2, Biology 1, pp. 271-324.

Glynn, P. W., R. H. Stewart and J. E. McCosker, 1972. Pacific coral reefs of Panamá: Structure, distribution and predators. *Geologische Rundschau* 61, 483-519.

Goreau, T. F. and W. D. Hartman, 1963. Boring sponges as controlling factors in the formation and maintenance of coral reefs. In R. F. Sognnaes (ed.), *Mechanisms of Hard Tissue Destruction*. Publication of the American Association for the Advancement of Science, no. 75, pp. 25-54.

Goreau, T. F. and W. D. Hartman, 1966. Sponge: effect on the form of reef corals. *Science* 191, 343-4.

Guida, V. G., 1976. Sponge predation in the oyster reef community as demonstrated with *Cliona celata* Grant. *Journal of Experimental Marine Biology and Ecology* 25, 109-22.

Halley, R. B., E. A. Shinn, J. H. Hudson and B. Lidz, 1977. Recent and relict topography of BooBee Patch reef, Belize. In D. L. Taylor (ed.), *Proceedings: Third International Coral Reef Symposium*. Miami: University of Miami, vol. 2, pp. 29-35.

Hartman, W. D., 1977. Sponges as reef builders and shapers. *American Association of Petroleum Geologists. Studies in Geology* 4, 127-34.

Hartman, W. D., 1979. A new sclerosponge from the Bahamas and its relationship to Mesozoic stromatoporoids. In C. Lévi and N. Boury-Esnault (eds.), *Biologie des Spongiaires, Colloques Internationaux du Centre National de la Recherche Scientifique*. no. 291, pp. 467-74.

Hartman, W. D. and T. F. Goreau, 1970. Jamaican coralline sponges: their morphology, ecology and fossil relatives. In W. G. Fry (ed.), *Symposia of the Zoological Society of London. The Biology of the Porifera*. no. 25, pp. 248-71.

Hein, F. J. and M. J. Risk, 1975. Bioerosion of coral heads: inner patch reefs, Florida reef tract. *Bulletin of Marine Science* 25, 133-8.

Hudson, J. H., 1977. Long-term bioerosion rates on a Florida reef: a new method. In D. L. Taylor (ed.), *Proceedings: Third International Coral Reef Symposium*. Miami: University of Miami, vol. 2, pp. 491-7.

Hunter, I. G., 1977. Sediment production by *Diadema antillarum* on a Barbados fringing reef. In D. L. Taylor (ed.), *Proceedings: Third International Coral Reef Symposium*. Miami: University of Miami, vol. 2, pp. 105-9.

Jackson, J. B. C., 1977. Competition on marine hard substrate: The adaptive significance of solitary and colonial strategies. *American Naturalist* 111, 743-67.

Johannes, R. E., 1967. Ecology of organic aggregates in the vicinity of a coral reef. *Limnology and Oceanography* 12, 189-95.

Jokiel, P. L., 1980. Solar ultraviolet radiation and coral reef epifauna. *Science* 207, 1069-71.

Lang, J. C., W. D. Hartman and L. S. Land, 1975. Sclerosponges: Primary framework constructors on the Jamaican deep fore-reef. *Journal of Marine Research* 33, 223-31.

MacGeachy, J. K., 1977. Factors controlling sponge boring in Barbados reef corals. In D. L. Taylor (ed.), *Proceedings: Third International Coral Reef Symposium*. Miami: University of Miami, vol. 2, pp. 477-83.

MacGeachy, J. K. and C. W. Stearn, 1976. Boring by macro-organisms in the coral *Montastrea annularis* on Barbados reefs. *Internationale Revue der gesamten Hydrobiologie* 61, 715-45.

McCloskey, L. R., 1970. The dynamics of the community associated with a marine scleractinian coral. *Internationale Revue der gesamten Hydrobiologie* 55, 13-81.

McLean, R. F., 1967. Measurements of beachrock erosion by some tropical marine gastropods. *Bulletin of Marine Science* 17, 551-61.

Moore, C. H., Jr. and W. W. Shedd, 1977. Effective rates of sponge bioerosion as a function of carbonate production. In D. L. Taylor (ed.), *Proceedings: Third International Coral Reef Symposium.* Miami: University of Miami, vol. 2, pp. 499–505.

Nassonow, N., 1883. Zur Biologie und Anatomie der Clione. *Zeitschrift fur wissenschaftliche Zoologie* 39, 295–308.

Neumann, A. C., 1966. Observations on coastal erosion in Bermuda and measurements of the boring rate of the sponge *Cliona lampa. Limnology and Oceanography* 11, 92–108.

Ogden, J. C., 1977. Carbonate-sediment production by parrot fish and sea urchins on Caribbean reefs. *American Association of Petroleum Geologists. Studies in Geology* 4, 281–8.

Otter, G. W., 1937. Rock-destroying organisms in relation to coral reefs. *Scientific Report of the Great Barrier Reef Expedition, 1928–29, British Museum (Natural History),* 1, 323–52.

Pang, R. K., 1973. The ecology of some Jamaican excavating sponges. *Bulletin of Marine Science* 23, 227–43.

Pomponi, S. A., 1979. Ultrastructure and cytochemistry of the etching area of boring sponges. In C. Lévi and N. Boury-Esnault (eds.), *Biologie des Spongiaires. Colloques Internationaux du Centre National de la Recherche Scientifique.* no. 291, pp. 317–23.

Randall, J. E., 1967. Food habits of reef fishes of the West Indies. *Studies in Tropical Oceanography* 5, 665–847.

Randall, J. E. and W. D. Hartman, 1968. Sponge-eating fishes of the West Indies. *Marine Biology* 1, 216–25.

Reiswig, H. M., 1971a. Particle feeding in natural populations of three marine demosponges. *Biological Bulletin* 141, 568–91.

Reiswig, H. M., 1971b. *In situ* pumping activities of tropical Demospongiae. *Marine Biology* 9, 38–50.

Russo, A. R., 1980. Bioerosion by two rock boring echinoids *(Echinometra mathaei* and *Echinostrephus aciculatus)* on Enewetak Atoll, Marshall Islands. *Journal of Marine Research* 38, 99–110.

Rützler, K., 1974. The burrowing sponges of Bermuda. *Smithsonian Contributions to Zoology* 165, 1–32.

Rützler, K., 1975. The role of burrowing sponges in bioerosion. *Oecologia* 19, 203–16.

Rützler, K. and I. G. Macintyre, 1978. Siliceous sponge spicules in coral reef sediments. *Marine Biology* 49, 147–56.

Rützler, K. and G. Rieger, 1973. Sponge burrowing: fine structure of *Cliona lampa* penetrating calcareous substrata. *Marine Biology* 21, 144–62.

Scoffin, T. P., C. W. Stearn, D. Boucher, P. Frydl, C. M. Hawkins, I. G. Hunter and J. K. MacGeachy, 1980. Calcium carbonate budget of a fringing reef on the west coast of Barbados. Part II — Erosion, sediments and internal structure. *Bulletin of Marine Science* 30, 475–528.

Smith, S. V., 1978. Coral-reef area and the contributions of reefs to processes and resources of the world's oceans. *Nature* 273, 225–6.

Smith, S. V. and D. W. Kinsey, 1976. Calcium carbonate production, coral reef growth, and sea level change. *Science* 194, 937–9.

Stearn, C. W. and T. P. Scoffin, 1977. Carbonate budget of a fringing reef, Barbados, In D. L. Taylor (ed.), *Proceedings: Third International Coral Reef Symposium.* Miami: University of Miami, vol. 2, pp. 471–6.

Stearn, C. W., T. P. Scoffin and W. Martindale, 1977. Calcium carbonate budget of a fringing reef on the west coast of Barbados, Part I — Zonation and productivity. *Bulletin of Marine Science* 27, 479–510.

Tunnicliffe, V., 1979. The role of boring sponges in coral fracture. In C. Lévi and N. Boury-Esnault (eds.), *Biologie des Spongiaires. Colloques Internationaux du Centre National de la Recherche Scientifique.* no. 291, pp. 309–15.

Vacelet, J., 1967. Description d'éponges Pharétronides actuelles des tunnels obscurs sous-récifaux de Tuléar (Madagascar). *Recueil des Travaux de la Station Marine d'Endoume,* Supplement 6, 37–62.

Vacelet, J., 1970. Les éponges Pharétronides actuelles. In W. G. Fry (ed.), *Symposia of the Zoological Society of London. The Biology of the Porifera.* no. 25, pp. 189-204.

Vacelet, J., 1977. Une nouvelle relique du secondaire: un représentant actuel des éponges fossiles Sphinctozoaires. *Comptes Rendus Hebdomadaire Des Seances de l'Academie Des Sciences. Serie D* 285, 509-11.

Vacelet, J., 1979a. La place des Spongiaires dans les systems trophiques marins. In C. Lévi and N. Boury-Esnault (eds.). *Biologie des Spongiaires. Colloques Internationaux du Centre National de la Recherche Scientifique.* no. 291, pp. 259-70.

Vacelet, J., 1979b. Description et affinités d'une éponge Sphinctozoaire actuelle. In C. Lévi and N. Boury-Esnault (eds.), *Biologie des Spongiaires, Colloques Internationaux du Centre National de la Recherche Scientifique.* no. 291, pp. 483-93.

Vincente, V. P., 1978. An ecological evaluation of the West Indian demosponge *Anthosigmella varians* (Hadromerida: Spirastrellidae). *Bulletin of Marine Science* 28, 771-9.

Vogel, S., 1977. Current-induced flow through living sponges in nature. *Proceedings of the National Academy of Science* 74, 2069-71.

Warburton, F., 1958. The manner in which the sponge *Cliona* bores in calcareous objects. *Canadian Journal of Zoology* 36, 555-62.

Ward, P. and M. J. Risk, 1977. Boring pattern of the sponge *Cliona vermifera* in the coral *Montastrea annularis*. *Journal of Paleontology* 51, 520-26.

Warme, J. E., 1977. Carbonate borers — their role in reef ecology and preservation. *American Association of Petroleum Geologists. Studies in Geology* 4, pp. 261-79.

Wilkinson, C. R., 1978. Microbial associations in sponges. I. Ecology physiology and microbial populations of coral reef sponges. *Marine Biology* 49, 161-7.

Wilkinson, C. R., 1979a. Nutrient translocation from symbiotic cyanobacteria to coral reef sponges. In C. Lévi and N. Boury-Esnault (eds.), *Biologie des Spongiaires. Colloques Internationaux du Centre National de la Recherche Scientifique.* no. 291, pp. 373-80.

Wilkinson, C. R., 1979b. Nitrogen fixation in coral reef sponges with symbiotic cyanobacteria. *Nature* 279, 527-9.

Wilkinson, C. and R. Garrone, 1980. Nutrition of marine sponges. Involvement of symbiotic bacteria in the uptake of dissolved carbon. In D. C. Smith and Y. Tiffon (eds.), *Nutrition in the Lower Metazoa*. Oxford: Pergamon Press, pp. 157-61.

Wilkinson, C. R. and J. Vacelet, 1979. Transplantation of marine sponges to different conditions of light and current. *Journal of Experimental Marine Biology and Ecology* 37, 91-104.

Wulff, J. L. and L. W. Buss, 1979. Do sponges help hold coral reefs together? *Nature* 281, 474-5.

Zhuravleva, I. T., 1970. Porifera, Sphinctozoa, Archaeocyathi — their connections. In W. G. Fry (ed.), *Symposia of the Zoological Society of London. The Biology of the Porifera.* no. 25, pp. 41-59.

Zhuravleva, I. T. and E. I. Miagkova, 1979. Comparaison entre les Archaeata et les Porifera. In C. Lévi and N. Boury-Esnault (eds.), *Biologie des Spongiaires. Colloques Internationaux du Centre National de la Recherche Scientifique.* no. 291, pp. 521-6.

Ziegler, B. and S. Rietschel, 1970. Phylogenetic relationships of fossil calcisponges. In W. G. Fry (ed.), *Symposia of the Zoological Society of London. The Biology of the Porifera.* no. 25, pp. 23-40.

Index